PERSPECTIVES IN RESOURCE MANAGEMENT
IN
DEVELOPING COUNTRIES

VOLUME 3
Ecological Degradation of Land

Details of the Series

Perspectives in Resource Management in Developing Countries

1. Resource Management : Theory and Techniques
2. Population, Resources and Development
3. Ecological Degradation of Land
4. Land Appraisal and Development
5. Soil Deterioration and Conservation
6. Water Supply and Quality
7. Water Utilization and Management
8. Groundwater Assessment and Development
9. Bioresource Depletion and Conservation
10. Environmental Hazards
11. Scenic Resources and Tourism
12. Energy Resources and Environment
13. Degrading Atmospheric Resources
14. Integrated Watershed Development

About the Editor

Professor B. Thakur, Formerly Vice-Chancellor of L.N. Mithila University, Darbhanga holds substantive position in the Department of Geography, Delhi School of Economics, University of Delhi. He has more than four decades of teaching and research experience in resource management, natural hazards and urban development at the Universities of Ranchi, Patna, Waterloo, Akron, Delhi and Mithila. He was a Commonwealth Scholar at the University of Waterloo (Canada), Fulbright Visiting Professor at the University of Akron (USA) and Shastri fellow at York University (Canada). He has successfully supervised 19 Ph.D. Theses, 35 M.Phil and 57 M.A. dissertations. He has published about 95 research papers in journals of India and abroad and has published 12 books. Professor Thakur has shouldered many responsibilities, such as Secretary General and President of National Association of Geographers, India; Recorder and President of Earth System Sciences of Indian Science Congress Association; Editor and President of Association for Geographical Studies, Delhi University; and Chairman of the Commission on Natural Hazards and Resource Management of NAGI.

Concept's International Series in Geography No. 5

PERSPECTIVES IN RESOURCE MANAGEMENT IN DEVELOPING COUNTRIES

VOLUME-3

Ecological Degradation of Land

Edited by
BALESHWAR THAKUR
Professor of Geography
Department of Geography
Delhi University, Delhi

CONCEPT PUBLISHING COMPANY, NEW DELHI-110059

Cataloging in Publication Data—DK
Courtesy: D.K. Agencies (P) Ltd. <docinfo@dkagencies.com>

Perspectives in resource management in developing countries / edited by Baleshwar Thakur.
v. 3 cm. — (Concept's international series in geography ; 5)
Contributed articles.
Includes bibliographical references.
Includes indexes.
Contents: v. 3. Ecological degradation of land.
ISBN-13: 9788180690365 (set)
ISBN-10: 8180690369 (set)
ISBN-13: 9788180694882 (v. 3)
ISBN-10: 8180694887 (v. 3)

1. Natural resources—Developing countries—Management. 2. Sustainable development—Developing countries. 3. Natural resources—Developing countries—Management—Technological innovations. 4. Developing countries—Population—Economic aspects. 5. Developing countries—Population—Environmental aspects. I. Thakur, Baleshwar, 1943- II. Series: Concept's international series in geography ; 5.

DDC 333.7 22

ISBN -13: 978-81-8069-036-5 (Series)
ISBN -10: 81-8069-036-9 (Series)
ISBN -13: 97881-8069-488-2 (Vol. 3)
ISBN -10: 81-8069-488-7 (Vol. 3)

First Published 2008

Published and Printed by
Ashok Kumar Mittal
Concept Publishing Company
A/15-16, Commercial Block, Mohan Garden
New Delhi-110059 (INDIA)

Phones : 25351460, 25351794
Fax : 091-11-25357103
Email : publishing@conceptpub.com

Dedicated to
Chanda, my wife, who gave me help,
encouragement and inspiration

Preface

Thinking minds in the developing world today are deeply concerned at the massive quantitative expansion in the population size and the concomitant pressure on natural resources. This population pressure has often exceeded the carrying capacity of the land as traditionally used. It has resulted in loss of cultivated lands, disappearance of forests, soil erosion, energy crisis, ecological imbalances, pollution hazards, environmental harassment, regional water shortage and deterioration of water quality. Little wonder that we cannot maintain a growing population at a high standard of living without adequate natural resources. Thus, natural resource studies have become a subject of great importance in the perpetuation and enrichment of civilized human existence and, ultimately, even our biological survival.

The purpose of this multi-volume series is, therefore, to contribute to and provide scientific knowledge on natural resources in developing countries for their rational and sustainable development. It creates an understanding of the critical and decisive role that resource development and management play in today's society by an intelligent garnering of natural resources, restocking of forests and the achievement of a long-term rotational agro-forestry, arresting of soil erosion and depletion, prudent management of range and preservation of fish reserves and mammals.

Identifying current issues and responses of renewable and non-renewable resources, it suggests scientific techniques to permit greater flexibility and sophistication in the assessment of alternative management schemes, and demonstrates the role of new technological developments in their inventory and appraisal. Assessing the impact of natural hazards in resource management, it develops a scientific basis for integrated resource management and probes deeply into the questions of future resource supply as well as adequacy.

B. Thakur

Publisher's Note

Concept Publishing Company has successfully launched its multi-volume series entitled "Concept's International Series in Geography." Till date four multi-volume series have already been published and well received among the academic world. The details of the series are as follows:

1. Perspectives in Agricultural Geography, edited by *Dr. Noor Mohammad* (in 4 volumes).
2. Perspectives in Urban Geography, edited by *Dr. C.S. Yadav* (in 16 volumes).
3. Perspectives in Geomorphology, edited by *Dr. H.S. Sharma* (in 4 volumes).
4. New Dimensions in Agricultural Geography, edited by *Dr. Noor Mohammad* (in 8 volumes).

The present series was initiated about more than a decade ago. We have received in encouraging response from the scholars around the world. Although it took long time to see the light of the day, the problems in utilization and management of Natural Resources are still the same and a great concern for the sustainable development in the developing countries.

This series is planned in fourteen volumes. Each volume covers one particular natural resource, in detail.

Ashok K. Mittal
Publisher

Acknowledgements

More than a decade has elapsed since the multi-volume project was launched by the Concept Publishing Company. Many geographers and earth scientists have contributed advice, comments and ideas in the course of preparing this book. Particularly, I have greatly benefited from the interaction with Professor P. Dayal, Professor A.K. Dutt, Professor L.N. Ram, Professor V.K. Verma and Dr. K.V. Sundaram. I am grateful to all of them for their generosity of ideas and sacrifice of time. I owe many thanks to my colleagues particularly Mrs. Preeti Joshi, Dr. P.K. Parihar, Dr. Ramashray Prasad, Ms. Punyatoya Patra and Ms. Ritu Ahalawat who offered helpful comments in reviewing the articles. I would also like to acknowledge the encouragement and moral support. I received over the years from late Prof. C.P. Singh, Professor Noor Mohammad, Professor S.K. Aggarwal, Dr. Anu Kapur, Dr. R.B. Singh, Dr. R. Thakur, Dr. Phool Singh Bhusan and Sri B.S. Dahiya.

Most importantly, the completion of this volume has been possible because of the tremendous support I received from the authors, particularly for their patience to cope with the delays on our part.

Some already published materials have also been reproduced, because these still have great relevance. I would like to offer my thanks to these authors and publishers who have allowed the use of copyright materials.

I express my deepest gratitude to Sh. Ashok K. Mittal of Concept Publishing Company for his encouragement, support, suggestions and full dedication in creating a special working environment, who made possible the successful completion of this book. I appreciate Ashok's powerful cover and beautiful text design. The computer age made this endeavour more efficient and expedient. I am thankful to Shri Bachi Ram, Cartographer, for his cartographic assistance, Mrs. Neeru Sachdeva, Rajeev Ranjan Kr. Sinha and Punam Tripathi for typing the manuscript.

I must thank my wonderful wife, Chanda, who gave me so much support, inspiration and practical help. I would specially like to thank my sons Sudhir and Rajiv, and daughter-in-laws Raj Rani and Madhuri, who took on onerous task of checking the completeness of the references for every chapter of the book.

Finally I hope that the book will stimulate and generate a new thinking on natural resource management in general and developing countries in particular and would be popular among the students and teachers of natural resource management, administrators and policy makers.

B. Thakur

Contents

List of the Contributors

Ahmad, Afroz, G.B. Pant Institute of Himalayan Environment and Development, Ministry of Environment and Forests, Government of India, Paryavaran Bhawan, Kosi-263643, Almora, U.P.

Awasthi, Rajesh, M. Phil, Student, Geography, Department of Advanced Liberal Studies, D.A.V.V., Indore (M.P.).

Bajpai, Manu, Geography, Department, Delhi School of Economics, University of Delhi.

Chand, R., Department of Geography, Kumaun University, Nainital-263002 (Uttarakhand).

Chandna, Vaneeta, Lecturer, Shaheed Bhagat Singh (Eve) College, Sheikh Sarai, Phase II, New Delhi-110017.

Daya Ram, Assistant Professor of Geography, Department of Soil Science, Haryana Agricultural University, Hissar-125004 (Haryana).

Jag Mohan, Department of Geography, Delhi, School of Economics, University of Delhi, Delhi-110007.

Joshi, D.C., Central Arid Zone Research Institute, Jodhpur.

Kapoor, Atul, C-6/90, Keshav Puram, Delhi-110035.

Kar, Amal, Central Arid Zone Research Institute, Jodhpur.

Kharkwal, S.C., Head, Department of Geography, Garhwal University Campus, Rauri, Garhwal-246001.

Lado, Cleophas, Senior Lecturer, Geography of Department, University of Transkei (Butterworth Beach), Private Bag, X 3031, Butterworth, Republic of Transkei, Southern Africa.

Maiti, Angur, Reader in Geography, S.S. Jalan Girls' College, Calcutta, 700073.

Marh, Bhupinder Singh, Department of Geography, Himachal Pradesh, University, Summer Hill, Shimla-171005.

Mehta, B.S., Department of Geography, Kumaun University, Nainital-263002 (Uttarakhand).

Mukhopadhyay, (Mrs.) Sutapa, Pool Officer, Department of Geography, Visva Bharati, Santiniketan-731235.

Murty, K.L. Narasimha, Lecturer in Geography, Department of Geography, Government College, Srikakulam-532001 (A.P.).

Pal, Saroj K., Professor, Physical Geography, Department of Geography, Delhi University, Delhi.

Pathania, Sanjay S., Department of Geography, Himachal Pradesh University, Summer Hill, Shimla.

Phanse, Vijaya, Head, Department of Geography, Government Arts and Commerce College, Indore (M.P.).

Prasad, Maheshwari, Professor of Geography, Ranchi University, Ranchi, Bihar.

Prasad, Ramashray, Senior Lecturer in Geography, Dr. Bhim Rao Ambedkar College (University of Delhi), Geeta Colony, Delhi-110031.

Saxena, S.K., Central Arid Zone Research Institute, Jodhpur, Rajasthan.

Sen, Amal Kumar, Former, Principal Scientists/Head, Division of Resource Survey and Momchoring, Central Arid Zone, Research Institute, L.C.A.R., Jodhpur, *Present Address:* 28, Lansdowne Teerace, Calcutta-700026.

Sharma, H.S., Professor, Department of Geography, University of Rajasthan, Jaipur-302004.

Singh R.B., Department of Geography, Delhi School of Economics, University of Delhi, Delhi.

Singh, Mehtab, Lecturer, University College, M.D. University, Rohtak.

Singh, Surendra, Principal Scientists, Central Arid Zone Research Institute, Jodhpur.

Szabolcs, I., Research Institute of Social Science and Agricultural Chemistry of the Hungarian Academy of Sciences, Budapest.

Thakur, B., Professor of Geography, Department of Geography, Delhi University, Delhi.

Trivedi, Venu, Professor, Department of Geography, Girls Post-Graduate College, Indore (M.P.).

Upreti, T.C., Department of Geography, Kumaun University, Nainital (U.P.).

Vangani, N.S., Central Arid Zone Research Institute, Jodhpur, Rajasthan.

Yadav, S.S., Associate Professor of Geography, K.R. (P.G.) College, Mathura (U.P.).

Introduction

Ecological Degradation of Land

BALESHWAR THAKUR

Introduction and Overview

Land degradation in the developing countries is an ancient phenomenon. However, recently, it has emerged as one of the most important environmental problems and has received considerable worldwide attention. It is by no limited to particular geographic zones, nor does it occur at a uniform rate; nor is limited to any particular ideology. Its distinctive and peculiar feature is that it exists virtually everywhere to one degree or another, thus, it continues to occur in diverse settings ranging from the Amazon basin to Australia, from Nile valley to Rajasthan and from Yellah basin in Jamaica to Zambia and Tanzania. About 5 per cent of the agricultural land in developing countries has been lost by degradation, and productivity has been appreciably reduced on a further 25 per cent and some 10 per cent of irrigated land is severely salinised (Young, 1998, 101). Its understanding is therefore essential to create sustainable land resource. Natural and social scientists alike have made great strides in understanding and explaining the process of accelerated land degradation.

Land degradation signifies loss or reduction in land productivity which restricts the land's productive capacity and reduces its ability to sustain yields or reduce economic returns (Chisholm and Dumsday, 1987, 7). It is a change in land that makes it less useful for human beings because it damages the physical, chemical and biological status of the land, loss of productivity of land is the issue that causes concern for millions of people in the Third World. Thus, land degradation is defined as the substantial decrease in either or both of an area's biological productivity or usefulness (Johnson and Lewis, 1995, 2). This may be either reversible in the short-term or non-reversible in the long-term. The complete loss of soil through erosion is, for practical practices, irreversible. When soil fertility decline is caused by loss of nutrients, these can be replaced, at a cost, by balanced fertiliser inputs; where the cause is degradation of soil physical properties brought about by lowering of organic matter, restoration will take a number of years (Young, 1998, 102). Two effects which are significant related to land degradation are those of on-site and off-site. On-site effects refer to lowering of productive capacity or reduced outputs and off-site effects refer particularly to changes in river flow, water quality and sedimentation produced by forest clearance and resulting erosion (Young, 1998, 103). There is increasing evidence that on-site effects are the concern of individual farmers, whereas off-site of the wider community. Special efforts are thus needed to control each of these since it is par excellence an interdisciplinary issue.

The central questions addressed about land degradation are : What are the ecological and human-induced processes of land degradation? How much land has already been lost and what are the current rates of degradation? What are the different methods of measuring degradation and what indicators can be chosen to detect change? Why has the society allowed extreme rates of degradation to continue? Is there not sufficient scientific understanding of processes of degradation to be prevented or at least minimised by policies which promote landuse and management? Analysis of why things are wrong in land degradation remains easier than prescribing remedies (Douglas, 1987, 617). The scientific challenge therefore is more exciting.

The causes of land degradation are complex and diverse. It is a complex set of processes and the rate of land degradation varies from place to place and from time to time. Generally, it is paralleled by growing population densities with overcrowded, unproductive and widespread small holdings without proper land reform and these are synonymous with developing world situations. But, population density is by no means its exclusive cause and it is not uniquely associated with densely populated land (Eden and Parry, 1996). It can also be stated that it is not the sheer number of people which account for land degradation rather mechanisation of land exploitation. Its occurrence is also paralleled by advancing technologies, which cause instabilities and environmental stress. However, it is argued that ultimately technology is neutral and no more than a tool of the land user who employs it. In fact, land degradation is the end result of any factor or combination of factors which damage the land, water or vegetation resources and restrict their use or productive capacity (Chisholm and Dumsday, 1987, 27). Johnson and Lewis (1995, 2) state that land degradation is a human-induced process and identify human interference as the basic cause of land degradation. Others state (United Nations Environment Programme, 1993, 133-34) that it may be caused by a variety of anthropogenic pressures including deforestation, overgrazing, unsustainable agricultural practices and industrial activities and they can be broadly termed as social in character. According to (Chisholm and Dumsday, 1987, 8), human influence through agricultural, pastoral and urban activities is critical in determining the rate of degradation.

An early approach providing a conceptual and institutional perspective and analysing in depth, the issues of degrading lands and soil conservation bearing directly on policy was brought together by Chisholm and Dumsday (1987) in their edited volume on "Land Degradation : Problems and Policies." This volume was specifically concerned with Australia. It focussed on issues of regional degradation ranging from soil erosion and salinisation to recreational damage and loss of habitats. The main emphasis was on socio-political aspect rather than bio-physical. The book presents a broad multi-disciplinary perspective on the challenge of problems of degrading land, including the on-site and off-site effects of soil erosion, nutrient loss and salinisation, and the conservation policies needed to meet the challenge. The book determines the socially important elements of the problem of land degradation; defines the biological and physical setting of the problem; sets out ways of pursuing land management in the national interest taking into account the legal, economic, social, political and organisational issues. A parallel volume edited by Blaikie and Brookfield (1987) on Land Degradation and Society has endeavoured to assess land degradation in relation to people, economies and societies using the land. The authors place land degradation in a social context, where degradation has been defined as a decline in the productivity of land and labour that is mainly caused by the exploitation of nature for short-term gain. Using an approach termed regional political ecology combining the concerns of ecology and political economy to cope with the constantly shifting dialectic between society and land resources, the authors have placed land degradation in a historical, political and economic context. Also, they have recognised not only the co-existence of human interference and natural degrading processes, but also parallel natural reproduction and restorative management. Barrow's (1991) volume on Land Degradation : Development and Breakdown of terrestrial Environments has attempted to produce a work which is global in perspective, and which stressed the integration of environmental and human elements of land degradation. Similarly, Johnson and Lewis (1995) in their volume on Land Degradation : Creation and Destruction have dealt with the causes and forms of land degradation and the means of managing and controlling it. Land degradation has been defined as a decline in natural and biological productivity which is either irreversible or which may not be recovered for at least one human generation. They have analysed linkages between natural and human systems and recognised two categories of land degradation : unintentional/unforecast and intentional/creative. This division provides a basis for analysing the economic and cultural causes of land degradation. Creative destruction and destructive creation are presented as different sides of the same coin (sustainable development). Eden and Parry (1996) in their edited volume on Land Degradation in the Tropics : Environmental and Policy Issues gave equal weighting to human and bio-physical processes of land degradation, and stressed the need and importance of looking at complex bio-physical variables and underlying social causalities, in order to aid the design of appropriate policy intervention. This is one of the most exciting and urgent tasks for today's geographers and resource managers because tropical countries are suffering from localised damaged and degradation due to legacy of exploitation and environmental mismanagement.

The present volume attempts to outline the salient aspects of ecological degradation of land to signify a reduction of land productivity as a result of a combination of physical, chemical and biological aspects under diverse environmental conditions. However, there is not sufficient scientific understanding of the ecological processes for degradation in the developing world. The ecological processes refer to physical and biological stresses threatening the stability of social-plant associations in agricultural ecosystems. The stresses cause serious instabilities produced by erosion, salinisation, waterlogging, deforestation, fertility loss, soil structure change, acidification and pollution of soil. Although there are linkages between some of the process, most of them work independently of each other. The net result is that large areas have been transformed into wasteland, the product of desertification, soil erosion, waterlogging, salinisation and industrial exploitation. Therefore, it is urgently needed to reverse this trend and to restore the productive capacity of wastelands in order to meet the demands of increasing population and other developmental needs (Nagaraja and Gautam, 1996, 144).

Wasteland is generally unproductive, ruinous and devastated land which is normally uncultivated and sparsely populated. Broadly, wasteland is defined as a land which is presently lying unused to its optimum potential, thus leading to national wastage (Government of India, 1990, 11). According to Stamp (1962), wasteland may be defined as that land which has been previously used but which has been abandoned and for which not further use has been found. In several other definitions, being attempted by the economists, the economic potential and the actual returns from the land seem to have formed the basis of identifying the wasteland. Accordingly, any land which gives less than 25 per cent of its potential may be considered as wasteland. Bhumbla and Khare (1987) have defined wasteland as the land which are ecologically unstable, have nearly lost their top soil, and have developed toxicity in the crops and trees. Wasteland is classified into two categories : cultivable and non-cultivable. Cultivable wasteland is capable or has the potential for development of agriculture, pasture and afforestation but is not being used to its optimum level owing to constraints, such as salinity, alkalinity, soil erosion, waterlogging, absence of vegetation cover, occurrence, occurrence of abundant deposit of sand, unfavourable physiographic positions and human negligence. All these lands are put to some productive use after proper treatment and reclamation measures. On the other hand, non-cultivable land is that which is barren and cannot be put to any productive use, e.g. snow-covered and glacier areas and barren out crop hill ridges. If unchecked, these wastelands can lead to irreversible loss of the natural resources. The important causes of wasteland formation are desertification, soil erosion, waterlogging, salinisation and shifting cultivation, whose satisfactory measurements are yet to be made.

The process of desertification resulting into land degradation is mostly prevalent in the tropical drylands, where 72 per cent of the world's drylands are distributed : 32 per cent in Africa, 32 per cent in Asia and 8 per cent in South America. Over the past thousand years, more and more land has been lost to the desert because there has been a slow but steady decline in land quality that turns forest into savanna, and savanna into barren, unproductive scrub. This progressive downgrading is known as desertification. It is further defined as land degradation in arid, semi-arid and dry sub-humid areas resulting from various factors including climatic variations and human activities (Kassas, 1994, 1). It is the spatial extension of desert-like conditions as a result of man's impact on ecosystem of semi-arid regions. It occurs mainly in desert-boundary regions and involves a complex of physical geographical processes having negative consequences on landuse, and ultimately leading to a strong disturbance of the normal ecological equilibrium. Thus, desertification is a process of impoverishment of arid, semi-arid and sub-humid ecosystems leading to reduced productivity of plants as well as it is also process of ecological degradation by which the productivity of land is lost or substantially diminished. Erosion (wind and water) and chemical degradation cover 95 per cent of processes of land degradation. Erosion always starts from the denudation of social surface for long periods of time, that makes it sensitive to the degradation forces, strong winds and running water (Houerou, 1994, 2). Consequently, grazing lands cease to produce palatable pasture, dryland agriculture fails and irrigated fields are abandoned due to salinisation, waterlogging and other forms of social deterioration. In sum, local edapho-climatic conditions and land management are of paramount importance in the process of land degradation. If land management remains unchanged these arid lands will be turned into man-made deserts in less than 150 years. This phenomenon, therefore, is one of the greatest pre-occupations faced by mankind today, and should receive due attention.

Soil erosion is occurring in developed and developing countries alike and is regarded as the most serious and widespread form of land degradation. The erosion has been severe from the earliest civilisations in the Middle East to the most recently cultivated lands in the America, Australia and South Africa. Although erosion is triggered by natural phenomena such as climatic change and organic movements, human activities involving the large-scale removal of the vegetation and exposure of the soil are the common agents (Ebisemiju, 1989, 136). In much of Asia, Latin America and Africa, change from forest to crop field is accompanied by an increase of surface run-off and an acceleration of soil erosion. Since 1940's, the continued deforestation in the Himalayas and north-eastern Thailand has created serious soil erosion problems in the lower Gangetic plain and Thailand. The Yellow river is depositing 1.2 billion tons of soil in each year. Worldwide farmers are losing an estimated 24 billion form of top soil from cropland each year (Brown, 1994, 1). The most catastrophic form of soil erosion is gullying, found as frequently on pastures as on cultivated land (Young, 1998, 104). Gullies are a common phenomenon in many parts of tropical world and gully erosion is probably the most serious problem. It is the extreme manifestation of the acceleration of erosion processes beyond the normal rate and is the indication of a deteriorating environment (Ebisemiju, 1989, 136). Ravines are advanced stage of gully erosion. They are noted for their enormous size and depth with vertical sides. Land degradation by ravines and gullies is widespread in India, Madagaskar, Lesotho, Malawi, Ethiopia, Haiti, Central Nigeria and the Himalayan foothills of Pakistan. Gully erosion due to river channel trenching has threatened vast tracts of agricultural land in these countries. The damage is greatest in the alluvial plains of the semi-arid and arid regions and most serious where it threatens precarious subsistence – oriented agricultural systems in the developing world (Haigh, 1984, 543). Zones of severe ravine trenching are found in the Yamuna-Chambal, Gujarat, Punjab Shiwalik Foothills and Chhotanagpur. There are other areas of substantial ravine erosions in India like Mahanadi valley, upper Son valley, upper Narmada and Tapi valleys, all on the southern fringes of the Gangetic plain. The widespread and conspicuous dambo gullying in northeast Zimbabwe is the result of complex interactions between human and physical factors including dambo clays, bed rock weathering, livestock increases, extension of dryland cultivation, and rapidly increasing human populations (Whitlow, 1992 and 1994; Whitlow and Campbell, 1989).

The resource value of soil in the agricultural productivity is its capacity to produce satisfactory yields of cultivated plants. Soil fertility is the ability of soil to provide all the essential plant nutrients in available form. It includes site characteristics of the soil surface and the internal chemical, biological and physical properties of the soil. Soil fertility decline covers changes in these properties which lower its productive potential. The chemical degradation refers to loss of plant nutrients and also acidification; physical degradation to adverse changes in social physical properties, including structure and water holding capacity; and biological degradation to lowering of soil organic matter, with associated decline in the activity of soil fauna (Young, 1998, 108). Plant nutrients are concentrated in the vegetation itself rather than in the soils from which nutrients tend to be rapidly leached by heavy rainfall in combination with high temperatures. Clearing of the forest vegetation breaks the normal cycling of plant nutrients. The ash resulting from the burning of branches and debris reduces soil acidity and makes some mineral immediately available for crop production; but, fertility tends to fall off rapidly and yields in the third year of cultivation are often only half of what is obtained in the first year. The decline in fertility and therefore of crop yields is the important reason for land degradation. Evidence is accumulating that fertility decline is extremely widespread in tropical world, particularly in those areas where shifting agriculture is commonly practised (Stoorvogel and Smaling, 1990).

Of particular importance to tropical agriculture are the two problems typically associated with canal-based irrigation, i.e. waterlogging and salinity. Waterlogging occurs due to progressive rise in groundwater table. On reaching the water near the root zone, the availability of oxygen to plant root system gets reduced. Ultimately, it results in increased salinity as it brings the dissolved salts and minerals of the soil close to the surface. The degradation is produced by increase and accumulation of salts in the soil. The salinised soil appears while on which stunted growth of crop takes place. Thus, land degradation due to salinisation and waterlogging is caused by the interaction of a number of factors such as : lack of provision of drainage in irrigated areas, over use of water in areas where only light irrigation can be sustained, seepage of water from canals and distributaries, siltation of drains, lack of land development before provision of irrigation, lack of lining of canals and distributaries, adoption of unsuitable cropping patterns, poor water use practices, difficulty in implementing landuse policies rigidly and poor conjunctive use of groundwater facilities (Bowonder, Ramana, Ravi and Srinivas, 1987, 331). Consequently, in most canal irrigated areas, soil deterioration through salt accumulation has assumed serious proportions. In some of the countries the problem is even worse. Area suffering from salinisation is 30-40 per cent in Egypt, 30-35 per cent in Syria, 25 per cent in Cyprus, 50 per cent in Iraq and 15-20 per cent in Austria (Navlawala, 1995, 38). Salinity affects productivity on about 20 to 30 in ha. or about 7 per cent of the world's irrigated land. Without exception the Near East countries from Libya to Pakistan are afflicted by salinisation problem. At risk are the Nile, Tigris, Euphrates and Indus basis. In India, about 2.5 per cent in ha. is under alkali soils and 4.5 in ha. under saline soils spread over Punjab, Haryana, Uttar Pradesh, West Bengal, Andhra Pradesh, Gujarat and Maharashtra. This indicates that the problem is widespread and exists practically all over the world, and both salinisation and waterlogging can be prevented through good irrigation design and practice (Johnson and Lewis, 1995, 79).

Because of critical role of minerals in development, large-scale mining operations are inevitable. Therefore, today, mining continues to be a key component of the economies of many Third World countries, and, in some cases, constitutes their economic mainstay. The expansion of mining has also been given renewed impetus by the World Bank, IMF and the World Trade Organisation in their Structural Adjustment Programmes and Policies for those countries under their tutelage in its drive to open up the South to the multinationals, including the big mining companies (Idris, 1998, Editors Note). The result has been a massive expansion in mining, especially in Africa and Asia. This has involved total removal of materials as in the case of sand and gravel extraction, low-grade oil, coal, phosphates, copper, slates, limestone and others. Unfortunately, this mining boom has been wreaking havoc in many countries of the South. Subsidence and collapse are common in most areas that have been actively mined. Subsidence alters drainage patterns and damages surface infrastructure, and collapse drastically alters the surface configuration of the affected areas and also degrades the affected areas. Land is altered as gravel pits and scars are formed due to the result of huge depressions here and there. Thus, mining leaves behind a degraded and pockmarked landscapes devoid of soil cover producing blighted soils and fouled water. The total land area affected by this process is of the order of 50-11 ha. in the world.

About the volume

The volume contains a total of twenty-two articles : four on wasteland, five on desertification, six on processes leading to land degradation and seven on land-man relationships and land resource depletion. They are arranged accordingly.

The paper by Chand, Upreti and Mehta deals with the identification and distribution of wasteland in the Kalpanigad basin of eastern Kumaun Himalaya. Here, the wasteland in general is reported to be 34 per cent of the total basin area. The cultivable waste, being the important feature of the wasteland, ranges above 61 per cent in the areas. However, there are wide variations at the village level ranging from a minimum of 13 per cent to a maximum of 75 per cent. In general, the majority of villages have 20 to 40 per cent wasteland.

The wastelands of Churu district in Rajasthan have been classified into seven categories which cover 61 per cent of the total area : sand desertic, degrading pastures/grazing lands, land affected by salinity and alkalinity, underutilised/degraded notified forest lands, undulating upland with or without scrub, mining/industrial wasteland and barren rocky/stormy waste/short rock areas. Sands-desertic cover the largest area (53.05 per cent of the district and 87.10 per cent of the wastelands in the district). The authors are of the view that much of the sandy terrain is not a waste in the strictest sense of the term, although the land is in a degraded state. Locally available plant species having some economic value have been identified and suggested for planting in different types of wastelands for their potential development. It has also been realised that rehabilitation of such degraded land holdings needs immense efforts and peoples participation.

The nature and extent of wastelands and factors affecting their development in the Kopai drainage basin area of Birbhum district have been studied by Mukhopadhyay. It is interesting to note that the upper and lowest most part of the basin have the highest percentage of wasteland. It has been observed that villages of the upper basin having the highest percentage of wasteland have also a high amount of land slope and ruggedness number. Though the percentage of wasteland is highest in the lower part of the basin, the slope and ruggedness number are not significant. Here, floods are the determinant factor for wasteland generation. Large patches of lateritic wastes occur in different parts of the mid-basin area. The wastelands have been divided into five parts : (1) stony wasteland, (2) laterite-covered wasteland, (3) river-borne wasteland, (4) marshy wasteland, and (5) anthropogenic wasteland. Their pattern of distribution as well as reclamation measures have also been suggested at the foothills of Ajodhya.

Maiti has attempted to study land and soil characteristics of Gajaburu micro-watershed Purulia for the development and management of its wastelands. Due to erosion, big gullies and ravines exist in this area. The soil is poor, undulating and excessively drained. The area is predominantly inhabited by the tribals where modern agricultural practices have not been adopted much. Low fertility of land, coarse texture of soil, lesser population density, relatively rugged terrain and soil erosion have been the sources of the determinants of wasteland in this area. The lands of this watershed have been divided into six capability classes, i.e. capability Class II, Class III, Class IV, Class VI, Class VII and Class VIII. The Class II, III and IV lands require conservational measures to make them suitable for growing crops after thorough management practices. The Class VI, VII and VII lands are not favourable for any crops. However, afforestation, agro-forestry or agro-horticultural programmes along with measures on gully pugging are required for development. The author is of the view that an watershed is the most scientific unit for efficient management of land and water resources as it is basically an agro-climatic unit with more homogeneity of land and other resources as compared to the revenue district.

The process of salinisation and alkanisation of both soil and water is the part and parcel of desertification. The paper by Szabolcs on Salinisation of Soil and Water as an Important Factor in Desertification mentions that a great part of existing deserts of the world are salinised and it contributes to the processes of desertification. The paper summarises the factors of salinisation such as deforestation, overgrazing, changing landuse pattern, depletion of freshwater layer and their environmental consequence. It is termed as secondary salinisation as developed alone to human activities. The paper also recommends different methods to combat the process of secondary salinisation and alkalisation.

Similarly, Singh in his paper on Assessment and Mapping of Desertification in Turkmenistan has highlighted the role of remote sensing in the identification and mapping of different types of desertification caused by natural and anthropogenic processes in the Karakum desert of Turkmenistan. It is suggested that anthropogenic processes are more dominant and widespread affecting 80 to 90 per cent of the different ecosystems in the desert than natural desertification processes. The suitable measures like crop rotation, judicious methods of irrigation like drip and sprinkles, rational grazing, and stabilisation of sand dunes and sand drift have been suggested for combating desertification.

The paper by Lado elucidates principles of desertification with reference to Arid and Semi-Arid Lands (ASAL) of Kenya and assesses the impacts of intensive crop-husbandry and existing farming systems on desertification. It includes irrigated agriculture, over-cultivation, shifting agriculture, intensive dryland farming practices, cultivation of cereals beyond the agro-ecologically set zones. The study also classifies existing crop area by type, including both tree and agricultural crops. But, in order to provide a sound scientific basis for rehabilitation and sustainable development of arid and semi-arid ecosystems, the study identifies the potential use of specific crops in desertification control and also proposes the appropriate monitoring mechanisms required for improved environmental management in crop production systems.

The paper on Problems of Desertification and Strategy for Combating it with Reference to Rajasthan by Sharma stresses on the causes of desertification in general and its relevance to Thar desert in particular as well as the strategy to solve them. On the basis of precipitation, a number of morphogenetic regions have been identified and theories concerning the causes of desertification in Rajasthan have been discussed, namely, (a) stationary theory, (b) sand-spread theory, (c) man-induced sand reactivation theory and (d) over utilisation of dry ecosystem through modern science and technology theory. The author stresses that alarming rise in human as well as livestock population in Rajasthan has led to over utilisation of the ecosystem like overgrazing, deforestation and exploitation of forest resources. The paper also deals with strategies to be adopted to check the outward growth of the desert, a few them being afforestation, groundwater utilisation, adoption of agricultural method apropos to the climatic environment, development of appropriate horticulture and animal husbandry along with a micro-watershed planning as a policy.

The impact of Desertification and overall Integrated Management of Resources in the Kuchch district of Gujarat has been taken up by Sen. The extent and form of desertification in the district is described with a special emphasis to the Banori region which is characterised as a barren, tireless vegetation low lying tract. The region has been divided into 13 geomorphological units on the basis of data procured through sun-synchronous satellite, namely, landsat of four lands and filters along with 5 per cent of ground truth verification. Based on this, the extent of desertification and its processes have been discussed and illustrated. Vegetational ecosystem along with geology, ground water resources and soil types have been described. Based on the limitation of resources and the constraints imposed by the arid environment, a few developmental planning strategy for the area have been suggested like proper and judicious utilisation and development of minerals, dairy, fishery, port, tourism and agricultural resources.

Prasad has studied land degradation caused by the shift of the Kosi channels within the two embankments. It is based on the configuration plans of the channels in the area and year to year variation in terms of area as well as channel movement. Area under study has been divided into four 'Channel Reaches' on the basis of varying fluvial processes and their characteristics are described. It has been indicated that the oscillation of channel bed is still continuing even after the engineering methods were adopted 40 years ago. In this process of Channel wandering, a large chunk of area is annihilated and, in this context, the paper highlights the problems of land degradation in the Kosi embankment area along with some suggestions to overcome.

The extent of waterlogging and consequent degradation of agricultural land in the Karnal district has been taken up by Singh and Singh. The study is based on secondary data on ground water level, different sources of its exploitation and the extent of area which has become saline due to excessive irrigation. The expansion of alkalinity and salinity has also been taken into consideration. The authors stress that unfavourable topography in bowl-shaped with depression with centre is the cause of the problem. The rise in ground water level in monsoon months leads to upward movement of sub-soil salt giving rise to salinity at the surface and affect the fertility of soil. The other cause of waterlogging is water accumulation on the surface through irrigation. The paper also demonstrates the genesis of waterlogging problem along with its spatial and temporal extent and reclamation methods. This problem has been observed to have started in fresh groundwater areas, whereas its increasing trend has been traced in marginal saline ground water belt of south western part of the district.

Prasad in his paper on "Impact of New Mining Technology on Land Degradation in Coalfields of Chhotanagpur" has highlighted the conspicuous aspects of the environmental impacts of coal production in the Chhotanagpur region. The region, covering the districts of Garhwa, Palamu, Hazaribag, Giridih, Chatra, Bokaro, East and West Singhbhum, Ranchi, Gumla and Lohardaga, is the treasure of minerals in Bihar. The study specifically deals with the CCL and BCCL (Central Coalfields Ltd. and Bharat Coking Coal Ltd.) which account for most of non-coking and the coking coalfields. They constitute the principal scene of intensive mechanised mining in both old and new coalfields. The region accounts for about 25 per cent of the total coal reserves of the country and has spearheaded industrial revolution not only in the region but in the whole country. The growth rate of production has increased after nationalisation in 1972 due to induction of new technology and resorting to open-cast method (higher level of mechanisation). However, it has proved to be more destructive to the environment and the devastation continues unabated. The basic open-cast equipments are shovels, dumpers, dozers, drills and draglines which have remarkably increased after 1972. The study highlights that the region has significantly gained economically but seriously suffered environmentally. The new technology of mining and associated operations has adversely affected the entire ecosystem of the coalfields. They are expressed as devegetation, topographic deformation, edaphic loss, drainage derangement, air, water and noise pollution and dereliction. The author warns that coal mining renders an area of 600 ha. of land per year biologically unproductive and may rise to 1400 ha. by the beginning of millennium. The effective control of the environment in such a large area is financially a gigantic task, although, complete restoration of the coal areas divasted by coal mining is likely to become a compulsion for the concerned mining companies.

Marh and Pathania have attempted to throw light on the problems of environmental degradation caused by state mining in Khanyara Mines in Kangra district of Himachal Pradesh. They have stated that the unscientific mining method and the tendency of mine owners to extract maximum possible materials has caused considerable environmental degradation. It has been created due to inappropriate way as well as over exploitation of the minerals, not suitable to the terrain and physical condition of the area. The main effects identified are landslides, land subsidence, accumulation of waste material on hill slopes, destruction of forest cover etc. It is stressed that the present level of production and supply can be reasonably achieved by restructuring the mining area to judiciously selected blocks, adopting scientific extraction procedures and evolving a more rational leasing procedure for mining rights.

The magnesite mining, located at Jhiroli village in Almora district (U.P.), was initiated in 1971 by the open-cast method using power shovels and bulldozers and processing it in shaft kilns to produce dead burnt magnesite. The main environmental impacts are on the physical and chemical environments, i.e. earth's atmosphere, water processes, flora and fauna, landuse recreation, aesthetic and human use value. However, one of the most serious impacts is soil impoverishment and loss which leads to a reduction in plant biomass, land's carrying capacity, crop yields and eventually to the intensification of sterile biological desert. The study by Ahmad shows higher rate of sedimentation than erosion which is deposited as silt in the drainage channel. The heavy amount of solid wastes is leading to the siltation of perennial water resources and agricultural lands leading to a decline in agricultural production. In view of this, strategies have been developed to restore damaged mined ecosystem using an ecological approach which takes into account the inter-relationship between land, water, air and all living things to ensure ecologically sustainable development.

Around cities and metropolitan cities, land is scarcely found. Due to growing population multi-storied buildings are the common features. High rise dwellings are mainly dependent on the supply of building materials and one of the most important is the availability of bricks prepared from loamy and silty soil. In this background Singh and Jag Mohan have analysed the status of brick kiln industry of Delhi and its impact on the deteriorating land quality as well as on the shrinkage of agricultural land associated with air pollution by dust and smoke. Increase in built-up areas of towns and cities is an ongoing process which involves acquisition of land by land developers and its conversion into urban uses. The study by Kapoor emphasises on land acquisition in Delhi Union territory during 1951-1989. The growth of Delhi and yearly influx of immigrants leads to large-scale acquisition of land for development purposes. The study is divided into 8 time periods, that is, 5 years each. The area of urban Delhi has increased several fold and new residential, industrial, commercial and recreational areas have come up on a scale never witnessed before. In the beginning land was acquired in small lots and there were not so many problems. But when the land was acquired in big lots, there were several problems. The problems have been analysed. The analysis drawn from the study shows that all those villages which were contiguous to Delhi Urban Area were fully acquired first and the process is continuous one. The land prices have shot up in and around the acquired area and an increase in land prices in an indication of DDA's failure to eliminate the scarcity of land.

The study by Chandna is an attempt to analyse and determine the loss of agricultural land in West Delhi for the time period 1981-91, especially the spatial pattern of encroachment and the type of land being converted. It is stated that many villages along the fringe of West Delhi have been swallowed by the city and their agricultural sites have been turned into residential sites. The encroachment follows a linear pattern along the Najafgarh drain and road. It has been found that the decade of 1981-91 is most important regarding the acquisition of land both in terms of number of villages and the amount of land acquired. It is also found that the highest quality land is being acquired maximum than the lowest quality.

The most serious problem associated with urban sprawl is the inefficiency of landuse. An attempt has been made by Phanse, Trivedi and Awasthi to highlight the problems associated with urban sprawl in a metropolitan city of Indore. It has resulted in loss of agricultural land which happens to be fertile black cotton soil. This loss amounts to 872 hectares. An equally serious matter is the loss of forest cover around the city. In 1991, Indore attained the status of a metropolitan city and every increase in population pressure will result in conversion of more agricultural land to residential and other urban user. The alternative lies in intensive use of arable as well as urban land to check further expansion of the city.

Because of population pressure on land resources, depletion has become a common phenomenon in the Third World countries. However, recently, our common land has also been in the process of being used, especially in the tribal dominated society. The study by Thakur and Prakash had dealt with the growth decline of village common land in the Jhabua district of Madhya Pradesh between 1971-91.

The spatial patterns of growth and decline in the distribution of common land have been mapped and analysed, and the association between socio-ecological variables and common land has also been determined. Narasimha Murty has analysed loss of agricultural land and changing utilisation of resource base in three selected villages of Srikakulam district (Andhra Pradesh). The author has suggested that agricultural land should not be transferred and put in for other than agriculture user. The changing resource base provides the farmer more benefit in a short period, the long-term disadvantages will be the gradual decrease of the soil subsurface. He also suggests that uncultivated land or barren land should be used for poultry farming and dairy farming. Ram and Yadav assessed the changes that took place in crop failure pattern between 1968-69 and 1987-88 of rainfed cropped area in Hissar tehsil of Haryana State. It is demonstrated that decrease in failure index of winter crops was higher than summer crops.

Saroj K. Pal analyses the Land-man relationship elaborately and he further discusses the development prospects for India in detail.

REFERENCES

Barrow, C.J. (1991), *Land Degradation : Development and Breakdown of Terrestrial Environments*. Cambridge : Cambridge University Press.

Bhumbla, D.R. and Khare, A. (1984), *Estimates for Wasteland in India,* Delhi : Society for Promotion of Wasteland Development.

Blaikie, P. and Brookfield, H. (1987), *Land Degradation and Society,* London : Methuen.

Bowonder, B., Ramana, K.V., Ravi, C. and Srinivas, C. (1987), "Landuse, Waterlogging and Irrigation management", *Landuse Policy*, 4: 331-341.

Brown, L.R. (1994), "Food", *Environmental Education Dossiers*, No. 9, Barcelona : Centre Unesco de Catalunya.

Chisholm, A. and Dumsday, R. (1987), *Land Degradation : Problems and Policies*. Cambridge : Cambridge University Press.

Douglas, I. (1987), "Book Review on Land Degradation and Society", *The Professional Geographer*, 615-618.

Ebisemiju, F.S. (1989), "Thresholds of Gully Erosion in a Laterite Terrain, Guyana", *Singapore Journal of Tropical Geography*, 10(2) : 136-143.

Eden, M.J. and Parry, J.T. (eds.) (1996), *Land Degradation in the Tropics : Environmental and Policy Issues*. London : Pinter.

Government of India (1990), Developing India's Wasteland, New Delhi : Ministry of Environment and Forests.

Haigh, M.J. (1984), "Ravine Erosion and Reclamation in India", *Geo Forum* 15(4) : 543-561.

Houerou, H.N.L. (1994), "Desertification and Desertilisation : Climatic Fatality or Human Mismanagement", *Environmental Education Dossiers, No. 7 on Desertification*, Barcelona : Centre Unesco De Catalunya.

Idris, S.M.M. (1998), "Editors Note on Social and Environmental Impact of Mining in the Third World", *Third World Resurgence*, 93.

Johnson, D.L. and Lewis, L.A. (1995), *Land Degradation : Creation and Destruction*. Cambridge : Blackwell.

Kassas, M. (1994), "Desertification: *Environmental Education Dossiers No. 7.* Barcelona" : Centre Unesco De Catalunya.

Navalawala, B.N. (1995), "Waterlogging : Problems and Solutions", *Yojana*, September, 38-40.

Nagaraja, R. and Gautam, N.C. (1996), "The Role of Remote Sensing and GIS in Wasteland Management in India", in M.J. Eden and J.T. Parry (eds.) *Land Degradation in the Tropics : Environmental and Policy Issues*, London : Pinter, 144-154.

Stamp, L.D. (1962), *The Land of Britain : Its Use and Misuse*, Third Edition, London : Longmans.

Stoorvogel, J.J. and Smaling, E.M.A. (1990), Assessment of Soil Nutrient Depletion in Sub-Saharan Africa : 1983-2000. Staring Centre Report 28, Wageningen, The Netherlands : Staring Centre.

United Nations Environment Programme (1993), Environmental Data Report, 1993-94. Oxford : Blackwell Publishers.

Whitlow, R. (1992), "Gullying within Wetlands in Zimbabwe : An Examination of Conservation History and Spatial Patterns", *South African Geographical Journal*, 74 : 54-62.

Whitlow, R. (1994), "Gullying within Wetlands in Zimbabwe : Morphological characteristics of Gullies", *South African Geographical Journal*, 76 : 41-48.

Whitlow, R. (1994), "Gullying within Wetlands in Zimbabwe : Gully Development and Factors Influencing Gully Growth", *South African Geographical Journal*, 76 : 41-48.

Whitlow, R. and Campbell, B. (1989), "Factors Influencing Soil Erosion in Zimbabwe : A Statistical Analysis". *Journal of Environmental Management*, 29 : 17-29.

Young, A. (1998), *Land Resources : Now and for the Future*, Cambridge : Cambridge University Press.

1

Wasteland Identification and Distribution in Kalpanigad Basin, Kumaun Himalaya

R. CHAND, T.C. UPRETI and B.S. MEHTA

The term wasteland has been defined in various ways by different workers. Traditionally, the land which was out of the choice of initial settler around his first land occupance unit might have been described by him as wasteland. The general opinion about the wasteland in native self-reliant communities may differ in accordance to their socio-economic pattern of life and dependence with local environment. The meaning and dimension of wasteland in a highly mechanised agricultural settlement or metropolitan societies will also be different. It is, therefore, a dynamic concept which is defined according to the contemporary landuse pattern and on an economic scale of the society at one given point of time. When viewed in the context of ever-increasing population pressure on land and declining capacity of land to support these populations, Stamp realised in principle that "no land is wasteland" taking into account the optimum use of all the available pieces of land.

While conducting landuse surveys in England he defined wasteland as that land which has been previously used but which has been abandoned and for which no further use has been found. Wasteland Survey and Reclamation Committee, Ministry of Food and Agriculture, India (1961), has defined wasteland as the land which is either not available for cultivation or is left out without being cultivated like fallows and culturable waste. The Natural Resource Division of the Planning Commission constituted in 1963 also agreed that the major types of wastelands include land not available for cultivation, barren and uncultivable waste. It also includes other uncultivated land excluding fallows and culturable waste, permanent pastures and land under miscellaneous trees and fallows. In the above definition wasteland is considered to be all those categories of land which are not under the use of forest pasture and cultivation. Shafi (1968) also defined wasteland for the purpose of extension of cultivation to other uncultivated land excluding fallows and fallows other than current fallow. Fallow land is, therefore, not considered under wasteland by him. The definition given by the Bureau of Economics and Statistics (1979) also followed Shafi's vision and defined wasteland. All land available for cultivation whether not taken up for cultivation or abandoned after five years for one reason or the other was considered as wasteland.

A more scientific outlook towards wasteland is propagated through the Society for Promotion of Wasteland Development, New Delhi, in recent years. Their definition of wasteland includes the following characteristics:

> (1) ecologically unsuitable, (2) whose top soil has been nearly completely lost and (3) which has developed toxicity in the root zones for growth of most plants, both annual crops and forests (Bhumbla and Khare, 1984). The National Remote Sensing Agency, Hyderabad (Gautam Narayan, 1988) has described wasteland as that land which is presently lying unused or which is not being used to its optimum potential due to some constraints. The National Wasteland Development Board set up by the Indian Government has finalised the definition of wasteland under the chairmanship of M.G.K. Menon on April 24, 1986. The board has recognised wasteland as that land which is degraded and is presently lying unutilised except as current fallows due to different constraints (Gautam and Narayan, 1988).

The above definitions have thrown sufficient light as to how the concept of wasteland was conceived and developed. It seems that the use factor including all use conditions in past, present and also future outlook had been dominant in the beginning. However, the wasteland gained new and much important interpretation through the appeal of environmental degradation and optimum potentiality of land in recent years (Bharat, 1987; Chaudhari, 1985; Patel, 1987; Prem Bhai, 1985; Singh, 1989; Shah, 1984; Yadav, 1986, 87).

The Himalayan Region

In spite of the above progress of work in the country, the nature of wasteland in the Himalaya is not clearly perceived and the relevant information is not available. Various schemes of Integrated Mountain Development including the World Bank projects are working checking soil erosion, afforestation, etc. But these programmes have been undertaken without conducting systematic surveys of landuse and inventory of land resource data. Some scantly attempts on the denuded wastelands are made by Gupta (1982, 84), Gupta Tejwani (1980), Shah (1982) and Gupta and Arora (1985). The land capability classification is also attempted by Jalal (1975, 76), Khybri (1980) and Shah (1985). These studies have not considered wasteland in particular. Afforestation of village community lands with the participation of village people has been already started by the Dasholi Gram Swarajya Mandal, Gopeswar, Chamoli and other sister institutions. But it can be followed only by the organisational will of that order. Wasteland problem has often received mention in the context of the degradation of the Himalayan environment. But there are no definite estimates of wastelands available. The wasteland identification project which has covered 146 districts in the country has not taken up any district from the Himalayan region. The information provided by the Revenue Department reveals a very old situation. In reviewing the overall situation, it is felt necessary to undertake a specific research programme on wasteland in the Himalaya. This paper is an abstracted version of the wasteland Management Project undertaken by the first author of this paper which is sponsored by the Department of Environment, New Delhi.

Selection of Study Site

The Kalpanigad, a tributary of the Ramganga (E) of river Kali in the eastern part of Kumaun Himalaya, is selected for the present investigation. It comes under the Lasser Himalayan zone of district Pithoragarh encompassing portions of Bin and Munakot blocks of tehsil Pithoragarh and Kanalichhina of tehsil Didihat. The catchment area of Kalpanigad is located between Latitude 29°30'28" and 29°43'28"N and Longitude 80°8'30" and 80°17'15" E covering an area of 120.04 sq. km. The basin boundaries are demarcated in the north by the Lori range and the Dwaj-Saurlekh range in the east. The southern boundary is shared by the Barma Dhura and Asurchula range separating it from the Sor valley. In the north-west Musgal Danda forms a watershed (Fig.1.1). The District Census Handbook of 1981 has an aggregate of 18,975 persons living in 106 villages in the study

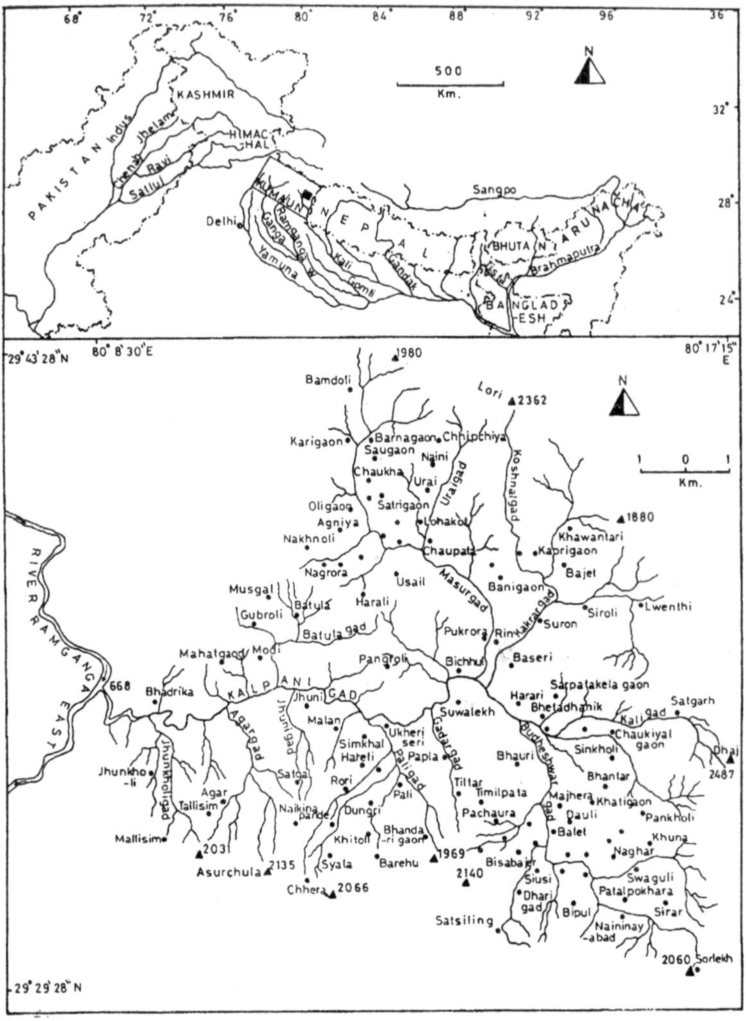

Fig. 1.1: Location of Kalpanigad Basin in Kumaun Himalaya (Small black rectangle).

area. The survey conducted during January-February, 1989, recorded 21,431 persons including 11,050 males and 10,381 females.

The basin is characterised by a dissected terrain and precipitous slopes very susceptible to erosion, landslides and other environmental hazards. Geologically, it negotiates the Berinag formation in north and overlying the Rautgara-Gangolihat, the Sor-Thalkedar (Mandhali) formation in the south. The Berinag thrust has caused an enormous damage, viz. displacement of rocks, slope failure, shifting of river terraces, mass movement of debris on slopes, ground cracking and, finally, the collapse of buildings and other properties in the area. The region has witnessed many tragic events even before the deforestation and mining activities started. The landslide of 1945 destroyed some 14 households and swept away and entire Ukheri hamlet of village Sheri. Many sites are experiencing continuous landslides year after year, even the houses in many threatened locations have been vacated.

In recent years, open-cast mining of magnesite and soap-stone in the Dewalthal area is being done in a very unsystematic way causing great damage to the terrain. The most sad part is that people of villages Lohakot, Usail, Chaupata and Joshigaon have started quarrying of soapstone in their agricultural fields. The result is that today the land affected by erosion and thus rendered barren is greater in extent than cultivated land. The extent of wasteland is, therefore, increasing at a faster rate. The consequences of a variety of development activities and other numerous processes initiated by human activity tend to disrupt the delicate balance of interacting natural processes. Human pressure in the basin is highest being 186 persons per sq. km. as per the 1981 census. The immediate mountain slopes around the villages are degraded. The forest area is only 11.47 per cent of the total reported revenue area excluding the reserve forest area of 20.20 sq. km. which are also not properly managed. There is an alarming situation regarding the fuel and fodder as about 65 villages in the area have absolutely no forest at all. They travel more than 15 km. to collect fuelwood. The availability of timber even for agricultural implements has been so scarce that 74 villages of the area have been reported to buy it from outside the region. About five villages buy it from even Jauljibi which is more than 100 km. away from the area. The pilot study has shown that more than 50 per cent families from the 19 villages have been permanently migrated to the Terai-Bhabar of district Nainital.

Identification Criterion

The identification of different types of wastelands depends upon the use factor or potentiality of such land. The estimates of wastelands also differ from one region to another depending upon the need of the society at a given point of time.

There are two fundamental ways of arriving at the wasteland

identification by including both culturable and non-culturable wastelands. Three categories of landuses are chosen here from the landuse classification of the Revenue Department.

Cultivable waste and area not suitable for cultivation are considered under culturable wasteland and rocky waste as unculturable wasteland. The basic idea behind this identification is to scrutinise the whole landscape in terms of all landuse systems so that the intensity or demand on wasteland may be related with all other components of the land systems. Thus, the classification of wasteland adopted here is in the following form:

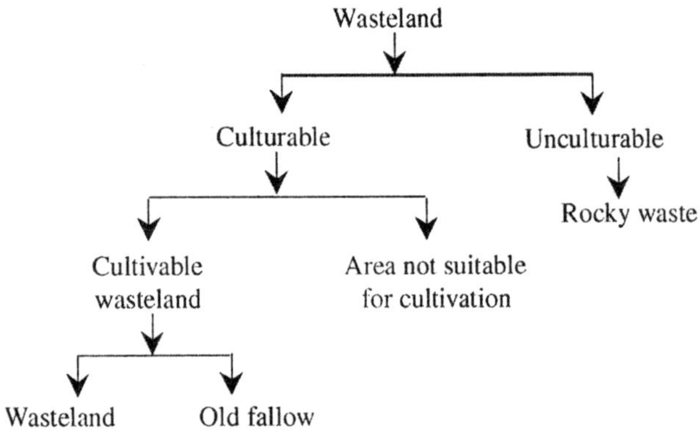

Composition of Wasteland

Table 1.1 shows that wasteland in the Kalpanigad basin was 3,400.74 hectare or 34.03 per cent of total reporting area in 1987-88. Under this, there are two categories, culturable wasteland formed substantially a high proportion of 3,133.76 hectare (92.15%) and the remaining 266.98 hectare (7.85%) came from the unculturable waste. The culturable wasteland includes 2,649.91 hectare of cultivable waste which is 84.56 per cent of total culturable wasteland. Cultivable waste is further divided into waste, fallow and current fallow. Waste is that category of land which is out of cultivation for more than five years. It is also known as *banjar* land by villagers and this category is the real wasteland. The marginal location and other socio-economic conditions of the farmers do not permit to bring back such lands under cultivation. The old fallow land which is also uncultivated for more than two years but less than five years also comes under the same conditions. Out of a total 2,649.91 hectare of cultivable waste, only 4.31 hectare (0.16%) is under old fallow. This shows almost negligible amount of fallow land which is not a common practice in the area. However, the current fallow is never included in the wasteland.

Area not suitable for cultivation is 483.85 hectare (14.23% of culturable waste) which cannot be utilised for agriculture unless a high cost of reclamation is involved. This is known as uncultivable banjar in land records. These are highly eroded uplands, steep slopes and undulating terrain combined with highly biotic interference. In most of the landuse classifications, this category is defined as unculturable land indicating least possibilities for agricultural use. But, personal verification of this land in village Sheri has revealed that at least these may be developed for vegetative cover. So this came under culturable waste in a broader sense. The rocky waste is unculturable in the true sense, that is, it cannot be put to any productive use. This category comprises 266.98 hectare (7.85%) of the total wasteland.

Table 1.1: Wasteland in the Kalpanigad Basin, 1987-88.

Patwari Circle	*Rocky Waste*		*Area Not Suitable for Cultivation*		*Cultivable Waste*		*Total Wasteland*		*% of Total Area*
	ha.	*(%)*	*ha.*	*(%)*	*ha.*	*(%)*	*ha.*	*(%)*	
Bisabajar	6.56	1.51	67.99	15.62	360.69	82.87	435.24	100	40.65
Jibi	0.99	0.58	2.92	1.70	167.39	97.71	171.30	100	54.75
Totanaula	58.14	10.31	81.15	15.19	424.86	75.31	564.15	100	38.43
Kharkdoli	-	-	-	-	7.01	100.00	7.01	100	22.63
Digtoli	3.51	2.28	-	-	150.42	97.72	153.93	100	27.19
Satgal	18.76	4.09	58.31	12.71	381.75	83.20	458.82	100	39.68
Chandak	7.24	16.72	-	-	36.07	83.28	43.31	100	42.01
Bangachhina	56.18	27.60	2.54	1.25	144.83	71.15	203.55	100	28.44
Dewalthal	20.59	7.68	9.76	3.64	237.83	88.68	268.19	100	25.81
Bamdoli	2.60	1.16	83.70	37.45	137.22	61.39	223.52	100	38.92
Bisonakhan	15.31	3.83	123.32	30.89	260.58	65.27	399.21	100	31.67
Siroli	68.61	30.09	53.56	23.50	105.83	46.42	228.00	100	18.96
Modi	8.48	3.47	0.60	0.25	253.43	96.29	244.51	100	49.75
Total	**266.98**	**7.85**	**483.85**	**14.232**	**2649.91**	**77.92**	**3400.74**	**100**	**34.04**

It is noteworthy that pasture land is not treated as wasteland in the revenue records. Most people have included pasture or village community land under wasteland. If these are included, the estimate of wasteland for the Kalpanigad basin exceeds (53.82%), more than half of the total reporting area. However, these are reserved for grazing purposes and also act as recreational ground for the village. Of course, these lands are required to be managed scientifically and can be taken for plantation only after the consent of the village community.

Figure 1.2 describes the composition of wasteland and their proportionate distribution at the Patwari circle level. One of the dominating features is that cultivable waste has contributed between 61.39 per cent in Bamdoli and 100 per cent in Kharkdoli with the exception of Siroli (46.42%). Area not suitable for cultivation is also found ranging up to 37.45 per cent in Bamdoli but in most of the cases it is found below 15 per cent. Kharkdoli, Digtoli and Chandak have no area under this category. Rocky waste has a lower proportion of 0.58 per cent in Jibi and 10.31 per cent in Totanaula. Chandak and Bungachhina have only high proportions being 16.72 and 27.60 per cent respectively.

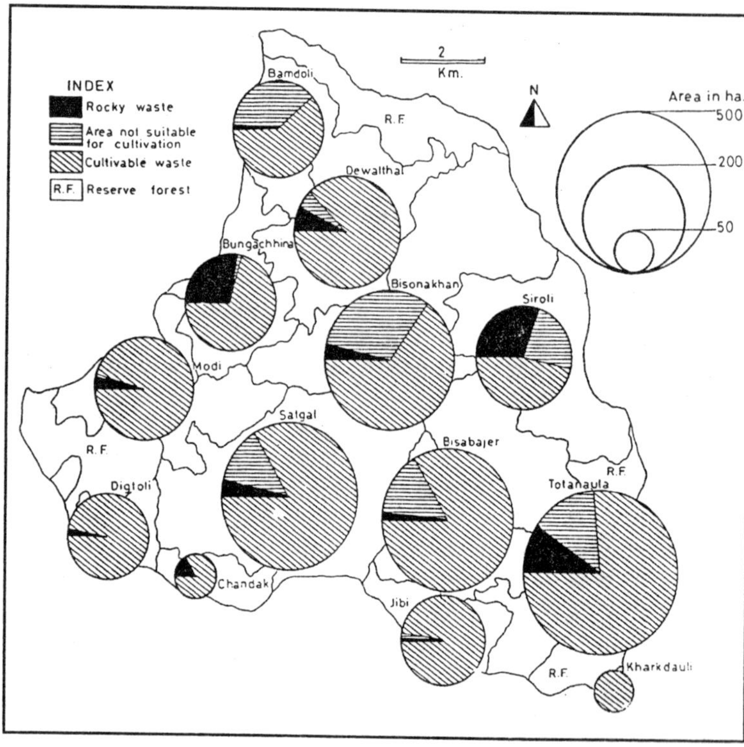

Fig. 1.2: Kalpanigad Basin: Patwari Circle-wise Composition of Wasteland. 1987-88.

Distribution of Wasteland

It can also be seen from Table 1.1 that the distribution of wasteland is observed as high as 54.75 per cent in Jibi while Siroli has the minimum, 18.96 per cent. There are seven Patwari circles which fall above the average of 34.04 per cent for the whole region. Modi (49.75%), Chandak (42.01%), Bisabajar (40.65%), Satgal (39.68%), Totanaula (38.43%) and Bamdoli (38.92%) are other six areas lying above the average.

Figure 1.3 is drawn to represent village-wise distribution of wasteland. There are wide variations in the distribution at village level from a minimum of 12.9 per cent in village Narkholi to the maximum 74.50 per cent in Suwalekh. The land of two uninhabited

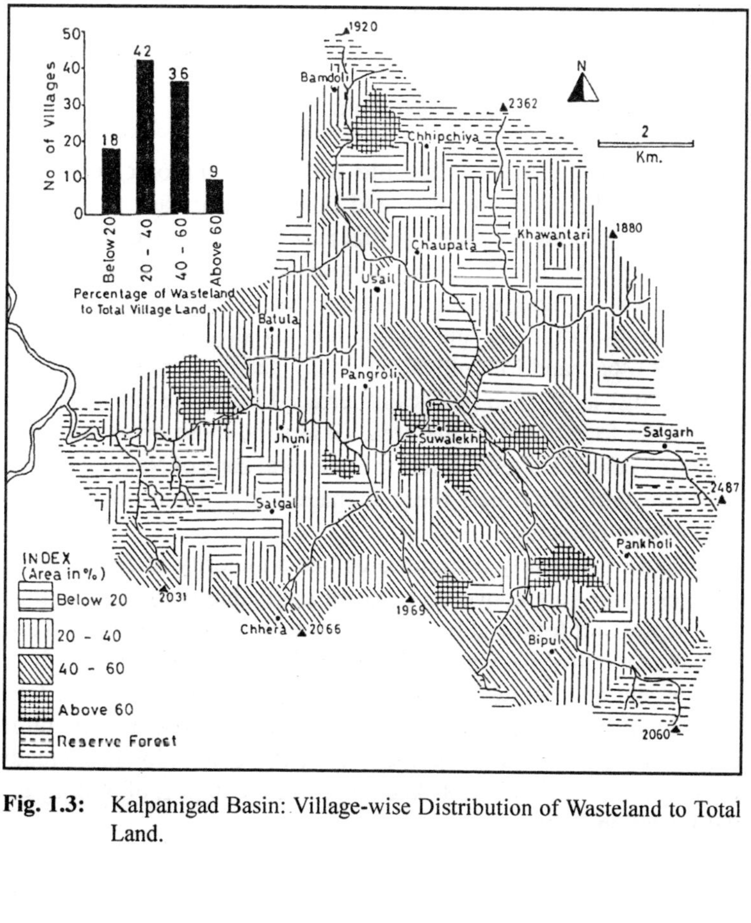

Fig. 1.3: Kalpanigad Basin: Village-wise Distribution of Wasteland to Total Land.

villages is found to be lying entirely waste. Nine villages have above 60 per cent wasteland and 36 villages also fall in the category of 40 to 60 per cent wasteland and remaining 18 villages out of a total 106 villages of the area fall below 20 per cent wasteland.

These estimates of wasteland are quite above the national averages of the wasteland data which is being derived from the remote-sensing method. Based on the satellite remote-sensing data of 1980-82, the National Remote Sensing Agency, Department of Space, Government of India, estimated 16.21 per cent wasteland for the whole country. The high percentage of area under Jammu and Kashmir (60.44%) is mainly due to a considerable area being under permanent snow. However, it is expected that the national wasteland identification project which is using the enlarged Landsat Thematic Mapper (TM) on the 1:50,000 scale would provide more accurate information.

Cultivable Waste a Real Wasteland

As mentioned earlier, cultivable waste is a major category of wasteland. The genesis of this land can be traced back to the food scarcity conditions prevailing in the past. Large areas of forest and scrub land even up to 45 degrees or more were taken up for cultivation as there was no other source of income due to less diversification in the economic structure and people had no alternative but to clear forest or other land for cultivation. Many people were granted lease in the forest land; such patches of terraced fields are still identified within the forest land in the area. Gradually, these areas were left barren as people started migrating in the terai zone of Nainital district or those who remained there managed to subsist on the non-agricultural pursuits.

Figure 1.4 clearly demonstrates that a majority of 56 (52.80%) villages contribute above 90 per cent cultivable wasteland to their total wasteland followed by 34 villages which have between 60 and 90 per cent cultivable waste. Villages falling in the category of 30 to 60 per cent and below 30 per cent of cultivable waste to their total wasteland are smaller being 9 and 7 respectively. It draws attention that the land rendered waste due to the extra-cultivation has already crossed the ecological limits keeping aside the impact of mining and deforestation in the area.

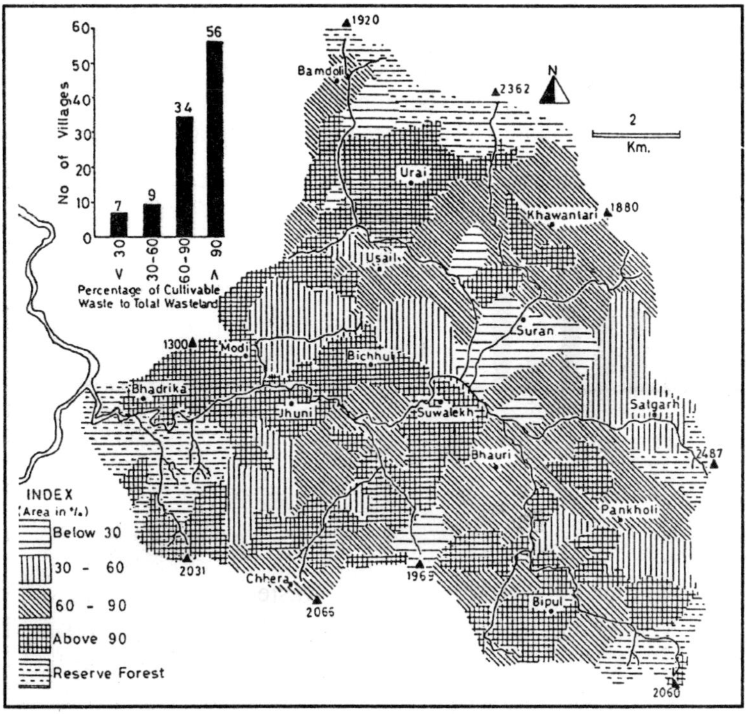

Fig. 1.4: Kalpanigad Basin: Village-wise Distribution of Cultivable Wasteland to Total Wasteland.

The share of the cultivable waste in the total cultivated land is shown in Table 1.2. A very high proportion of villages (68.08%) represent cultivable waste more than double its cultivated land.

Table 1.2: Cultivable Waste to total Cultivated Land, 1987-88.

Percentage of Cultivable Waste to Cultivated Land	*No. of Villages*
Below 50	10
50-100	23
100-150	19
150-200	18
200-250	08
Above 250	26
Uninhabited	02
Total	**106**

The spatial pattern of cultivable waste has also been depicted in Fig. 1.5. The wasteland is found extensive mostly on the upper slopes and ridge tops as these surface locations are less attractive for agriculture. Although some portions of wasteland are now converted into good forest land, most of these are still degraded lands.

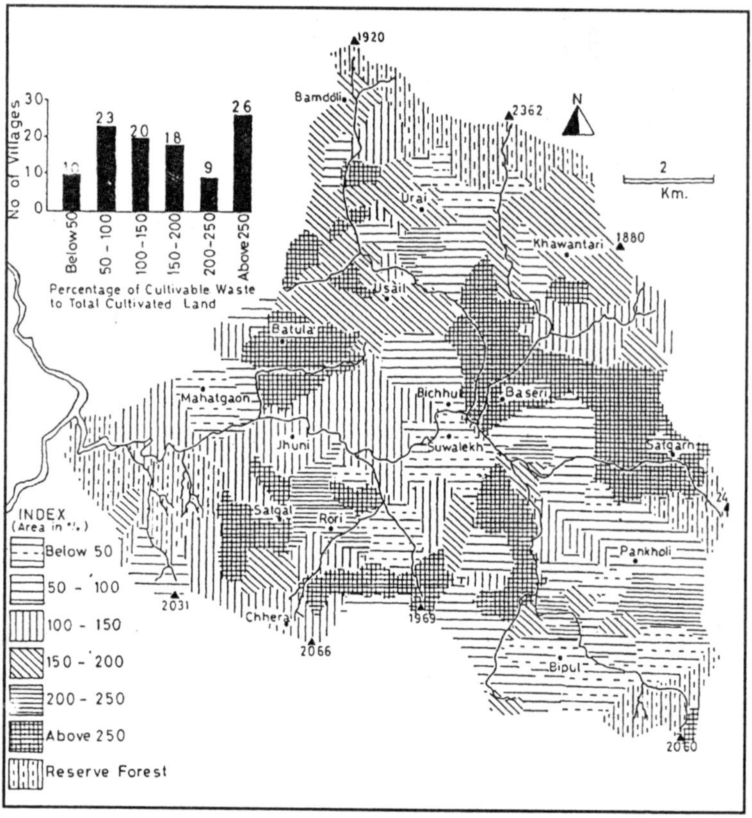

Fig. 1.5: Kalpanigad Basin: Village-wise Distribution of Cultivable Wasteland to Total Cultivated Land.

Area not Suitable for Cultivation

These are the lands mostly situated in very rough and fragile surface conditions and show no definite landuse features. These are also not commonly observed in all the villages. About 43.40 per cent villages have no such lands. Out of 60 villages with this land category, exactly

50 per cent have below 15 per cent share of this land out of their total wasteland. Villages with 15 to 30, 30 to 45, 45 to 60 and above 60 per cent of this category land to total wasteland are 14, 8, 2 and 6 respectively (Fig. 1.6). The ever increasing demand for fuel and fodder can be met through the plantation in suitable pockets of these lands. There has been no strong basis for putting land under this category.

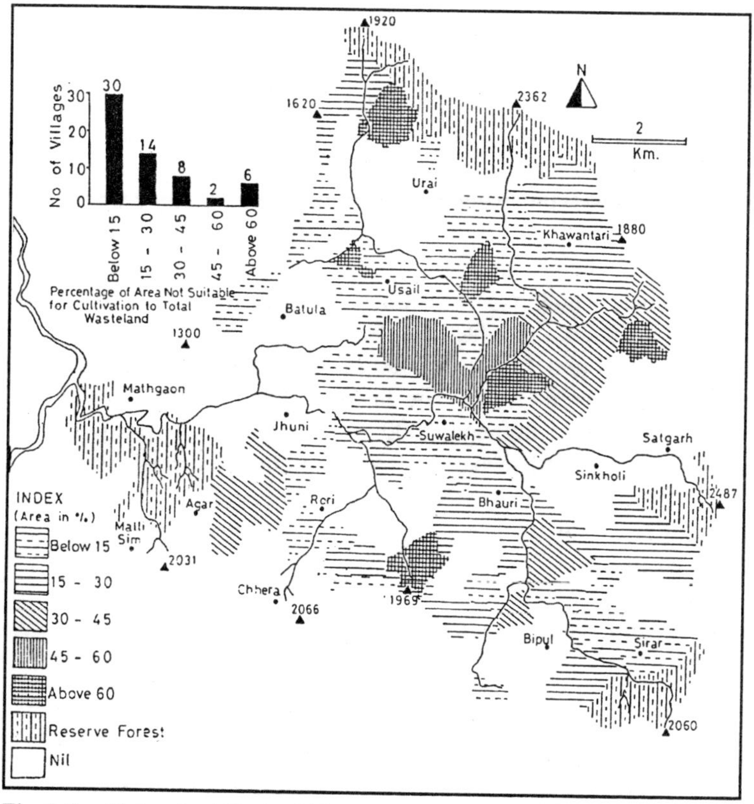

Fig. 1.6: Kalpanigad Basin: Village-wise Distribution of Area not Suitable for Cultivation to Total Wasteland.

Geographic Distribution of Wasteland

The wasteland identification project has already included upland and steep sloping areas as major wasteland categories at national level. In fact, each component of landscape has a considerable effect on the distribution of wasteland. Although these geographical relationships appear stronger at large scale, however, some of the influences can be perceived even at the micro level. The distribution of wasteland in relation to height, slope aspects and drainage is considered here for interpretation.

Wasteland and Altitude

The wasteland patches were delineated in toposheets of 1:50,000 scale according to their location in various height groups. A total of 43 patches were identified and the area measurements in each height group were noted. The height data has been classified with an interval of 400 m. Wasteland patches occurring in each height group are presented in Table 1.3. A maximum of 18 wasteland patches are found to occur in the height group of 1,200 to 1,600 m. with the minimum of one in below 800 m. The maximum patches accounting for 76.76 per cent of the total recorded number occur in two height groups of 1,200 to 1,600 and 1,600 to 2,000 m. The area estimates follow a somewhat different trend. The highest area is (45.81%) though recorded between 1,200 and 1,600 m. while the height group of 800 to 1,200 m. recorded second largest area in the total wasteland. The lowest area (0.54%) also recorded from above 2,000 m. height group. The spatial extent of thee patches along various height groups is presented in Fig. 1.7.

Table 1.3: Distribution of Wasteland Patches in Various Height Groups.

	Wasteland Patches		*Wasteland Area*	
Height in m.	*Number*	*% of Total Patches*	*Area (sqkm)*	*% of Total Area*
Below 800	1	2.33	0.81	2.07
800-1200	5	11.63	14.46	36.96
1200-1600	18	41.86	17.92	45.81
1600-2000	15	34.88	5.72	14.62
Above 2000	4	9.30	0.21	0.54
Total	**43**	**100.00**	**39.12**	**100.00**

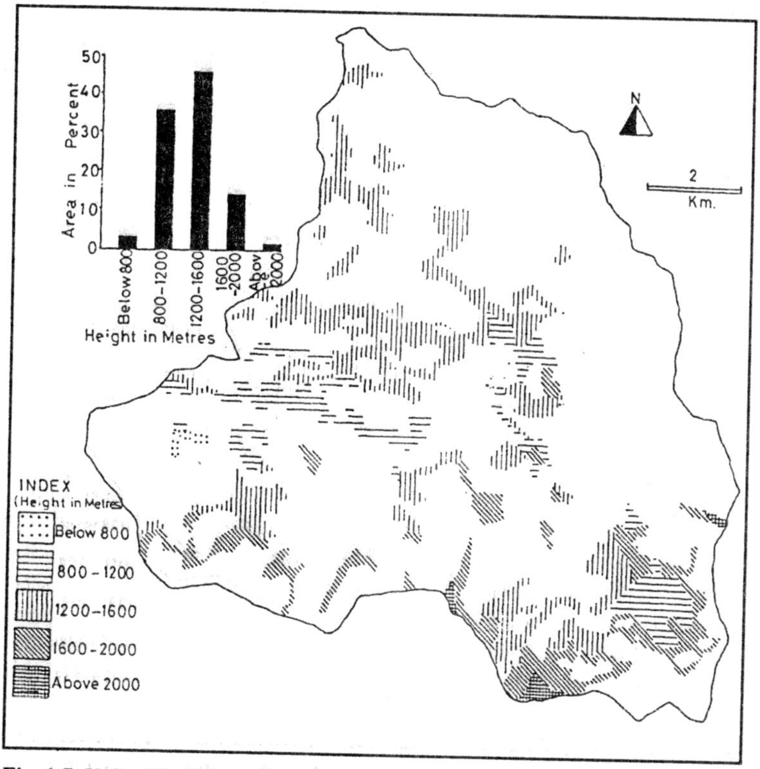

Fig. 1.7: Kalpanigad Basin: Distribution of Wasteland in Various Height Groups.

Wasteland and Slope

The slope data has been categorised into five groups with an interval of four degrees. The number of wasteland patches along with the area measurement are presented in Table 1.4.

The wasteland patches delineated from the average slope map are 65. The slope category of 24 to 28 degrees concentrates maximum patches (35.38%) followed by 30.77 per cent patches in 20 to 24 degrees. The minimum number of patches (6.15%) are found in the slope group of above 32 degrees. The slope group of 20 to 24 and 24 to 28 degrees together account for 66.15 per cent of total wasteland patches. Taking area into consideration, again the highest area is recorded (44.07%) from 24 to 28 degrees of slope but the second largest area belongs to 28 to 32 degrees of slope group. The minimum area is (2.25%),

Table 1.4: Distribution of Wasteland patches in various slope groups.

	Wasteland Patches		*Wasteland Area*	
Slope in Degrees	*Number*	*% of Total Patches*	*Area (sq.km.)*	*% of Total Area*
Below 20	6	9.24	1.44	3.68
20-24	20	30.77	5.13	13.11
24-28	23	35.38	17.24	44.07
28-32	12	18.46	14.43	36.89
Above 32	4	6.15	0.88	2.25
Total	**65**	**100.00**	**39.12**	**100.00**

however, found above 32 degrees of slope. Figures 1.8 represents the spatial extent of wasteland in the above-mentioned slope categories.

Wasteland and Slope Aspects

The impact of slope aspect in the general landuse pattern in the hills has been widely emphasised. It enables easy identification of such wastelands as need to be diverted to any kind of landuse. A total of eight possible aspects of slope have been demarcated through the help of relief and drainage map and wasteland patches in all these aspects are counted. A total of 74 patches were identified and their distribution is presented in Table 1.5.

Table 1.5: Aspect-wise distribution of wasteland.

	Wasteland Patches		*Wasteland Area*	
Slope Aspect	*Number*	*% of Total Patches*	*Area (sq.km.)*	*% of Total Area*
1. East	6	8.11	2.82	7.21
2. West	8	10.81	5.59	14.29
3. North	12	16.21	5.20	13.29
4. South	12	16.21	6.25	15.98
5. North-east	9	12.17	2.30	5.88
6. North-west	6	8.11	5.66	14.47
7. South-east	7	9.46	3.13	8.00
8. South-west	14	18.92	8.17	20.88
Total	**74**	**100.00**	**39.12**	**100.00**

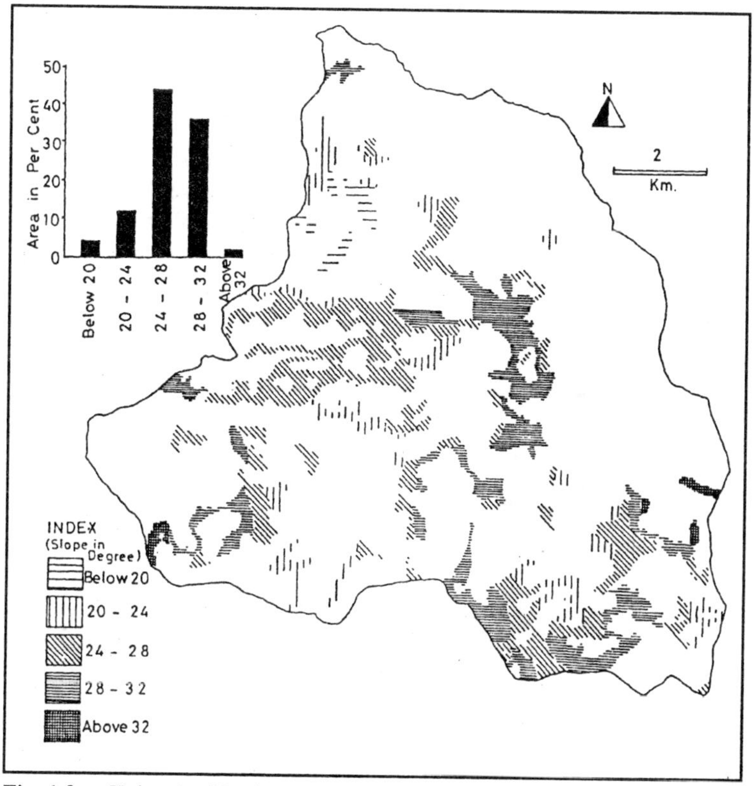

Fig. 1.8: Kalpanigad Basin: Distribution of Wasteland in Various Slope Groups.

The wasteland patches located in south-west are found to rank highest (18.92%) of the total identified patches. North and south faces of slope rank next concentrating equal amount of patches (16.21%). On the other hand, the least number of patches occur in east (8.11%), north-west (8.11%), west (10.81%) and south-east (9.46%) categories. This indicates that the east and west aspects which get maximum sun hours are utilised most and thereby render least wasteland areas. As this, north and south-facing slopes are found under maximum expansion of wasteland. The forested north slopes are becoming barren and the south-facing slopes have been extensively used for agriculture. As a result, the percentage of cultivable wasteland is also increasing on these slopes.

Wasteland and Drainage Density

It is presumed that drainage must have a role in the location and development of wasteland. The excessive drainage or the scarcity both create poorer conditions for the landuse. The drainage density is worked out as total length of streams per sqkm and four drainage density groups from below 1.9 km./sq. km. to above 3.8 km/sq. km. were identified in the area. These categories were plotted in the map (Fig. 1.9). The wasteland patches in these drainage density groups were counted and are presented in Table 1.6.

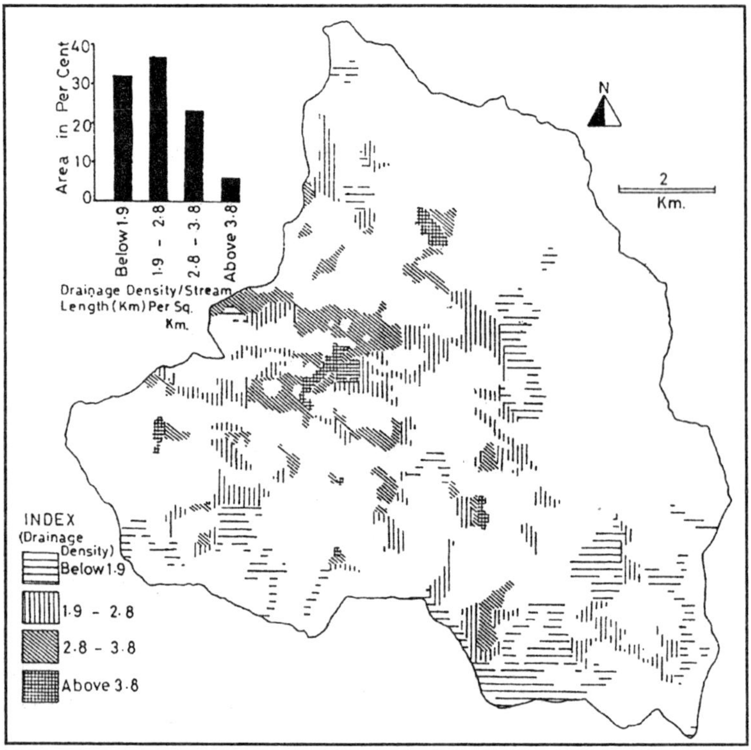

Fig. 1.9: Kapanigad Basin: Distribution of Wasteland in Various Drainage Density Groups.

The wasteland patches are found to occur highest (45.45%) in the drainage density group of 1.9 to 2.8 followed by (27.27%) in the next group of 2.8 to 3.8. The minimum number of patches (6.82%) is observed above 3.8. However, if the area distribution is considered, the second largest category is shifted to the drainage density group of 1.9 keeping the largest and smallest concentration in the same groups.

Table 1.6: Distribution of Wasteland in Various Drainage Density Groups.

Drainage Density/ Stream Length (km.) per sq. km.	*Wasteland Patches*		*Wasteland Area*	
	Number	*% of Total Patches*	*Area*	*% of Total Area*
Below 1.9	18	20.46	12.66	32.36
1.9 to 2.8	40	45.45	14.62	37.37
2.8 to 3.8	24	27.27	9.68	24.75
Above 3.8	6	6.82	2.16	5.52
Total	**88**	**100.00**	**39.12**	**100.00**

These distributions, particularly in relation to slope and drainage, have led to some important generalisations. However, they also contribute to the need of geographical database in the wasteland studies. In fact, the frequency pattern has been governed more by the topographical nature which is relatively moderate in the area. Consequently, all the distributions are accumulated mostly in the moderate size classes in each case. The area dos not present extreme slope, height and drainage characteristics so the wasteland patches occurring in extreme conditions are comparatively few. The second important point is the increasing human interference that has altered dramatically the actual limit set by the physical processes.

REFERENCES

Bharat, B. (1987), *Wastelands for Industry, Ecology*, Vol. 2, No.1, pp. 21-29.

Bhumbla, D.R. and Khare, A. (1984), *Estimates of Wastelands in India*, Society for Promotion of Wasteland Development, New Delhi.

Chaudhari, Kamala (1985), "Wastelands and the Rural Poor", Paper presented in Symposium on Forestry for Rural Development, FAO Regional Office, Bangkok, October 2, 1985.

Gautam, N.C. and Chennaiah, G.C.H. (1985), "Landuse and Land cover Mapping and Change Detection in Tripura using Satellite Landsat Data", Inst. *Journ. of Remote Sensing*, Vol. 6, No. 3-4, pp. 517-528.

Gautam, N.C. and Narayan, L.R.A. (1988), *Wastelands in India*, Pink Publishing House, Mathura, p. 96.

Gopal Krishnan, K. (1982), *India's Wet Desert, Earth Scan Feature*, London.

Gupta, R.K. and Arora, Y.K. (1985), "Technology for the Regeneration of Degraded Lands in the Himalaya", in J.S. Singh (ed). *Environmental Regeneration in Himalaya-Concepts and Strategies*, Gyanodaya Prakashan, Nainital, pp. 169-85.

Gupta, R.K. (1982), "Sisal Plantation for Wastelands", *Indian Farming*, March issue, 1982.

Gupta, R.K. and Trjwani, K.G. (1980), "Management of Denuded Lands in the Himalaya", *Indian Journal of Soil Conservation*, Vol. 8 (2), pp. 146-56.

Gupta, R.K. (1984),"Restoration of Wasted Community Lands for Silvi-pastoral management under Social Forestry in U.P. Himalaya", in S.C. Joshi *et al.*, (ed), *Rural Development in the Himalaya: Problems and Prospects*, Gyanodaya Prakashan, Nainital, pp. 163-75.

Jalal, D.S. (1975), "Classification and Potential Productivity of Arable Land in Kumaun Himalaya with special reference to district Pithoragarh", *Uttarakhand Bharati*, Vol. 1(3), pp. 65-75.

Jalal, D.S. (1976), *Land Utilisation in Pithoragarh*, Chaitanya Publication, Allahabad.

Khybri, M.L. (1980), "Land Capability Classification for the Himalaya Region", in *Proceedings, National Seminar on Resource Development and Environment in the Himalayan Region*, D.S.T., Govt. of India, New Delhi.

Patel, U. (1987), "Wasteland Regeneration: Our Priority", *Yojana*, Vol. 30. (24), pp. 4-7.

Prem Bhai (1985), "Problems of Opportunities of Wasteland Development", Paper presented in the Seminar on Opportunities and Problems on Wasteland Development, August 3, 1985, Society for Promotion of Wasteland Development, New Delhi.

Stamp, L.D. (1948), *The Land of Britain: Its Use and Misuse*, p. 433.

Shafi, M. (1968), "The Problem of Wastelands in India", *Geographer*, Special Number 21, Aligarh Muslim University Society, Aligarh.

Singh, S. (1989), "Wasteland Monitoring using Remote Sensing Techniques", *Wastelands News*, Vol. 4(4), pp. 2-8.

Singh, A. (1989), "Geographic Information System for Wasteland Development Planning", *Wastelands News*, Vol. 4 (4) pp. 9-11.

Shah, R. (1984), "Economics of Wasteland Development Projects undertaken by Gujarat State Rural Development Corporation" in Kamal Sharma (ed), *Economics of Wastelands Development*, Society for Promotion of Wasteland Development, New Delhi.

Shah, S.L. (1982), *Socio-economic, Technological, Organisational and Institutional Constraints in the afforestation of civil, Soyam, Usar and Wasteland for resolving fuelwood crises in the Hill district of U.P.*, Vivekanand Parvatiya Krishi Anusandhan Shala, Almora.

Shah, S.L. (1985), "Some Methodological and Pragmatic Issues in Land capability classification in Himalaya", in J.S. Singh (ed), *Environmental Regeneration in Himalaya-Concept and Strategies*, Gyanodaya Prakashan, Nainital, pp. 115-29.

Yadav, H.R. (1986), *Genesis and Utilisation of Wastelands — A Case Study of Sultanpur District*, Concept Publishing Company, New Delhi.

Yadav, H.R. (1987), *Wasteland Diagnosis and Treatment*, Concept Publishing Company, New Delhi.

2

Wastelands Mapping in Churu District, Using Remote Sensing and Suggestions for Development

S.K. SAXENA, AMAL KAR, D.C. JOSHI and N.S. VANGANI

The deserts are generally considered as vast wastelands where nothing grows, or where the land is underutilised and degraded. According to the National Wasteland Development Board (NWDB, 1987) wastelands are defined as those lands which are degraded, but could be brought under vegetative cover with reasonable effort, and which is currently underutilised, as also the land which is deteriorating due to lack of appropriate soil and water management, or on account of natural causes. Wastelands can, therefore, result from inherent/imposed constraints, such as location, environment, chemical and physical properties of the soil, or financial or management constraints (NRSA, 1986; NWDB, 1986, 1987). With such a broad definition it is often very difficult to judge the land of the Thar desert for wasteland mapping, as much of it is now teeming with population, although the climatic and terrain constraints are severe, as compared to those in the wetter parts of the country, and the land is highly vulnerable to degradation processes.

The authors participated in a nationwide wasteland mapping project, using Landsat TM data at 1:50,000 scale and mapped the wasteland categories in Churu district of Rajasthan—a highly degraded and vulnerable sand-covered district in the north-eastern Thar. The wasteland categories were demarcated as per the guidelines and definitions of NWDB (1987), while the characteristics of each unit and its development potential were worked out on the basis of remote-sensing, field survey and a knowledge of the local environment.

Environmental Setting

The Churu district with an area of 16830 sq. km. is located in the northern part of Rajasthan State, between 27°24', and 29°0'N Latitude and 73°46' and 75°41'E Longitude. The district is extensively covered by deep sandy undulating plains having variable density of sandy hummocks, sand dunes and other loose sandy features. Isolated hills of metamorphic rocks occur mainly around Gopalpura, Dungras, Bidasar and Biramsar in the south. These are flanked by gravely pediments and buried pediments. The flat plains with widely spaced dunes occur near Sahawa, Taranagar and to the east and south of Rajgarh. A narrow strip of younger alluvial plains occurs near Navrangpura in the north-east where a small stream, Kantli, flows. This stream rises in the Aravalli hills near Khandla in Sikar district and gets lost in sand after entering Churu district. Major saline depressions (Ranns) occur near Sujangarh, Chhapar and Parihara villages.

Climatic Features

Mean annual rainfall in the district is 363.8 m.m., but the western part covering Dungargarh tehsil receives 261.2 m.m., whereas the south-eastern part covering Sujangarh tehsil gets 372 m.m. Nearly 85 per cent of the annual rainfall is received during July to September. Winter rainfall contributes less than 6 per cent of the total. Mean maximum and minimum temperatures are 32.7° and 16.8°C respectively. May and June are the hottest months when the temperature sometimes crosses 45°C. December and January are the coldest months, when the minimum temperature often goes down to –5°C. Aridity index varies from 75 to 83 per cent. Mean annual wind speed is 8.5 kmph, but in summer months wind speed of 16.4 kmph or more is prevalent. High wind regime and intense solar radiation along with spells of duststorms become common during the summer.

Human and Livestock Population

The 855 villages and 11 towns of the district record a human population of 1,539,470 (1991) which is nearly three times the figure for 1951 (5,23,650). The population density is 91 persons per sq. km.

The livestock population is also fairly high (15,74,522) and has doubled during the last 30 years. Sheep and goats form the bulk (67.3%) of the animal population and these have tremendous impact on the district's scanty vegetation resources.

Material and Methods

The procedure laid down by the National Remote Sensing Agency (NRSA, 1986) was followed for identification and mapping of wastelands. Standard false colour composites (FCC) of Landsat TM bands 2, 3 and 4 at 1:50,000 scale were used for interpretation in conjunction with the Survey of India topographical sheets at 1:50,000 scale and available aerial photographs. The identification and delineation of different types of wastelands were, in the first instance, carried out with the standard visual interpretation techniques (Bowden and Pruitt, 1983). The provisional boundaries of each category of wastelands were demarcated. A ground verification survey was then carried out to finalise the wasteland type and their locational accuracy. Additional information on landform attributes, type of land degradation processes (water erosion, wind erosion/deposition, salinity-alkalinity, etc.), soil characteristics, status of natural vegetation and hydrological conditions were also collected for assessing the conditions of wastelands and to recommend suitable measures for their rehabilitation.

Wasteland Categories and Their Characteristics

Interpretation of TM data and ground truth survey revealed that out of the 14 categories of wastelands outlined by NWDB and NRSA (1986), only seven could be recognised in Churu district (Table 2.1). The characteristic of each category is described below.

Wasteland Categories

Sand-desertic (8927.58 sq. km.) : The most dominant wasteland category within Churu district is sand-desertic, covering 53.05 per cent of the district area and 87.10 percent of the total wasteland area. This unit usually appears on TM FCC in a light-to-dull-yellow tone. In areas of intense biotic reactivation the tone become whitish. The sand dunes and sandy hummocks could be easily separated from the flat sandy terrain because of their shape. The unit occurs in almost all the parts of the district, in the form of high sand dunes and narrow sandy undulating interdunal plains. All the three major types of dunes, viz. longitudinal, parabolic and transverse, are quite stabilised but excessive biotic pressure on the dune body through regular ploughing, overgrazing and removal of scrubby vegetation have very much loosened the structure of sand and reactivated the dune slopes at many places, resulting in the piling of sand on dune crest, slopes and leeward side. At places *barchans* or hummocks are formed in the adjoining fields of interdunes or near habitation sites. The problem has been accentuated due to four years of continuous drought (1984-87). The dunes and hummocky interdunes are widely cultivated for *Kharif* cultivation, but because of high vulnerability to soil erosion hazards and mismanagement, these have been categorised as wastelands.

Table 2.1: Wasteland Categories in Churu District.

Sl.No.	*Wasteland Category*	*Area (sq. km.)*	*Percentage of District Area*	*Percentage of Total Wasteland Area*
1.	Sands-Desertic	8927.58	53.05	87.10
2.	Degraded pastures/grazing land	1264.75	7.50	12.34
3.	Land affected by salinity/alkalinity	15.17	0.09	0.15
4.	Underutilised/degraded notified forest land	32.03	0.19	0.31
5.	Upland with and without scrub	3.35	0.02	0.03
6.	Mining/industrial wasteland	5.70	0.03	0.06
7.	Barren rocky/stony waste/sheet rock area	1.20	0.01	0.01
	Total	**10249.78**	**60.89**	**100.00**

The southern and western parts of Sujangarh and Dungargarh tehsils are infested with high and stabilised longitudinal and parabolic dunes, dominated by *Prosopis cineraria* community, with a density of 25-30 plants/ha. The central part of the district covering Ratangarh, north of Sujangarh, south of Sardarshahr and the eastern part of Dungargarh tehsil have very high concentrations of transverse and other dunes with high sand activity. Only scrub vegetation, mostly dominated by *Calligonum polygonoides–Leptadenia pyrotechnica* community, may be recorded on majority of the dunes. The northern parts of Sardarshahr and Ratangarh tehsils have more of the parabolic dunes with Prosopis-Calligonum community. In Rajgarh tehsil the sand activity is comparatively less and dunes are widely spaced out. Salvadora-Maytenus-Capparis are supported on the dunes of this zones.

Formation of *barchan* dunes, which are free of any vegetation, are mostly threatening many habitations of the district. They require immediate attention for their stabilisation.

Degraded Pastures/Grazing Lands (1264.74 sq. km.) : This unit, covering the second largest area of the district (7.50% of the total area) occurs on almost all the landform types. The major occurrences are, however, on low dunes and adjacent interdunal plains. Invariably a scrub community of Leptadenia pyrotechnica-Calligonum polygonoides predominates such grazing land. Due to prolonged drought, the dry and damaged clumps of Lasiurus sindicus and Cymbopogon jwarancusa are encountered in a few cases. Most of the pastures/grazing lands appear on the enlarged TM FCC in distinct geometric shapes. The degraded units are usually light grey to light brown in colour with a coarse to mottled texture.

In a few cases the associated shrub species are C. decidua, Calotropis procera, Lycium barbarum and Ziziphus nummularia. Many lands on dunes are devoid of any ground cover, except Cyperus arenarius. In the northern and eastern parts of the district (Rajgarh and Sardarshahr), the degraded grazing lands are mostly located on the flat or sandy undulating alluvial plains. Salvadora oleoides–Acacia nilotica community prevails on heavy but shallow soils with a hard *kankar* pan below. Capparis decidua is the main associated shrub. Dichanthium annulatum seems to be the main fodder grass of these grazing lands, but during the drought of 1985-87 the clumps were virtually levelled with the ground. In the southern part of the district a few grazing lands occur on pediments and buried pediments. Here Prosopis cineraria dominates, whereas common shrubs are C. decidua, Euphorbia caducifolia, Prosopis juliflora, Lycium barbarum and Ziziphus nummularia. Amongst the grasses Dactyloctenium sindicum and Eleusine compressa were recorded in a very poor state of health.

Land Affected by Salinity-Alkalinity (15.17 sq km) : In Churu district the salt-affected wastelands are the saline *ranns* or depressions, covering 0.09 per cent of the district area. These are located at Sujangarh, Tal Chhapar, Parihara, Absar, Charanwasi, etc., and are characterised by a flat, heavy-textured saline/alkaline soils. On the TM FCC these lands appear in a milky white (dry) to light blue or light greenish blue (wet) tone, having a fine to mottled texture. The presence of halophytic vegetation along the peripheries of the *ranns* and in the less saline parts within the *ranns* provide a light brown tone. They provide hard surface during the dry season, but get flooded during the monsoon period. The slope varies from 0° to 1° and the dominant particle size is 0.06 to 0.07 m.m. These depressions are the gathering ground of evaporates, especially common salt and gypsum, and are linked with the buried former stream courses. Almost all the depressions are presently enclosed by high stable dunes. Exploitation of gypsum or salt is carried out at certain parts. The Chhapar area has been declared as a protected forest. Prosopis juliflora is the dominant tree of this saline flat. Sporobolus marginatus, Desmostachya bipinnata, Aleuropus logopoides make a grass cover in the protected zone. Other halophytic species encountered are Tamarix articulata and Suaeda fruticosa.

At Sujangarh, the halophytic shrubs and grasses like Suaeda fruticosa, Salsola baryosma and Sporobolus marginatus dominate the saline patch, whereas only Capparis decidua with a fairly good density (150/ha.) dominate the Parihara rann. Small interdunal saline flats with gypsiferous or calcium-rich soils are locally exploited, due to which they are recorded in highly degraded conditions. Mutilated or depauperated cushion-shaped vegetation is represented by species like Prosopis, Capparis and Calotropis procera.

Underutilised/Degraded Notified Forest Lands (32.03 sq. km.): The notified forest areas within the district occur around Chhapar, Ratangarh, Churu, Dungargarh, Sardarshahr and Rajgarh. These appear on the TM FCC in a greyish brown to light brown or reddish brown tone, depending on the nature and extent of vegetation cover, and hence distinct tonal boundaries with the surrounding units. At Chhapar the forest area is nothing but the saline depression where soils are heavy-textured and support Prosopis juliflora as the main community. The other notified area covers the sand dunes and interdunal plains as contiguous patches. Prosopis cineraria with 30-40 trees/ha. dominates. Other associated species in the reserved forest areas are Calligonum polygonoides, Lycium barbarum, Calotropis procera, Ziziphus nummularia and

Leptadenia pyrotechnica. Plantation of Acacia tortilis has been carried out in some parts of these notified areas. Rajgarh degraded forest land with deep to very deep sandy soils supports the Salvadora oleoides-Acacia senegal-Maytenus emarginata community. Associated shrub species are Capparis decidua, Lycium barbarum and Leptadenia pyrotechnica.

Undulating Upland With or Without Scrub (3.35 sq. km.) : This unit, covering 0.02 per cent of the district area, occurs mainly in the southern part where the isolated low hills of metamorphic formations are exposed near Gopalpura, Biramsar and Dungras. Here skeletal debris cover the slopes of the hills/uplands. The slope varies between 10° and 30° and supports scanty and widely spaced scrub vegetation on shallow gravely soils. The unit appears on the TM FCC in a medium to dark grey or greenish blue tone and has a mottled texture. Main species encountered are Acacia Senegal, Maytenus emarginata, Euphorbia caducifolia and Capparis decidua. The areas having a shallow to moderately deep soil support a well-defined community of Euphorbia-Salvadora-Capparis. Some of the associated species are Ziziphus nummularia, Leptadenia pyrotechnica and Lycium barbarum. These shrub species generally occupy the ephemeral water courses.

Mining/Industrial Wasteland (5.70 sq. km.) : The major mining activity within the district is near Chhapar where the saline depression is partly exploited for the expansion of common salt. Small stone quarries are located in the rocky areas near Bidasar for construction material. Disturbed mining sites do not support any vegetation, except a few annual species. These appear on TM FCC in a white or dark blue tone.

Barren Rocky/Stone Waste/Sheet Rock Areas (1.20 sq km) : The isolated quartzite hill at Randhisar is barren. The hill slope does not have appreciable colluvial debries cover and hence is unable to support any perennial vegetation. Dark grey to greenish blue tone with an irregular shape are their characteristic features on the TM FCC.

Suggestion for Rehabilitation of Wastelands

For successful rehabilitation of the wastelands the following priorities need to be considered for all the wasteland categories.

Protection from Biota

Development of enclosure around the rehabilitation sites is the foremost requirement, which may be carried out by (i) barbed wire fencing of three to four strands all around the area; (ii) developing a live, tall hedge of non-palatable species like Euphorbia caducifolia, Prosopis juliflora, Parkinsonia aculeate, etc., and (iii) digging a 1-2 m. deep and 1 m. wide trench all around the area. The raised bund of excavated soil may be utilised for planting or seeding by non-palatable species.

Development of Water Harvesting Sites

For the successful establishment and growth of plants permanent water bodies are required to provide water to the transplanted top feed species.

People's Participation

Involvement of the local people is essential for the success of the programme. Their collective effort can only provide a check to the encroachment on protected lands and help in raising the plants and thus rehabilitating the wastelands.

Nursery Development

Six months to one year before developing the wasteland, a programme should be undertaken to raise the desired tree and shrub species in polythene tubes/bags. The seedlings should be at least six months old for better establishment. Species suitable for different wasteland categories are described below:

Sand-desertic

In this unit the movement of sand is to be checked. To achieve this goal it is necessary first to grid the area with available brushwood in a chessboard design. Sand-binding and drought-resistant perennial grass species like Cenchrus cillaris (Dhaman), Lasiurus sindicus (Sewan), Panicum turgidum (Murut) and Cymbopogon jwarancusa

(Bui ghas) may be planted through seeding or rooted slips methods during the monsoon. Indigenous top species like Prosopis cineraria, Acacia Senegal, Tecomella undulata, Calligonum polygonoides, Maytenus emarginata, Mimosa hamata, etc., may be transplanted during July-August. When developed, this silvipastoral system can provide sustainable production of forage and feed for a longer period of time and minimise the wind erosional hazard. Another important plant, Citrullus colocynthis (Tumba), a creeper, needs to be encouraged on sandy hummocks, as it is a very good sand-binder. Its fruit has immense utility in medicine, as well as in the preparation of non-edible oils for soap-making.

Degraded Pasture/Grazing Lands

Most of the common village grazing lands in the district are in the sandy terrain and are highly degraded. These degraded pastures may be developed in a phased manner. The whole area may be allowed for open grazing. The grass species suitable for the Churu region are: C. cillaris, C. setigerus, L. sindicus, Panicum antidotale and Cymbopogon jwarancusa. A mixture of these grasses shall be ideal for re-seeding. Rooted slips and pelleted seeds may also be tried. The top feed species suitable for sandy wasteland, together with exotic top feed shrub species like Dichrostachys nutan and Colophospermum mopane may be tried. A density of 120 shrubs and 30-50 tree species per hectare may be maintained.

Land Affected by Salinity/Alkalinity

The first requirement is to reduce the salinity level and create conditions suitable for the establishment of halophytic species. This is possible in the peripheries of the saline depressions. To achieve this goal, trenching and bunding may be taken up first and carried out at an interval of 5-6 m. It will help in the leaching/reduction of salinity. Salt-tolerant species of shrubs and trees like Suaeda fruticosa, Haloxylon salicornicum (Lana), Tamarix articulatae, Salvadora oleoides (Jal) and Prosopis juliflora may be raised on the bunds, either by direct seed-sowing or by sapling transplantation. In the silvipastoral system seeds of salt-tolerant grasses like Sporobolus marginatus (Kharagas), Dichanthium annulatum (Karad) and Desmostachya bipinnata (Dab) may be tried by hand-dibbling or rooted slips for successful establishment.

Underutilised/Degraded Notified Forests

Seed nucleus already exists in this unit. Therefore, grazing should be completely stopped here as a first step. Soil conservation measures like contour furrowing, trenching, bunding, etc., should then be taken up to improve the soil moisture. The natural flora will then colonise the furrows and trenches. Trees and shrub species suitable for the sites, like Acacia senegal (Kumut), Anogeissus pendula (Dhokra), P. cineraria, P. juliflora, A. tortilis, Albizzia lebbek (Siris), Salvadora oleoides (Jal), Calligonum, Ziziphus and Capparis, raised in polythene tubes (6-9 months old), may also be transplanted, keeping a density of 100-150 per hectare.

Undulating Upland With or Without Scrub

To spread the water evenly on sloping land, soil conservation measures like contour furrowing and trenching should be carried out at a regular interval of 3-5 m. This will increase the moisture regime along each contour furrow and help in natural regeneration of vegetation. In order to hasten the process of development, the saplings (6-9 months) suitable for such rocky habitat should also be planted and reseeding of grasses like Sehima nervosum, C. setigerus and Chrysopogon fulus should also be undertaken.

Mining/Industrial Wasteland

The abandoned sites of salt-mining and stone-mining need separate treatments like land-levelling, filling of shallow pits, shaping of mining muck heaps, etc., as the first step of development. Salt-tolerant species should be planted on saline patches and species suitable for shallow soil areas on stony/rocky areas, after providing suitable amendments to the planting pits. The species have already been identified and mentioned.

Barren Rocky/Stony Waste/Sheet Rock Area

The area around the quartzite hill at Randhisar may be utilised to harvest maximum run-off, especially at the distal end of the pediments. Some catchment treatments are required to achieve this goal.

Conclusion

The Churu district, with its high human and livestock population density, has a high requirement for fuel, feed and fodder, besides crops. Yet, the present vegetation resources within the district are insufficient to cater to the needs of the increasing demands. Since the terrain is dominantly sandy, a high biotic pressure on its vegetation resources leads to widespread land degradation and gradual wasteland formation, in spite of the fact that the average annual rainfall at Churu, the district headquarters, is 363.8 mm. Deep ploughing of the sandy terrain, including the dune slopes, encroachment on the marginal lands and economic compulsions to dispense with the traditional systems of long fallow and deferred rotational grazing, is adding to the problems. Remote-sensing plays a key role in the identification and mapping of these lands.

In spite of such a scenario, much of the area, including the dune-covered terrain, is divided into individual land holdings, where rainfed Kharif Cultivation is practised. In other words, much of the sandy terrain is not a "waste" land in the strictest sense of the term, but is in a highly degraded state. Rehabilitation of such a degraded sandy terrain with individual land holdings is a difficult, but not impossible, proposition. Choices and decisions are to be made at the individual farmer's level and, hence, many of the species suggested for development through vegetative propagation had to be of economic utility and locally available. Environmental education and people's constructive participation are required for their adoption.

Acknowledgements

The authors are thankful to the Director and the Head, Division of Resource Survey and Monitoring, Central Arid Zone Research Institute, Jodhpur, for providing facilities to carry out the study. We are also grateful to NRSA and NWDB for providing financial assistance and satellite imagery. Shri Ganga Ram, Shri Basha Ram, Shri S.S. Dave and Shri Abhay Singh provided the necessary technical help and deserve special thanks.

REFERENCES

Bowden, L.W. and Pruitt, E.L. (eds.), (1983), *Manual of Remote Sensing.* Virginia: American Society of Photogrammetry, False Church, pp. 882-91.

NRSA, (1986), *Manual of procedure for wastelands mapping using remote sensing techniques*. Hyderabad: National Remote Sensing Agency, p. 52.

NWDB, (1986), *Wastelands definition and classification.* New Delhi: Technical Task Group Report, National Wasteland Development Board, p. 14.

NWDB, (1987), *Description and classification of wastelands.* New Delhi: National Wasteland Development Board.

3

Determinants and Reclamation of Wasteland
A Study of Kopai River Basin in Birbhum District

SUTAPA MUKHOPADHYAY

A wasteland is defined as land which is uncultivable or is presently lying unutilised due to different constraints but has been used previously, which is giving a very low actual return compared to its economic potential, which is ecologically unstable, or whose top soil has completely lost its fertility status, which has developed toxity for the growth of crops and trees due to environmental or anthropogenic problems (Yadav, 1985). Among the five main landuse categories of the census book culturable wasteland is defined as the land within cultivated holdings which has not been cultivated during the current year and the past five years or more in succession. Such lands may be either fallow or covered with shrubs, jungles, bamboo bushes and are not included under orchards or forests. Another category of the landuse type is the area not available for cultivation, it stands for barren and uncultivable land which cannot be brought under cultivation without incurring a high cost. It also includes lands occupied by buildings, roads and railways or are under water, e.g. rivers, canals, etc. So for an analysis of wastelands in the present context both culturable wastelands and the barren, uncultivable lands are grouped together. The current fallows are not considered as wastelands because such lands are left fallow to maintain the fertility status of the soil and within a year same lands are recultivated.

The demand for agricultural crop is increasing day by day with the increase in population. But a proportion to increase in this cultivated land has not occurred. A large area remains as wasteland throughout the country. So the Government of India has given emphasis to the reclamation of the wasteland. In this paper the nature and extent of the wastelands along the Kopai drainage basin area of Birbhum district of West Bengal has been discussed. Some reclaiming measures have also been suggested for the optimum utilisation of these lands. Though agriculture is the mainstay of economy in this river basin, due to lack of proper management the rural economy is suffering greatly, and there is a tendency among the rural people to migrate to the nearby urban centres in search of work as daily wage labourers. As a result about 50 per cent land of the total basin area has remained as wasteland. So it is high time to assess the wastelands of this area and suggest measures for their economic utilisation.

Geo-environment of the Kopai River Basin

Kopai is an important river of the southern part of Birbhum district and covers an area of 436.5 sq. km. This river originates from Khajuri village of Santal Pargana district of Bihar and after crossing some parts of six police stations, namely, Dubrajpur, Khoyrasol, Rajnagar, Illambazar, Bolpur and Lavpur (See Table 3.1), it joins the river Bakreswar 3 km. south of the Labhpur urban centre. With a total length of 110 km. the river flows in west-east direction in between the river Mor in the north and the Ajay in the south. The altitude of the basin varies from 157 to 25 m. Topographically, the whole river basin may be classified into 3 parts:

High Land of the Plateau Fringe

Granite and gneissic covered eastern part of the Chhotanagpur plateau is the source region of this river basin. Its altitude varies from 100 to 157 m. It occupies about 17.88 per cent of the total basin area.

Undulating Plain Land

The altitude of this part varies from 60 to 100 m. Laterite, upper tertiary and lower quarternary alluvial are the main components of this tract. It occupies about 25 per cent of the total basin area.

Lower Plain Land

The area below 60 m. altitude is of this category where laterite and alluvial of lower quarternary age are the main constituents. Swampy lands are the prominent features of the confluence zone. This region occupies about 57.12 per cent of the total area.

Climatic condition of the basin is of the hot and dry continental type. The average temperature of the hot season is 40°C and of the winter season 18°C. Average annual rainfall is 1,400 mm.

Table 3.1: Police Station-wise Wasteland Distribution Pattern.

Name of the Police Station	*Number of Mouzas of each P.S.*								*Total Number of Mouzas*
	% of wasteland to the total area								
	<1.1	1.1-5.09	5.10-10.09	10.10-15.09	15.10-20.09	20.10-25.09	25.10-30.09	>30.09	
Khoyrasol	—	6	6	9	5	2	4	14	46
Rajnagar	—	—	—	1	—	—	—	—	1
Dubrajpur	—	19	30	9	7	4	6	5	80
Illambazar	23	28	6	5	4	1	—	—	67
Bolpur	22	18	17	8	6	6	6	4	87
Lavpur	—	1	1	3	3	8	4	5	25

Wasteland Distribution Pattern

Mouza-wise wasteland of the basin has been calculated (1990-91) on the basis of the report of the Agriculture Department and of the field observation. It is interesting to note that the upper and the lowest most part of the basin have the highest percentage of wasteland (Fig. 3.1). In the case of the Khoyrasol police station 18 mouzas of the total 46 have more than 25 per cent wasteland of the total area. In some villages like Dhasunia, Faridpur, Kendukari, Aniyat and Haripur more than 50 per cent land remains as wasteland. In the Dubrajpur police station though 30 mouzas are under 5 to 10 per cent wasteland category still 11 villages have more than 25 per cent wasteland. Moreover, in some villages like Kapastari, Dubandha, Ashatganj, Aulia, Khosmahammadpur 35 to 40 per cent land is treated as wasteland.

In the Bolpur police station 16 villages have more than 20 per cent wasteland, among them Ballavpur, Kobimohanpur, Taltor, Mahula, Amdahara, Sitapur, East Islampur, Surul and Shyambati have more than 26 per cent wasteland to the total area.

In the case of the Lavpur police station of the 25 villages, in 17 villages more than 20 per cent land may be categorised as wasteland. Some villages like Patharghata and Talbona the percentage of

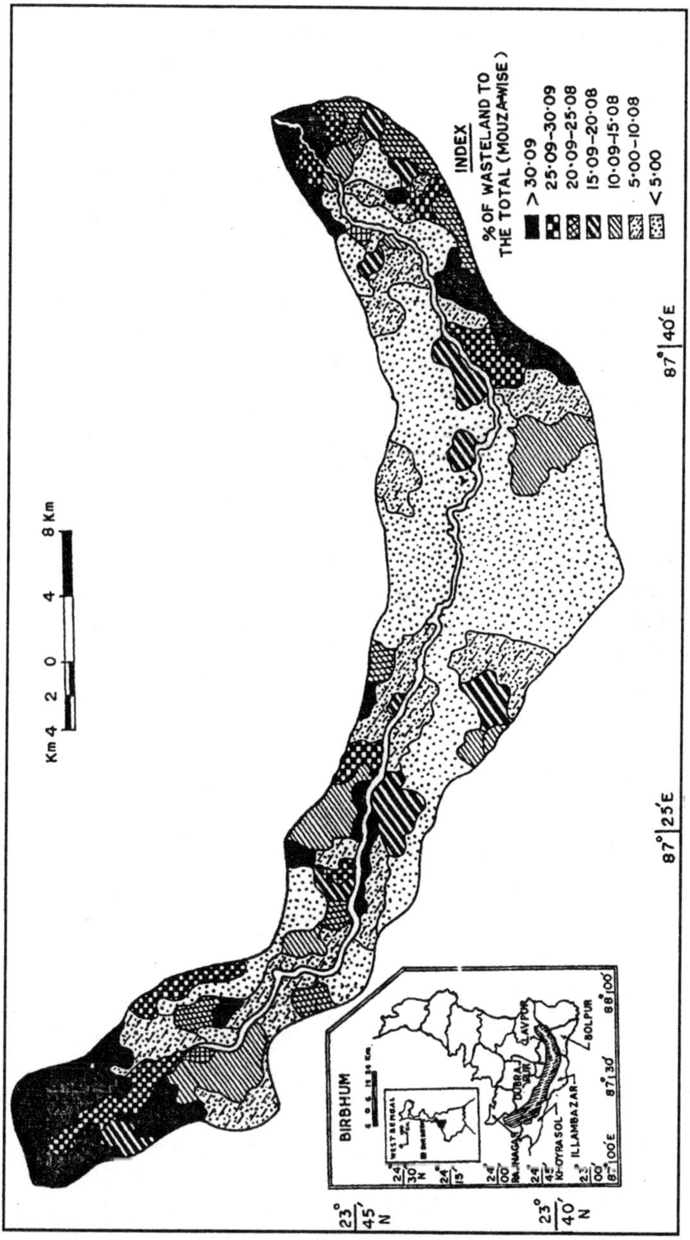

Fig. 3.1: Wasteland Distribution of Kopai Basin.

wasteland to the total area is 39.14 per cent and 36.90 per cent respectively.

Determinants of Wasteland

The factors which are responsible for the growth of the wasteland are morphometric, climatic, hydrologic, pedogenic and anthropogenic. Among all other morphometric factors slope and ruggedness have the prime importance. These variables are interrelated with each other, because the areas with high altitudes account for a greater amount of land slope and ruggedness number. It is observed that the villages of the upper part of the basin having the highest percentage of wasteland have also a high amount of land slope and ruggedness number. The slope of the wasteland of this tract varies from 3° to 4° and the average ruggedness number is 0.28. The high slope and ruggedness number induce a high rate of erosion, as a result fertile top soil is washed away. It is also observed in the field that the applied fertilisers and irrigated water move downslope in this sloping land and accumulate in the low-lying areas. It is also observed in the field that there is much overland flow through drains either natural or artificial, due to which the fertility status of the soil decreases on account of erosion by surface drainage.

It is also observed that though the percentage of wasteland is highest in the lower part of the basin, the slope and ruggedness number are not at all significant here. The hydrological factor, i.e. flood, is the determinant factor for this wasteland generation. From the field observation it is found that the flood may occur almost every year in this part and in some areas flood water may stand for even 4-5 months after the rainy season. These sand-crusted low lands are not at all suitable for agriculture. So large areas are found as wasteland in this tract.

But among all other factors pedogenetic factor is the most important one for such a distribution pattern of the wasteland. Granite and gneissic-covered stony wasteland are the predominant feature of the upper part of the basin, so in some villages the percentage of wasteland is even 30 to 50 per cent of the total area of the villages. Large patches of lateritic waste occur in different parts of the mid-basin area. These bear nodules, pebbles and cobbles, and are either largely unvegetated or carry only sparse scrub jungle. Water-holding capacity of this type of soil is low, it has high pH and lack of NPK (Roy Chaudhury, 1966).

Moreover, in some areas of the Bolpur and Dubrajpur police stations a hard, black, indurated horizon of the laterite appear close to the surface which is not favourable for the growth of the crops.

Moreover, some of the wastelands may develop as a result of either misuse or overexploitation of land by man (Jha, 1987) and this type may be categorised as anthropogenic wastelands. Removal of vegetation and overgrazing have often led to increased run-off and sediment yield, thus causing impediments in natural drainage.

Wasteland Types

On the basis of the above discussion wastelands can be divided into following groups:

Stony Wasteland

Granite and gneissic-covered upper and middle part of the river basin have such stony wasteland where cultivation is almost impossible due to low fertility status. About 5 per cent land of the total river basin area is under this category.

Laterite-covered Wasteland

Lateritic undulating land with a veneer of boulders, pebbles and cobbles remains as wasteland in various parts of the river basin. It is estimated that about 20 per cent of the total basin area is under this group. Iron and calcium carbonate mixing ghooting are generally available in the lateritic soil and its amount gradually increases with the increasing of the depth. This type of pebbly and boulderly soil is found in a vast area of Prantik near Santiniketan where excessive erosion occurs. This type of land can be compared with the *banjar* (Yadav, 1985) land of upper India.

Lateritic uplands with iron bands are found in various parts of the Dubrajpur and Bolpur police stations. In Ballavpur, Kobimohanpur and Goalpara mauzas of the Bolpur police station a 80-99 cm. thick black iron band is observed on the crust. These lateritic wastes are underlain by a bed of clay. In Dubrajpur the lateritic layers are underlain by gneisse.

These patches of lateritic wastes are either largely unvegetated or carry only sparse scrub vegetation. The Forest Department has taken only 125 acres of land of the Bolpur police station for the afforestation programme and most of the land is laid as wasteland.

River-borne Wasteland

Kopai river in its middle and lower parts has changed its course at various places. The old river courses in some parts of the Bolpur police station remain as wasteland. After a detailed survey of these parts the present researchers has considered those lands as a separate unit.

Such type of land is observed near Bhallavpur at the north-eastern part of Bolpur-Suri State highway which occupies about 30 sq. m. area. Further downward near Kobimohanpur mouza another such remnant course covering 70 sq. m. area is also marked. From the field survey it is found that the river changed its course during 1974-75. Water is flowing through these old courses in the rainy season but it remains dry throughout the year. The intermediate area between the old and the new course has also a good sandy clay cover but the cultivation cannot be practised here regularly as these river-borne lands have no definite owner. During winters, in some parts, sporadically mustard and winter vegetables are grown, otherwise the lands are considered as barren.

Marshy Wasteland

In the lower part of the river basin vast tract is flooded every year. The rainwater stands in some low-level land therefore these waterlogged areas cannot be used for agriculture or any other purposes.

Anthropogenic Wasteland

This type of land is the result of either unscientific deforestation by man or the result of overgrowing, etc. It is difficult to separate this type of land from the other types and in Table 3.2 this type is included as other in the last column.

As a detailed analysis of the wastelands of the whole basin is not possible within the framework of this paper, 8 villages, 2 from each wasteland type have been selected for sample studies. Sample villages have more than 25 per cent wasteland of the total area. In Table 3.2

Table 3.2: Wasteland Types and Landuse Pattern of Sample Villages.

Name of Village (P.S.)	*Av. Altitude (mt)*	*Landform Pattern*	*Landuse (% to the total area)*			*Wasteland Types*				
			Forest	*Cultivable Land*	*Wasteland*	*Stony Wasteland*	*Lateritic Wasteland*	*River-borne Wasteland*	*Marshy Wasteland*	*Others*
Dhasunia (Khoyrasol)	130	Granite & gneissic covered, steep slopping land	—	39.36	55.37	50.00	—	—	—	4.37
Kenduakari (Khoyrasol)	111	Granite & gneissic covered plateau fringe zone	—	44.93	50.87	45.52	—	—	—	5.35
Surul (Bolpur)	44	Lateritic undulating land	—	56.75	40.90	6.98	32.62	—	—	3.30
Shyambati (Bolpur)	40	Pebbly lateritic soil, old alluvium covered level land	—	45.17	50.30	—	40.00	—	—	10.30
Ballavpur (Bolpur)	47	Undulating land with laterite band, alluvium along the river side	24.74	41.03	26.26	5.00	15.61	5.00	—	1.38
Kobimohanpur (Bolpur)	45	Pebbly laterite & sandy alluvium covered level land	5.56	63.20	25.00	—	12.38	10.00	—	2.66
Talbona (Lavpur)	37	Sand & silt mixed soil in low land	—	43.69	36.90	—	—	10.01	29.95	7.64
Mangalgras (Lavpur)	36	Sand & silt mixed soil in low land	—	62.20	26.92	—	—	6.21	15.75	4.96

the physiographical condition, landuse pattern and the types of wasteland ecology have been given in detail of each village.

Suggested Measures for Reclamation of Wasteland

Agriculture is the mainstay of economy in the Kopai River Basin area and there is little scope for the development of mineral-based industries in the region. Also a vast tract of wasteland has caused ecological imbalances. The resultant effects are soil erosion, increase in surface run-off and sediment yield. If it remained unchecked, it will affect the existing cultivable land.

Some of the wasteland reclamation methods were unsuccessful after some time, because the lands reverted to their original condition due to mismanagement of the reclamation measures. The economics of reclamation of various wasteland type is closely linked to the management aspect (Yadav, 1990). The cost of reclamation depends upon the nature and extent of degradation the wasteland concerned has undergone. The following measures are suggested for making better use of the wastelands by restoring their ecological balance:

(a) Environmental factors are not the same throughout the basin area for the generation of all types of wastelands. So a micro level analysis of each type be made to obtain existing and optimal conditions of varying wastelands and the future programme must be planned on the basis of this scientific analysis.

(b) To prohibit unscientific cutting of trees from the riverside, sloping land and erodable area and to develop awareness of tree protection among the rural folk.

(c) Akashmoni, eucalyptus, etc., the quick-growing varieties of trees, are planted in the wasteland by the Forest Department but along with them some indigenous varieties like Sal, Segun, etc., can also be planted as mixed forestry. These trees will bring economic benefits.

(d) Conversion of some wastelands into pasture ground by planting some leguminous trees like Humilis, Stylothansis, Pensilusa, Kubabul (Subabul), etc. This is highly practised in different parts of Bankura and Purulia, and other drought-affected districts of West Bengal (Mukhopadhyay, 1989).

(e) Removing the lateritic morrum-covered top soil of about 1 m. the underlying clay bed is exposed which is suited for agricultural practises. This method was introduced in 1990 in one-acre land of Kamalakantapur village of the Bolpur police station by one family. The main operation involved in this reclamation is the levelling, bunding and land preparation. The total capital investment for this programme was about Rs. 5,000 which included wages to the labour for levelling, land preparation, fertiliser application, irrigation, cultural practices and also the cost of inputs like fertiliser, hiring of irrigation pump, seeds, etc. Last year (1991) about 4,000 pieces of cauliflower and nearly 10,000 kg. of potato were produced from this field which fetched a considerable profit to the family.

This encouraging result may allure the villagers to reclaim the vast *danga* land remaining along both sides of the river Kopai in the middle part of the basin.

(f) The Government must initiate steps to distribute plots of land among the villagers from the river-borne wastelands for their economic utilisation. This sandy soil is suitable for the growth of potato and mustard. It is estimated that the capital required for the preparation of this land is Rs. 3,500 and the estimated yield of potato and mustard will be 7,000 kg. and 600 kg. per acre respectively. This no doubt is a profitable proposition.

(g) Due importance should also be given to the production of vegetable crops in the newly reclaimed land instead of the traditional paddy cultivation, as vegetables have a ready market in the nearby urban centres like Dubrajpur, Bolpur, Illambazar, etc.

(h) Pisciculture can be practised in the old river courses by making bunding at its two ends.

(i) Drainage facilities must be improved to drain out the water from the low waterlogged areas.

(j) Calcium carbonate rich *ghooting* and iron mixed laterite band can be used as mineral wealth for the establishment of cottage-level industries.

(k) Settlements and industries can be established on some wastelands which cannot be reclaimed for agricultural use or pasturage.

(l) Recently the Soil Conservation Department has introduced a new irrigation system in the depressed area of the undulating lateritic land where polythene sheet is used to store the rainwater. This programme has started in Purulia village of Dubrajpur in June 1989, and by this process 8 hectares of land is irrigated. It is a hopeful measure for the proper utilisation of the *danga* land.

In fine it can be suggested that the involvement of the local people in the reclamation programme is most essential. Adequate arrangements should be made to organise demonstrations of reclamation measures in each village. As once the reclamation projects are initiated, the villagers will have the responsibility to follow up the action.

REFERENCES

Jha, V.C. (1987), "Wasteland types and their effective utilisation in Birbhum District", *The Deccan Geographer*, Vol. XXV, No. 2 & 3, Secunderabad, pp. 231-242.

Mukhopadhyay, M. (1989), *Resource planning and its impact on rural development*. Allahabad : Vohra Publishers, pp. 158-160.

Roy Choudhury, S.P. (1966), *Land and Soil*. New Delhi : N.B.T., pp. 25-39.

Yadav, H.R. (1985), "Determinants and utilisation of Banjar land : A Case Study of District Sultanpur", *Geographical Review of India*, Vol. 47, No. 3, Calcutta, pp. 6-9.

Yadav, H.R. (1990), *Reclaiming Wastelands*. New Delhi : Concept Publishing Company, pp. 100-105, 117-19.

4

An Integrated Approach for Development of Wasteland on Watershed Basis

A Case Study in Purulia, West Bengal

ANGUR MAITI

Purulia, the drought-prone lateritic tract of West Bengal, lies between Latitudes 22°40' and 23°45'N and Longitudes 85°45'E and 86°45'E and covers a total area of 6,259 sq. km. Physiographically, the entire area is an eastern most extension of the descending tableland (475 m. above sea level) of Chhotanagpur and represents typical transition between the hilly Chhotanagpur plateau to the west and alluvial flat plains to the east.

Land degradation has created a serious ecological and socio-economic crisis. Apart from degraded forest land it has been estimated that out of a total geographic area of 6,259 sq. km. about 1,050 sq. km. area is suffering from some form of degradation. These lands are named as "Wastelands". For the sustainable use of the "Wasteland" as well as to check land degradation Purulia district finds its place under "National Wasteland Development Programme". The Wasteland Development Programme has recently been included in the Technology Mission.

The basic objective of this programme is to start projects at the village level aiming at integrated approach to land management and wasteland development based on micro watershed level plans. The watershed approach provides the ideal means for integrated development. The progress achieved by the watershed development programmes in Purulia is a case in point. This has been studied with special reference to "Gajaburu Watershed" (KBI-107B).

Wasteland a Serious Issue

The problem of wasteland has become a serious issue and it has become more so with the development of technology for increasing agricultural production. The development of technology and growth of human and cattle population have been involved as a factor for the growth of wasteland. The development of wasteland, according to the Davision concept, may be explained according to time, process and stage.

There is no uniform definition of wasteland. It may be defined as ecological constraints and economic potential of land. Wastelands are regarded as ecologically unstable with almost complete loss of topsoil and with appreciable toxicity of life forms. On the basis of economical potential and actual return, the lands which give less than 20 per cent of this economic potential are "wastelands". The Wasteland Survey and Reclamation Committee defined wasteland as those land which are either not available for cultivation or are left out of cultivation as fallows or cultivable waste. Strictly speaking, these are marginal lands and partly belong to long fallow. The wasteland development programme in India has been given priority in order to protect environment from further degradation. Wastelands produce much below their potential and the area is increasing each year at an alarming rate. They need careful identification, inquiry, classification, conservation and planning for proper reclamation.

Nature and Cause of Wasteland Formation

Wastelands of Purulia district are generally very poor in soil depth due to an excessive soil erosion. Most of the wastelands are situated at or near the ridges, i.e. at higher altitudes, presence of a quartz-feldspar layer in the horizon of the wastelands makes the soil excessively drained and creates hindrance to penetration to the roots of agricultural crops to lower depths of the soil. Due to soil erosion the quartz-feldspar layer gets exposed and makes the land unsuitable for agriculture.

The main reasons for soil erosion besides the physiographic and climatic factors are unregulated and uncontrolled removal of vegetation, excessive grazing and upland cultivation without soil conservation measures. Cropping pattern on upland is also such that these areas remain fallow for two or three years. These factors have resulted in deterioration of quality, composition and density of vegetation.

Table 4.1 shows the area under different types of land designated as wastelands in Purulia.

Table 4.1: Different Types of Land in Purulia.

(In Hectare)

Sl.No.	*Type of Land*	*Area*
i)	Total forest land	87600
ii)	Gullied and ravenous area	5100
iii)	Undulating uplands	12200
iv)	Degraded forest land	8600
v)	Barren rocky/stony wasteland	2400
vi)	Steep sloping area	1700

Source: Project Report on Wasteland Development, Purulia, West Bengal.

These lands constitute about 34.25 per cent of total forest land of the district. Moreover, population explosion has resulted in denudation of the areas once covered with dense canopy.

Wasteland Development on Watershed Basis

Wastelands as a type of landform can be attributed to a complex process interacting at varying intensities. Watershed management is a comprehensive term meaning the rational utilisation of land and water resources of a watershed for optimum production with minimum hazard to natural resources. Disaster due to drought is not only a meteorological phenomenon, but basically an economic one. About 80 per cent of the population of Purulia suffer due to drought because they do not have the capacity to withstand even a small setback in the agricultural production. The watershed development will reduce the frequency of disaster caused by drought by restoring the ecological balance and improving the water resources in the watershed areas. Watersheds, being natural hydrological entities respond more effectively to the various engineering, biological and cultural treatments designed to maximise production. It essentially relates to the soil conservation in the watershed and means proper landuse, protecting land against all forms of deterioration, building and maintaining soil fertility, conserving water for farm use and proper management of local water for drainage.

With the objective of fulfilling the aforesaid necessities the integrated wasteland management plans at the local level on watershed basis have been taken up with the following main components:

Goals

A. To check land degradation.
B. To put wasteland to sustainable use.
C. To increase biomass availability.
D. To restore ecological balance.

Strategy

A. People's participation at all stages specially through Panchayats.
B. Integrated landuse planning on watershed basis.
C. Preparation of mouza-wise action plans.
D. Emphasis on conservation and natural regeneration.
E. Production of fuelwood, fodder, timber.
F. Extension of appropriate technology for wasteland development.

With this view a large number of watersheds have been identified in the district. Amongst these the Gajaburu Watershed has been selected for study with reference to wasteland development.

Gajaburu Micro-Watershed Project (KBI-107B)

The Gajaburu Watershed falls geographically within the south-west corner of the Purulia district at Arsha Police Station at the foothill of Ajodhya. This area is drained by "Jamuna Jore" which is a tributary of the river Bandhu and ultimately is drained by the main river Kangsabati. The rainfall in this area is erratic with an occasional dry spell in between rains. The mean annual average rainfall is 1,255 mm. The watershed comprises villages Rajpati, Pathardi and Chatubhasa covering an area of 712 hectare. The distance of the watershed is about 32 kms. from the district town.

The shape of the micro-watershed is elongated with 250 to 350 m. height. According to the physiographic position, lands are situated at different top sequences:

1. The undulated uplands with common features of rock outcrops. Relief is normal with heavy run-off.
2. Gently to very gently sloping lands with medium run-off.
3. Lowland valleys. Relief is concave with slow run-off.

The topography is rolling with varying land slope. Due to erosion, big gullies and ravines exist in this area. The soil is poor, undulating and excessively drained. The area is predominantly inhabited by the tribals where modern agricultural practices have not been adopted much. On the marginal areas of forest cover, low fertility of land, coarse texture of soil, lesser population density, relatively rugged terrain and soil erosion during the rainy season are the sources of the determinants responsible for the presence of wasteland in this area. The slope in cultivated uplands varieties from 1 to 4 per cent, whereas near the ridges it exceeds 10 per cent. The lower reaches have terraced paddy fields. The slopes along the nullah are eroded and are engulfing the cultivated land converting it into gullies.

Land Capability

The land capability classification is a systematic arrangement of different kinds of lands according to those properties that determine the ability of land to produce on a virtually permanent basis. The following land capability classes, sub-classes with their slopes and soil units, existing and proposed landuse were identified, recognised and suggested in Gajaburu micro-watershed for an effective development plan in this area (Table 4.2, Fig. 4.1). There are five main land capability classes designated as Class II to VI, grouped in accordance with progressive increase of degree of limitations. Land capability classes are further sub-divided into land capability sub-classes based upon the dominant limitations or hazards in their use. Three sub-classes such as "e", "w", "s" are recognised involving (e) risk of erosion, (w) wetness, (s) root zone limitation. Here, the land capability classification is mainly based on the environmental parameters of morphometric (slope and ruggedness number), climatic (rainfall), hydrological (drainage density and behaviour of water table), inherent soil characteristics and personal field observations (Table 4.3 and Table 4.4). On the basis of this the following six land capability classes have been ascribed to various wastelands:

Wastelands Capability Classification

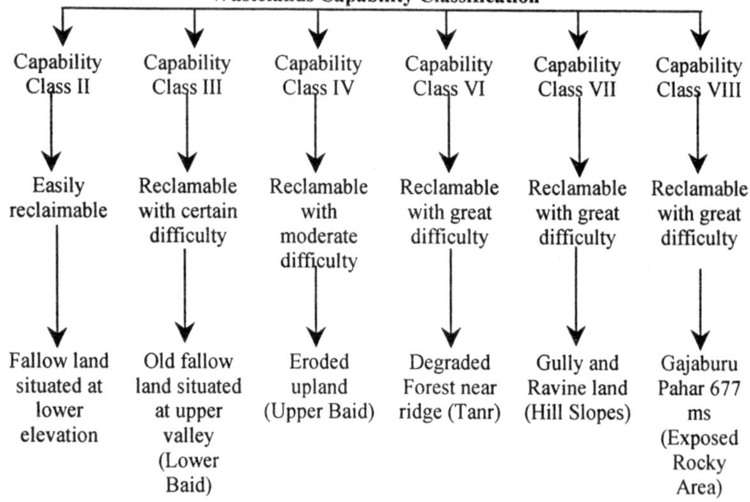

From the land capability classification it is clear that the watershed has got variegated landscapes and the entire watershed has got no land without any limitation for crop production. The Class II, III and IV lands require conservational measures to make them suitable for growing crops after thorough management practices. The Class VI, VII and VIII lands are not favourable for any crops, however, afforestation, agro-forestry or agro-horticultural programmes along with measures on gully pugging are required for development work in this type of land.

Revenue Land Classification

On the basis of the productivity, its situation and texture of the associated soils, the lands have been classified by the Revenue Department into following classes:

TANR : These lands are found on the highest part of the sloping surface composed of gravelly soil with bare outcrops which remain as fallow land, especially after deforestation of the area.

BAID : According to descending order next to the "Tanr" lies the "*Baid*" land with sandy and acidic soil.

Due to inadequate water supply and drainage condition these lands are mainly used for minor crops, oilseeds with poor yields or under village pasture.

KANALI: The next lower land with sandy loam soil is "*Kanali*" which receives drainage water from above. These are bunded and are mainly used for paddy crops with good yields.

BAHAL : "*Bahal*" land with sandy loam to sandy clay soil is situated in the valley bottom. They are exclusively used for paddy cultivation with good yields.

Table 4.2: Land Capability Classification in Gajaburu Micro-Watershed.

Capability Class	*Relief and Slope*	*Limitation*	*Existing Landuse*	*Proposed Landuse*
II-1	Valley bottom very gentle slope (1-2%)	Low permeability	Paddy cultivation	Land treatment and cropping
II-1	Valley plain (1-3%)	Low permeability	Paddy cultivation	Land treatment and cropping
IIw-1	Mid-valley plain (1-3%)	Low permeability poor drainage	Well bunded for paddy cultivation	Land treatment and cropping
IIIe-1	Gently undulating (1-3%)	Moderately eroded prolonged dryness	Terraced paddy field and barren lands	Soil conser-vational work to put under dry-land cultivation
IVe-1	Upper valley slopes (3-5%)	Severely eroded stoney, gravelly and rocky surface	Wasteland	Suitable for agro-forestry/ agro-horti-culture
IVe-1	Upland above 300 m. (5-10%)	Severe gully erosion, stoney gravelly and rocky surface	Uncultur-able wasteland	Work on gully plugging and afforestation

Table 4.3: Characteristics of Different Soil Series in Gajaburu Micro-Watershed along with Geomorphic Unit and Slope.

Sl.No.	*Name of the Soil Series with Symbol*	*Taxonomic Class at Great Group Level*	*Parent Material*	*Geomorphic United with Slope in %*	*Solum Depth in cm*	*Colour* *1. Surface* *2. Sub-surface*
1.	Chakra (C)	Ochraqualfs	Granite-gneiss and schist	Valley bottom (1-3%)	100-150 cm	1. DBR to DGR 2. YBR to GBR
2.	Kanchanpur (K)	Ochraqualfs	-do-	Toe-slopes (1-3%)	88 cm	1. LTYBR to YBR 2. YBR to BR
3.	Palpal (P)	Haplustalfs	-do-	Mid-valley plain (1-3%)	34 cm	1. YBR to DYBR 2. YBR to RBR
4.	Salaidahal (S)	Ochraqualfs	Alluvium	Valley plain and levees	152 cm	1. GR to YBR 2. G to Olive-Gray
5.	Rajdih (R)	Storthents	Granite-gneiss and schist	Upper convex (3-10%)	20 cm	1. DYBR to DBR 2. DBR to DRBR
Sl.No.	Texture 1. Surface 2. Sub-surface	Structure 1. Surface 2. Sub-surface	Permeability	Surface PH	Organic Carbon Surface in %	Basic Infiltrating Rate in cm/m
1.	1. S to SL 2. CL to C	1. SBK 2. Mid SC to SBA	Moderately Rapid	5.8 to 6.2	.39%	9.30
2.	1. SCL to CL 2. CL to C	1. Big to Mod. SBA 2. Mod. Coarse SBK	Very slow	5.4	.40%	1.80
3.	1. SL 2. SCL to CL	1. Mod. SBK 2. Mid. SBK	Mod.	5.8 to 6.2	.57%	1.90
4.	1. CL 2. CL to C	1. Mid. SBK 2. SCBK to ABK	Slow	5.5	.70%	2.10
5.	1. GSL to SL 2. GSCL to SC	1. Big SBK 2. Mid. SC to SBA	Mod. Rapid	6.5	.35%	NA

N.B.: Abbreviations and notations used are as per Munsell Colour Chart and Soil Survey Manual (USDA, 1952).

Table 4.4: Hydrological Parameters of the Watershed.

Rainfall	*Water Table*	*Drainage Density km./sq. km.*	*Ruggedness Number (RG)*
1. Pre-monsoon 168.57 mm	1. Pre-monsoon below 1 m	1. Below 280 m contour line 0.750	1. In plains very low (VL) RGVL Range 0.20 to 0.25
2. Post-monsoon 1056.59 mm	2. Post-monsoon 1 to 3 m	2. Above 280 m contour line 1.500 in upper catchment	2. In undulating upper catchment Moderate (M) RGM Range 0.075 to 0.100

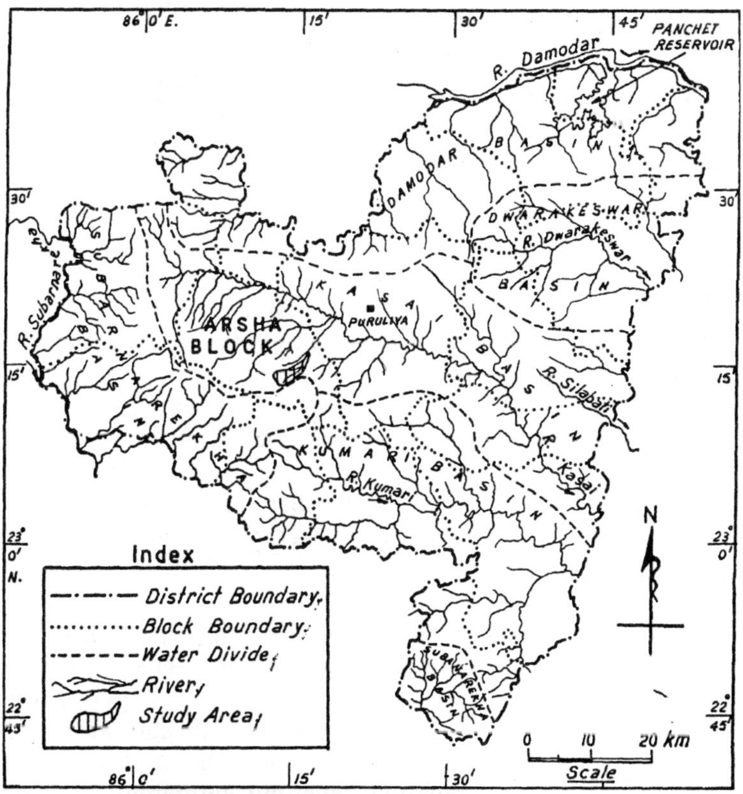

Fig. 4.1: Location map Gajaburu Micro-Watershed, Arsha Block, Purulia, West Bengal.

The aspects of micro top sequence as well as catenial associations are identified to study the land forms, erosion hazard including drainage status on the landuse. The lands belonging to "*Kanali*" and "*Bahal*" are found only along the lower valley bottom of Jamuna Jore. Except this the entire watershed is characterised by Tanr and *Baid* lands. Contour farming should also be used in conjunction with contour strip cropping, terracing, bunding, etc., in controlling erosion by surface flow in the cultivated land.

Soil

The soils are the result of the bio-chemical and physical processes operating upon the uppermost earth crust under various topographic and moisture conditions over a period of time.

The soils have developed from granite as the parent rock. The area is covered mostly with sedentary soils and portions of it are occupied by alluvial and colluvial soils. There are five soil series identified in the Gajaburu Micro-watershed. The land capability units are defined as soil-mapping units that have the same limitations in their management response and practices required. Table 4.2 exhibits the salient features of different soil series present in this area. A detailed soil and land capability classification of the watershed has been shown in Figure 4.2. The following five soil series are found in the watershed:

CHAKRA(C) : These soil series are well protected from erosion by sustainable bunds and are well adopted to paddy with very good yields.

KANCHANPUR (Kc) : They are well protected from erosion by suitable terracing and bunding. These are under extensive paddy cultivation with good yields.

PALPAL (P) : Lands under these series are mostly well protected from erosion by suitable terracing and bunding and intensively cultivated with paddy.

SALAIDAHAR (S) : These soils are susceptible to sub-soil erosion with formation of surface cracks. They are poorly drained with high water-holding capacity. These are cultivable wastelands which under good management (i.e. bunding, levelling, terracing) can grow paddy and other crops.

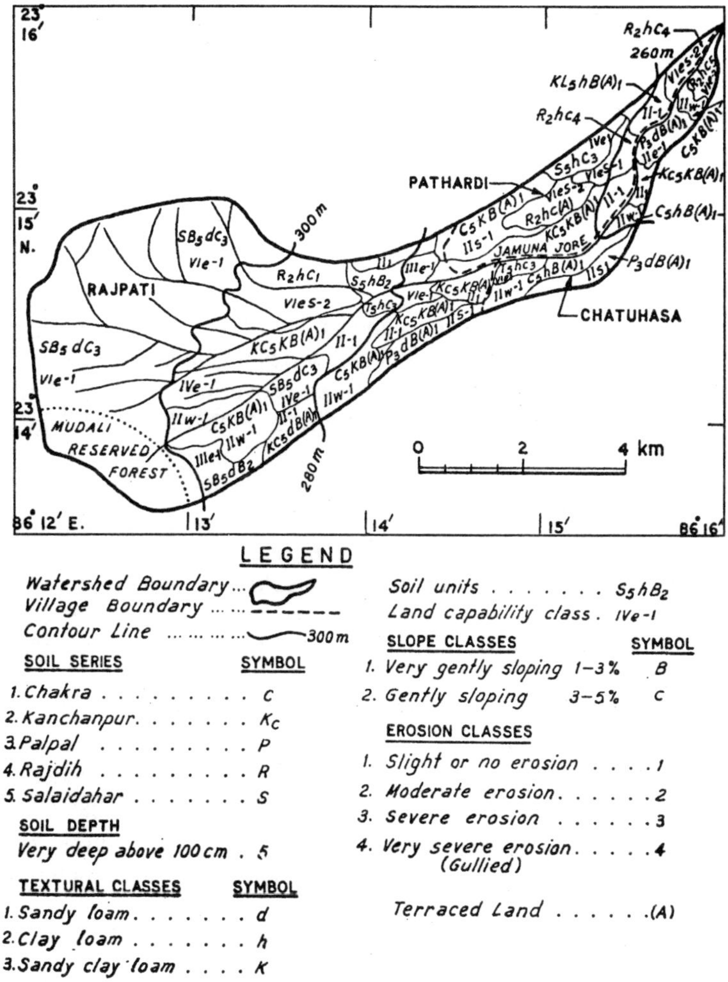

Fig. 4.2: Detailed Soil and Land Capability Gajaburu Micro-Watershed, Puruliya, West Bengal.

Source: All India Soil and Landuse Survey Organisation.

RAJDIH (R) : Rajdih series have developed upon granite-gneiss and schist complex. These are very severely eroded with gullies and ravenous lands. The permeability is rapid. The soils are excessively drained. These are under uncultivated wastelands under grass and new plantations of accacia and also other forest vegetations.

The Rajdih soil series cover about 62 to 70 per cent area of Rajpati village while Chakra and Palpal soil series together and Sirkaband soil series are found in marginal areas covering 25 per cent and 5 per cent of the watershed respectively. In Pathardi and Chatuhasa both Chakra and Salaidahar soil series cover more than 60 per cent of the area while Palpal-Rajdih-Kanchanpur soil series share rest of the area. Lands under these soil series are mostly well protected from erosion by proper soil conservational measures.

Soils grouped in Class II to IV are cultivable and crops can be grown under proper and scientific management. Soils under Class VIII are neither suited to crop husbandry nor to pasture and silviculture but are left for reserved forest, recreation, etc.

The micro-watershed suffers from acute soil erosion problems. The area being undulating and sloppy the velocity of run-off water gets accelerated and carries a very high quantity of eroded materials to the fertile land situated at valley bottom. Severe gullies have been formed in the micro-watershed. Faulty agricultural practices, overgrazing and deforestation are the main reasons of increased hazards.

Landuse

Landuse is an expression of man-nature interaction. Optimum use of the limited land resources of a region essentially depends upon a judicious and comprehensive landuse and agricultural planning considering all relevant factors to get maximum benefit from the land. Land capability classification is the most potent indicator for landuse planning in any region. On the basis of resources, socio-economic survey and land capability classification, detailed soil analysis and nature of wasteland a proposed landuse pattern was prepared along with existing landuse. Tables 4.5 and 4.6 and Figs. 4.3 and 4.4 show the present and proposed landuse of the entire watershed under planning respectively.

Tanr lands occur at high-level surface, partly forested and partly bare. *Baid* lands occupy the next lower level covering about 30% of the total area. *Kanali* lands (15% to 20% of the total area) are moderately fertile and are usually covered by sandy soil. The *Bahal* lands are low-lying areas in pockets of fertile transported alluvium.

Table 4.5: Existing Landuse.

Sl.No.	*Purpose*		*Area (in hectare)*
1.	Cultivated paddy land		164.00
2.	Culturable wasteland		
	(a) Fallow land	80.00	
	(b) Gullied land	20.00	100.00
3.	Horticultural plantation		2.00
4.	Unculturable wasteland (mostly forest land)		420.00
5.	Other landuse		20.00
		Total	**712.00**

Table 4.6: Proposed Landuse.

Sl.No.	*Purpose*	*Area (in hectare)*
1.	Additional area can be brought under paddy cultivation from gully plugging measure	16.00
2.	Arjun plantation for Tasar culture with dry-land cropping will be done from Compartmental Field Bunding Scheme	30.00
3.	Additional area to be brought under dry-land cultivation from Land Levelling Scheme	20.00
4.	Additional vegetable cultivation for Field Channel Scheme	2.00
5.	New area will be brought under Fruit Plantation	10.00
6.	Upper catchment of gullied area will be covered by grass	10.00
	Total	**88.00**

Source: Soil Conservation (Extension), Purulia, West Bengal.

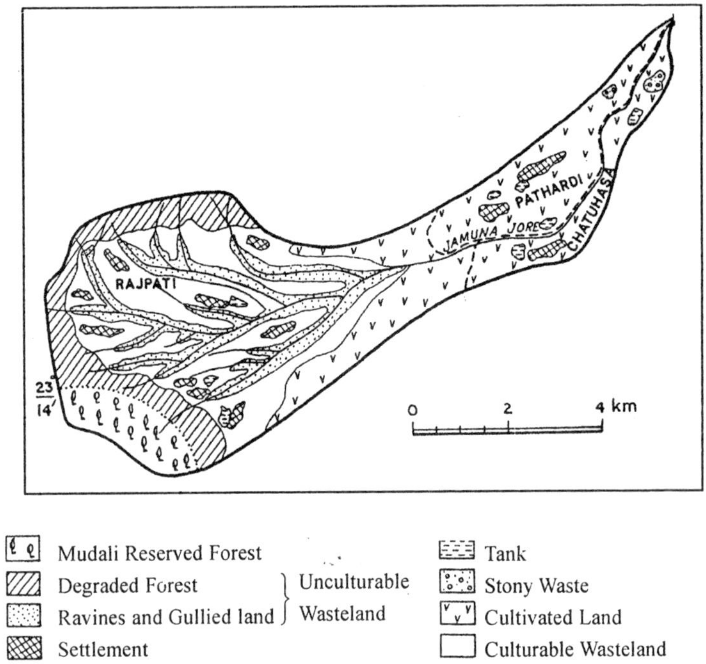

Fig. 4.3: Present Landuse: Gajaburu Micro-watershed, Puruliya, West Bengal.

There are both scientific and socio-economic parameters in the complex matrix of proper soil conservation and landuse. Disregard for this matrix is the major cause of land degradation and present level of soil erosion. Such a colossal waste can be controlled by altering the existing landuse to a pattern more comfortable with the terrain conditions.

The preservation of natural forest is a primary necessity for ensuring the genetic and biological diversity for posterity. Forests are treated as a revenue resource. With the removal of forests the livelihood crisis of the forest-dependent community has become acute. Rain-fed farming in slopes and uplands with repeated failures and low output is continuing without a break. This process can be checked by properly motivating the people and ensuring their participation at every level of rehabilitation and restorative programmes of wasteland development. Extension of soil forestry with suitable tree-mix and inter-mixing the tree crops with quickly harvestable crops and rotationally-mixed leguminous vegetables, grams, pulse, arhar shall

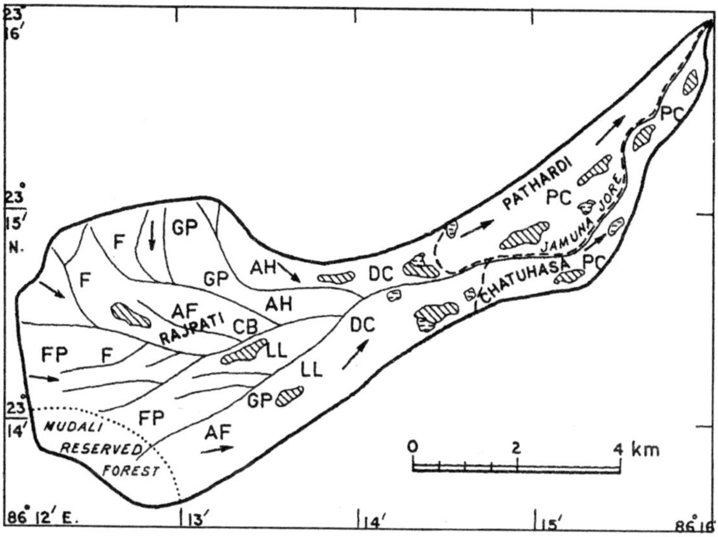

▧	Settlement		CB	— Contour Bunding
▤	Tank		LL	— Land Levelling
FP	— Forest Protection		AH	— Agro-Horticulture
F	— Forestry		DC	— Dry-Land Cultivation
AF	— Agro-forestry		PC	— Paddy Cultivation
GP	— Gully Plugging			Slope of Watershed

Fig. 4.4: Proposed Landuse: Gajaburu Micro-watershed Puruliya, West Bengal

be the likely option for rehabilitation of fallow lands. This shall provide an alternative livelihood for the impoverished land owners.

Further, the new approach, i.e. the application of right quantity of soil mix with atmospheric water-binding bio-degradable "POLYMERS", bio-incubators and bio-fertilisers which are capable of removing the periodic requirement of chemical fertilisers and irrigation will ensure biological reclamation of wastelands. Afforestation with extensive pastures and horticultural development in the relatively steeper slopes of the basin catchments can ensure soil and water conservation and a reduction of overgrazing problems. It would be an effective check on the major component of the soil erosion.

Extension of irrigation facilities by constructing water-harvesting structures, Jore bunds, etc., by arresting run-off water at the time of the rainy season is the most urgent need of the tribal cultivators. Along with the land capability limitations, acute land shortage and constraints to the increase of irrigation facilities may finally determine that the best alternative to further intensification of agriculture would be through raising of fuelwood, timber, fodder and horticulture. On a rough estimate one hectare of wasteland through scientific land management practices can produce about 10 tonnes of biomass beyond the fifth year. Most essential needs of the cultivators are fuel, fodder and timber. Wastelands could be used for pastures, fodders and horticulture. The obvious solution is to switch over to the forest-pasture-dryland farming pattern of landuse. This would promote both soil and water conservation.

Conclusion

It may not be possible to reclaim every bit of wastelands in the whole expanse of the area under study. The main objective is to bring the available wasteland to its optimal landuse after considering the land capability, site conditions and local needs. This would result in checking further land degradation, in increasing of biomass availability and also help in restoring ecological balance. Further, this will create employment opportunities within the village and thereby check the migration of labour during the working season.

Reviewing the objectives, strategies and options, it can definitely be said that watershed development is conceptually a two-fold management activity. It is simultaneously a programme of defensive as well as ameliorative mechanism in the developmental effort. Development without human face is tantamount to destruction.

Acknowledgements

The author is grateful to Shri Pranab Kumar Sengupta, Chief Soil Survey Officer, Soil Conservation Office, Government of West Bengal and Shri Anadinath Panda, Agricultural Development Officer, Soil Conservation (Extension), Purulia, for their valuable guidance and various suggestions in preparing this paper.

REFERENCES

All India Soil and Landuse Survey Organisation, Ministry of Agriculture and Irrigation, Govt. of India, Report No. Agri. 476, July 1979, pp. 16-30.

Bali, J.S. (1979), "Watershed management to fight drought", Souvenir, International Symposium on Hydrological Aspects of Droughts, New Delhi, p. 61-66.

Jaiswal, N.K. and Kottee, N.V. (1981), "Watershed approach", *Development of Drought-Prone Areas*, pp. 366-422, National Institute of Rural Development, Hyderabad.

Maiti, A. (1985), "A Case Study of Social Forestry in Purulia District", *Impact of Development on Environment*, Vol. 1, pp. 146-51.

Maiti, A. (1990), "Water Balance — An approach for a climatic study of drought in Purulia", West Bengal, *Indian Journal of Landscape Systems and Ecological Studies*, Vol. 13, No. 1, pp. 86-94.

Maiti, A. (1991), "Irrigation Need in Purulia — A climatic approach", *Indian Journal of Landscape Systems and Ecological Studies*, Vol. 13, No. 1, pp. 86-94.

National Wasteland Development Board (1986), *Technical Task Group Report on Wastelands — Definitions and Classification,* Government of India.

Project Report on Wasteland Development, National Mission of Wasteland Development, District Purulia, West Bengal, 1990, pp. 1-20, Government of West Bengal.

Sharma, K.C., "Watershed Management : An integrated approach" (in *Proceedings of the Summer Institute on Resource Inventory and Landuse Planning*, Jodhpur, Central Arid zone Institute), pp. 210-30.

USDA, (1984), "A manual on conservation of soil and water", *Handbook of Professional Agricultural Workers*, Oxford Publication Company.

Yadav, H.R. (1986), *Wastelands Capability Classification: Genesis and Utilisation of Wastelands*, pp. 172-81, Concept Publishing Company, New Delhi.

Yadav, H.R. (1989), *The Concept of Wasteland: Dimensions of Wasteland Development*, pp. 3-7, Concept Publishing Company, New Delhi.

5

Salinization of Soil and Water as an Important Factor in Desertification

I. SZABOLCS

The process of desertification which is a result of different environmental and human effects includes several essential part processes. Not all these processes are indispensable for desertification but most of them play smaller or bigger roles whenever and wherever desertification takes place.

One of such processes is the salinization of soil and water which is widely extended in the arid and semi-arid regions of the world. More than 50 per cent of the approximately 10 million km^2 of existing salt-affected soils can be found in arid and semi-arid areas mainly in deserts. According to the rules of geo-chemistry, aridity is favourable for the accumulation of water-soluble salts in the crust of weathering in the ground layers and soil horizons as well as in waters which leads to the formation of salt-affected soils. This rule is valid not only in relation to the environmental conditions developed without human interference in the past but also to man-made deserts which are a typical phenomenon of our days.

The accumulation of water-soluble salts, mainly sodium chlorides and sodium sulphates and partly sodium nitrates as well as some soluble calcium and magnesium salts, not only results in the lowering of soil fertility but also has an adverse effect on the environment in general. Saline conditions prevent not only plant growth but also the biological activity of micro-organisms, impair the purity of drinking water, and paralyse the metabolism between the compounds of soil, atmosphere and hydrosphere which would supply the biosphere with nutrients.

Salinization is not only a part process of desertification but also a direct influence on its development because if the soil and/or waters are saline the whole environment becomes less resistant to desertification than it would have been under conditions perfectly identical except for salinity. One we know this, it is evident that in combating desertification the combat against soil and water salinization has an essential significance.

As it has been mentioned above, a great part of existing deserts of the world are salinized but salinization also contributes a lot to the recent processes of desertification. Therefore, when man-made deserts are at issue, the man-made salinization of soil and water should also be taken into consideration.

Salinity and Secondary Salinization in Desert Agriculture

Particularly in arid and semi-arid regions, irrigation is as old as agriculture itself. The lack of rainfall made imperative the application of irrigation in many ancient agricultural systems. The problem stemmed from the fact that such systems developed mainly in such arid regions where, according to the landscape geochemistry of deserts and semi-deserts, this promotes salt accumulation. This problem was closely related to nearly all of the ancient cultures employing irrigation.

The effect of irrigation on salinization during the whole history of using this method in dry countries has never been fully elaborated in a comprehensive volume, though many books and papers describing this adverse effect and its consequences have been published all along the years.

It is well known that, e.g. in the valleys of the rivers Tigris and Euphrates in old Mesopotamia, fertile soils supplied abundant quantities of grain and other produces for a long time, feeding large populations in places covered by bare deserts in modern times. It is also well known that in ancient China, the Indus Valley and South America vast territories turned into deserts affected by salinity during irrigation by ancient societies. The problem of secondary salinization runs through the whole history of mankind. Evidently, there was neither sufficient knowledge nor technical means to predict, explain and combat salinization for many thousand years and, in consequence, the degradation of the fertility of soils and other adverse effects were recognised too late to do anything against their development. The process forced people to leave the land that became saline and others to cease production or to shift the irrigation to another place which, in many cases, also became salinized later. As long as new territories were available the shifting of irrigated agriculture temporarily solved the problem but either the growing density of population or the exhaustion of new land led to tragic consequences. More than one such example is known from history.

The Main Aspects of Secondary Salinization and Alkalization

With the development of science, step by step, the causes and regularities of salt accumulation have been disclosed. Modern soil sciences and other branches of science explain nearly completely the processes and hazards of salt accumulation in irrigated agriculture. Unfortunately, in spite of a comprehensive knowledge in this respect, the processes of salinization and alkalization have not been arrested, not even substantially diminished; on the contrary, they appear in new territories causing an enormous harm to a growing number of countries even today.

It is necessary to briefly summarise the main aspects of salinization and alkalinization caused by irrigation, and the agricultural as well as environmental consequences.

There are only a few chemical elements and compounds which play a decisive role in the salinization and alkalization of grounds, soils and water. They are as follows:

Cations:	Anions:
Ca^{2+}	No_3^-
Mg^{2+}	$C1^-$
Na^+	SO_4^{2-}
K^+	HCO_3^-
	Sio_3^{2-}

The ions and compounds, the migration and accumulation of which lead to the formation of saline and/or alkali soils, behave diversely in the weathering processes taking place in the earth's crust.

It is evident that in arid and semi-arid areas weathering processes result, as a final product, in water-soluble compounds which, due to the lack of precipitation, do not move from the place of their formation.

As it can be seen above, these final products are mainly responsible for the salt accumulation in rocks, soils and waters.

The geochemistry of salts in a certain place is determined by the mobility of the compounds formed and by the sequence of the precipitation of weathering products. The mobility of the rock-forming elements depends on the following factors:

(a) the stability of the crystalline network,
(b) the radius of ions formed during weathering, and
(c) the charge of the ions formed during weathering.

It is evident that the possibilities of translocation for weathering products depend mainly on their mobility.

Accumulated salts may be found in many places, in soils and waters of dry areas and constitute a potential hazard for irrigated agriculture.

The source of salts can be the salt content of surface waters which may be used for irrigation purposes. In this respect the picture is rather multi-coloured in arid and semi-arid regions. While in many places, particularly in big rivers, the water is of good quality, i.e. it has a low salt concentration, in other places, particularly in small tributaries, lakes, lagoons and swamps, often saline water exists. Evidently if this water is utilised for irrigation it will result, sooner or later, in secondary salinization and/or alkalinization. Here should be mentioned that on sea shores and in related areas the salt content of the sea water has a direct or indirect effect on the salinity of adjoining territories.

In arid and semi-arid regions the main cause of secondary salinization is irrigation; more precisely and application of improper methods of irrigation and drainage in the given area. If the goal is to harvest ample amounts of agricultural products irrigation is indispensable in deserts but if it entails salinization it will, by triggering a chain reaction, result in further desertification. The main causes of secondary salinization are:

(1) High salt content of irrigation water, and
(2) Rise of groundwater table bringing up salts to the soil profile.

Besides the two factors mentioned above, inadequate drainage systems also result in salt accumulation in the irrigation system and in its environment because the outflow of saline water is prevented or hindered.

Tremendous amounts of soluble salts can be found in the water of some rivers traversing deserts and semi-deserts. The water quality problems must be studied individually in such cases in order to control the chemical composition of irrigation waters.

Table 5.1: Annual Salt Accumulation in the Catchment of Two Rivers.

River	*Catchment area, km²*	*Airborne salts, t/km²*	*Discharge to sea, t/year*
Volga	1.401.949	2	8.000.000
Amu Darja	308.804	10	226.000.000

Table 5.1 also indicates that in many regions the quantity of airborne salts, too, should be taken into consideration. However, when comparing this amount with that of the salts collected and transported by waters, we shall find that, with the exception of some seashore districts, the quantity of airborne salts is negligible.

In arid regions, the ash of halophytes may contribute to the salinity of soils and waters. However, it is difficult to say whether it is a cause or a consequence when speaking in terms of landscape geochemistry, because halophytes grow as a result of an intensive salinity in their environment.

The main reservoir of saline water, in deserts and semi-deserts, is the groundwater. As a result of the geochemical processes described above, the bulk of water-soluble salts causing salinity during irrigation accumulates in the groundwater. As long as the groundwater table is deep and the moisture cannot come up through the capillary flow to the soil profile, even saline ground waters do not involve immediate salinization. As an effect of irrigation, the groundwater may rise so high that it can reach the surface layers and cause salinization even in the case of irrigation with good quality water. Unfortunately, this simple rule is often ignored during the planning and exploitation of many recent irrigation systems. The hazard of the rising ground water table by irrigation is underestimated because if this table is in great depths (at 10-20 m.) before the construction of the irrigation system it can easily be elevated to 1-2 m. depth below the surface, particularly in the case of lack or imperfection of the drainage systems.

It can be concluded that with the exception of places with good natural drainage, which is rather rare, in deserts and semi-deserts the lack of artificial drainage leads, sooner or later, to secondary salinization. Here should be noted that in this paper the term "secondary salinization" will be used in its conventional meaning, as differing from primary salinization which develops without human interference. Secondary salinization develops due to human activities, mainly due to irrigation.

Processes other than Irrigation causing Secondary Salinization

Improper irrigation and drainage are responsible for more than three quarters of secondary salinization in arid and semi-arid areas, and there are other mistreatments meted out by man to nature which also lead to the intensification or initiation of desertification. The most important instruments are as follows:

Deforestation

In arid and semi-arid conditions, but even sometimes in non-arid conditions, intensive deforestation often results in changes in the water economy of the landscape by turning a forest into bare land or plantation. It results in changes in the flow and metabolism of compounds, partly of soluble salts, in soils and waters, and often in considerable salt accumulation by the elevation of groundwater table or in alterations in the biological processes in plants and soil micro-organisms, e.g. in the 1970s and 80s in north-east Thailand large territories have been salinized even without irrigation following deforestation. Cutting of forest is dangerous when the hazard of desertification exists even without salinization but in many cases these processes are interrelated.

Overgrazing of Pastures

Overgrazing of pastures is another factor which often leads to the intensification of both salinization and desertification. Again, desertification caused by overgrazing is not always associated with salinization but, like in the case of other processes described above, interrelation is frequent. In consequence of overgrazing the original balance and metabolism of compounds between the natural plant cover and the soil alter. Concurrently, often salinization develops due to the decomposition of organo-mineral compounds and is followed by the accumulation of water-soluble salts. As yet another consequence of overgrazing, the ground water table often rises and elevates great amounts of salts into the soil profile.

Changing the Type of Cultivation

Similar problem may arise in the wake of changing the type of cultivation, e.g. by turning natural meadows into arable land or utilising them for plantation. The latter method of cultivation often drastically changes the water and nutrient balance in soils depleting them and initiating the chain reaction of desertification.

In several cases the annihilation of the natural plant cover of pastures and meadows contributes to the decomposition of the organic matter in the soil which again furthers the processes of desertification. Parallel with such processes the elevation of the table of mineralised ground water may be another source of salt accumulation.

Depletion of Fresh Water Layers

Depletion of fresh water layers near the surface of the soil and their replacement with saline ground water from deeper layers may happen due to irrigation (tube-wells) or by the utilisation of fresh-water reserves for drinking, animals, household purposes, etc. The effect of these courses of action is adverse in two respects, i.e. they result in salinization of soil and water as well as in the promotion of desertification processes.

Utilisation of Wood and Shrub Reserves in Arid Areas for Fuel

It is well known how widespread is this practice in many dry countries of Africa, Asia and Latin America. By burning the organic matter of plants growing on arid soils further plant development is hindered and the surface of the land is exposed to effects initiating and furthering desertification processes. Due to the mineralization of organic matter considerable amounts of water-soluble salts accumulate in soil layers and cause salinization in many places. The extensive practice of utilising of manure for fuel in many arid and semi-arid countries results in the same consequences because the burning of manure also adversely changes the balance of organic compounds in the given territory and leads to the accumulation of water soluble salts.

The cases described above only represent the main processes where salinization and desertification are closely interrelated and which develop in vast territories due to human actions which are improper for long-term development and contribute to the extension of desertification in many dry areas.

The above-mentioned and other processes are closely interrelated and their joint effect contributes more adversely to the development of desertification than their arithmetical sum. They are closely interrelated with minerological, hydrological and other factors of desertification increasing their influence.

In combating, preventing and predicting desertification, the combating, preventing and predicting of salinization is always essential. In the following pages some aspects of the prediction and prevention of man-made salinization will be described.

Combat against Secondary Salinization and Alkalization Methods and Recommendations

The extension of irrigation is and remains a major prospect for increasing yields and for nourishing the world's population. If the exploitation of irrigation systems is done carefully as also during the planning and construction phases of irrigation systems the necessary preliminary surveys and precautions are carried out and taken into consideration carefully, the production of food and raw materials can be multiplied in the future on a worldwide basis.

It is a must to intensify both the theoretical and technical activities relating to the hazard of secondary salinization and alkalization. Up to now the influence of the geochemical and hydrogeochemical processes has been underestimated in many places which, sooner or later, result in salt accumulation by irrigation in the given territory. While an abundance of studies and quality requirements for irrigation water are available, the effect of ground water on soil salinity is often left out of consideration. That is why a comprehensive preliminary study is necessary before planning new and extending existing irrigation systems, not only of soils and surface waters, but also of underground waters and layers (Darab-Ferencz, 1969).

In relation to the influence of ground water on salt accumulation, first of all the so-called critical depth of ground water table should be determined. This level means the depth below which, owing to natural or irrigated conditions, leaching prevails while above this level salt accumulation takes place in the soil profile. In other words, the salt regime of the given territory is in equilibrium.

Salt Regime

The salt regime of soils denotes accumulation and leaching of soluble salts, i.e. the periodical changes of these processes in the soil. The examination of the salt regime, the periodical changes in the quantity and quality of salts reveal the prevailing direction of their movements, whether the accumulation or the leaching of salts takes place under the given conditions.

Salt Balance

Although the study of the salt regime supplies important data on the dynamics of a salt-affected soil, it does not present in itself any further information. Therefore, the next step is to compare the salt content of the soil measured at given times and to express them in salt balances. Salt balances reveal the effects, or the effects to be expected, of amelioration or agro-technical measures on the changes in the degree of salinity and alkalinity of the soil as well as those of irrigation on its soluble salt contents. The establishment of salt balances requires an exact knowledge of the cause, speed and degree of accumulation or leaching.

The following data are needed for the establishment of the salt balance, regardless of the extent of the area concerned:

(a) Total amount of soluble salts at the beginning and the end of the observation.
(b) The increase in the soluble salt content during the observation.
(c) The decrease of the soluble salt content during the observation.

Three types of salt balances may be distinguished:

(a) Stable salt balance.
(b) Balance of salt accumulation.
(c) Balance of leaching.

The salt balance of soils may be established for both larger, e.g. a drainage basin, and smaller territorial units.

A salt balance may be established either for the whole soil profile or for given genetic horizons.

The tendency of the salt balance of a soil depends on the joint effect of numerous factors. From among these factors the following must be specifically mentioned.

Depth and Chemical Composition of Groundwater

When examining the water table, we must take into account the fact that it fluctuates periodically. The depth of the water table is influenced by local factors, e.g. the amount of precipitation in the given place, the micro- and meso-relief of the area. It is also affected by the amount of water getting into the catchment area from various sources supplying the groundwater and by the time when its effect manifests itself in the place examined. The depth of the ground water also changes when the natural conditions are artificially altered due to the effect, e.g. of inland drainage of irrigation.

The analysis of the chemical composition of the groundwater must be carried out simultaneously with the determination of the depth of the water table.

The following qualitative data of the groundwater must be given:

(a) The total salt content of the ground water in mg/1.
(b) The relative amount of sodium salts in the groundwater.
(c) The relative proportion of magnesium salts within the alkali earth metal salts.
(d) The distribution of soluble salts in the groundwater according to the anions.

The physical and water regime properties of a soil also affect its salt regime to a considerable extent. Of these two should be specifically mentioned: the water conductivity of the soil profile and, if present, the depth of the accumulation horizon of heavier mechanical composition or of disadvantageous water regime properties.

The Relief of a Given Area Influences the Salt Regime of Soils by Determining:

(1) The infiltration of rainfall or irrigation water into the soils of different relative height;

(2) The wetting of the soil profile; and
(3) When the downward movement of rainfall or irrigation water is followed by the upward movement of groundwater or the moisture content of deeper layers.

The Techniques of Irrigation

Under irrigated conditions the salt regime of the soils also depends on the method and technical level of irrigation.

In the case of surface irrigation, as a function of the amount of the applied irrigation water, leaching on a larger scale may take place. We must, however, take into account that part of the applied irrigation water gets into the groundwater and this may result in a rising of the water table, decreasing the possibilities for the leaching of soluble salts or even in shifting the salt balance of the soils toward the positive side, if the drainage conditions of the irrigated area are unsatisfactory.

Calculation of Salt Balance

Given the knowledge of the factors influencing the increase and decrease of the salt content of soils, the salt balance may be established on the basis of the following equation:

$$b = a +/d + \frac{CV}{Mt_{fs}} . 10^{-5} /$$

where

b = soluble salt content of the soil at the end of the observation in mg/100 g soil

a = soluble salt content of the soil at the beginning of the observation in mg/100 g of soil

C = the salt concentration of the irrigation water in g/1

V = the quantity of the irrigation water applied during the observation period in m^3/ha

M = the thickness of the soil layer for which the salt balance was established, in m

t_{fs} = the bulk density of the soil

d = the salt regime coefficient of the soil in g/100 g of soil

The change that has occurred in the salt content of a soil during the observation period is expressed in the salt regime coefficient. It gives the difference between the amount of the salts leached from the soil and the amount of the salts which got into the soil from sources other than the irrigation water.

As it is rather difficult to calculate the value of "d", we better determine the change in the salt content of the soil and calculate the salt regime coefficient on the basis of the following correlation:

$$d = b - /a + \frac{CV}{Mt_{fs}}.\ 10^{-5}\ /$$

Having completed the calculations for several irrigated soils, we can distinguish the following characteristic cases:

(a) The salt regime constant is of a negative value.
(b) The salt regime constant is of a negative value but the salt content of the soil increased during the observation period. This means that although leaching took place to a certain extent, more salt was introduced with the irrigation water than could be leached out.
(c) The salt regime constant is of a positive value and the soluble salt content of the soil increased during the observation period. This indicates that more salt got into the soil — not only from the irrigation water but also from the ground water and perhaps from other sources, too — than could be leached out.
(d) The salt regime coefficient of the soil may be zero. In this case the salt balance of the soil is stable and the current depth of the ground water table is practically tantamount to the critical depth.

To calculate the salt regime coefficient with the above simple equation requires the regular determination of salt movements in the given territory.

Where even such determinations are difficult, Kovda (1980) recommends a simple empirical estimate of the critical depth of the ground water:

$$L = 170 + 8t \pm 15$$

where:

L = the critical depth in cm

t = the mean annual temperature in Centigrade

Clearly, this equation only has a very general validity because the only variable is the mean annual temperature of the region.

Salinity Hazard

For the control of the possible hazard of salinization and/or alkalization in irrigated areas, or areas to be irrigated, the following factors should be studied and determined:

(1) Climatic factors, such as temperature, rainfall, humidity, vapour pressure, evaporation and their fluctuations and dynamics.
(2) Geological, geomorphological, geochemical, hydrological, hydrogeological and hydrochemical factors, such as: natural drainage, depth and fluctuation of the water table, direction and velocity of horizontal groundwater flow, salt content, composition of the groundwater, etc.
(3) Soil factors, such as soil profile, texture, structure, saturated and unsaturated water conductivity, soluble salt content, salt composition and salt profiles, exchangeable cations, pH, etc.
(4) Agrotechnical factors, such as landuse, crops, cultivation methods, etc.
(5) Irrigation practices, such as the amount of irrigation water, method, frequency and intensity of irrigation, salt content and composition of irrigation water, natural and artificial drainage, etc.

The above-mentioned factors determine the aims and methods of the preliminary survey of soils made in order to define the degree or existence of potential salinity and/or alkalinity.

Evidently, the environmental conditions on the one hand, and the methods of the utilisation of the territory in question, on the other, should be taken into consideration when an area is evaluated in this respect. Due to this fact, different limit values and different methods— based on uniform principles — should be selected in the course of this procedure.

In Table 5.2 a scheme of methods, recommended for the control of salinity and alkalinity in irrigated areas, is given. This table shows that the prediction of secondary salinization and alkalinization of the soils to be irrigated should be based on a preliminary survey of the landscape and soils before the construction of the irrigation system.

In this way, it is possible to take the necessary steps for the prevention of adverse processes.

Table 5.2: Scheme of Methods Recommended for the Control of Salinity and Alkalinity in Irrigated Areas.

A.	*Before construction of irrigation system*	
	Preliminary Survey	
	Landscape	*Planned Irrigation*
	climate	available irrigation water quantity and quality
	hydrology	groundwater depth and quality
	hydrogeology	technology of irrigation
	geomorphology	cropping pattern tolerance
B.	*During irrigation*	
	Monitoring	
	salinity and alkalinity of soil and groundwater table	
	chemical composition of groundwater	
	chemical composition of irrigation water filtration	
	physical soil properties	
	toxic elements, if any, in soil and water	

During irrigation, a well-organised monitoring of the soil and water properties is to be conducted in order to record changes, if any, and to take precautions, if necessary. Monitoring methods as well as the timing and location of sampling depend upon the local conditions.

In the course of making the survey and monitoring, in order to develop a reliable method for the prediction of salinization and alkalinization, the following problems must be solved:

(1) The main sources of water-soluble salts (irrigation water, ground water, surface waters, salty deep soil layers, etc.) must be identified.
(2) The main features of the salt regime must be characterised (salt balance); and the whole range of natural factors influencing the salt regime must be analysed.
(3) The effect of irrigation on the water and salt regimes of the soil must be determined.

Consequently, an exact salinity and/or alkalinity prognosis must be based on the evaluation of many natural and human factors and a thorough knowledge of the existing soil processes.

Numerous handbooks and recommendations are available on quality requirements for both soils and waters as the subject and means of irrigation respectively. We have to note that any of them must be adjusted to local conditions – (Szabolcs-Darab, 1982). It is beyond the aims of this paper to compare and evaluate different irrigation water quality systems of the different countries and regions. As a general rule, however, the following properties can be listed as the subjects of determination for all systems:

(1) Total salt content of irrigation water.
(2) Sodicity (Na^+) of irrigation water.
(3) Alkalinity of irrigation water.
(4) Mg content of irrigation water.
(5) Boron content of irrigation water.

The mapping of the results of preliminary and subsequent surveys constitute not only a good display of soil and environmental conditions of the irrigated areas, or areas to be irrigated, but also guidelines for proper irrigation and land protection. Such systems elaborated by various authors, including the author of this paper, for different places and conditions are also available in technical literatures (Szabolcs-Darab-Varallyay, 1969).

The monitoring system of irrigated areas must be elaborated and/or adapted also to closely related local circumstances. Soils, irrigation water and groundwater must be studied regularly and whenever discrepancies occur with the predicted salt regime, the necessary measure should be taken either by diminishing the acreage or the intensity of irrigation or by improving the drainage of the land.

Here only a few, rather random examples are furnished of the methods of studies aiming at the prediction and prevention of secondary salinization and alkalinization of irrigated soils. Our weakness is not the lack of such studies, but rather the lack of sufficient knowledge of how to adopt properly the correct methods. One of the main tasks for the future is to intensify the study of the possible hazard of increasing irrigation on soil and environment and to elaborate up-to-date and economical methods for the combat against he growing danger of the development of harmful processes.

REFERENCES

Darab, K., Ferencz, K. (1969), *Soil mapping and control of irrigated areas*, Hungary, OMMI Publ. Ser. 1, No. 10. Budapest.

Kovda, V.A. (1980), *Problem of combating salinization of irrigated soils*. UNEP.

Szabolcs, I. and Darab, K., (1982), Irrigation water quality and problems of soil salinity. *Acta Agronomica Scientiarum Hungaricae*. Tom. 31 (1-2), pp. 173-94.

Szabolcs, I., Darab, K. and Varallyay, Gy. (1969), Methods of predicting salinization and alkalinization processes due to irrigation on the Hungarian Plain. *Agrokémia és Talajtan* 18. Suppl. 351-76.

6

Assessment and Mapping of Desertification in Turkmenistan

SURENDRA SINGH

Desertification is one of the serious global environmental problems which has attracted the attention of environmentalists and the common people. Natural and anthropogenic processes are responsible for the development of the degradation of vegetative cover, wind erosion, water erosion, salinity and waterlogging types of desertification which have affected the biological productivity of the arid, semi-arid and dry sub-humid ecosystems. In order to combat desertification, the information on the present status of desertification is an essential prerequisite. Accordingly, the investigations on the assessment, monitoring and mapping of the desertification have been done by various workers in different countries of the world (Anon, 1976; Paylore and Hanry Jr., 1976; Anon, 1977; Jaiswal (ed.) 1977; Singh *et al.*, 1978, 1992; Dregne, 1991; Rozonov, 1992).

In the Karakum desert of Turkmenistan, the assessment and mapping of desertification were done using the remotely sensed and ground truth data by Kharin *et al.*, (1985) and also the methodology developed by FAO (1984). The findings of the above studies and also the study done by the author in the field and laboratory during his visit to Turkmenistan in the year 1990 are discussed in this paper.

Methodology

The study on the assessment and mapping of desertification in the Karakum desert of Turkmenistan was done using a three-tier system, viz. air, space and ground (Kharin *et al.*, 1985). Aerial photographs and Meteor. Salyut-4 and Soyuz-22 space imagery were interpreted and the areas affected by different types of desertification identified, delineated and mapped using the methodology developed by FAO (1984). The existing information in the form of maps and reports on different natural resources and their degradation was also used. The field studies were also carried out to check the information collected from the remotely sensed data and existing maps. The desertification hazard (DH) classes such as slight, moderate, severe and very severe were also determined by taking into account the sum of the values of the indices of desertification status (DS); desertification rate (DR); inherent risk (IR); domestic animal pressure (AP) and population pressure (PP). Based on these aspects the desertification status and risk maps of Turkmenistan were prepared (Figs. 6.1 and 6.2).

Results and Discussions

Based on the evaluation and the assessment of the present status of the natural resources of Turkmenistan by using aerial photographs, space photographs and ground truth, two types of desertification processes, viz. natural and anthropogenic, responsible for the present-day desertification in the Karakum desert of Turkmenistan have been identified (Babaev, 1982). In the Thar desert of Rajasthan similar types of desertification processes have also been identified (Ghose *et al.*, 1977).

Natural Processes

The manifestations of the climatogenic or natural desertification caused by natural processes are sand ridges and sandy plains, rock outcrops, gravels and stones, saline areas, clay surfaces devoid of vegetation and water erosion (Babaev, 1982). Slight to moderate wind erosion type of desertification occurs in the western, northern and eastern parts of the republic as manifested by sand mounds and sand sheets. Severe to very severe wind erosion type of desertification in the form of barchan dunes and sand ridges has been observed in the Karikul, Nebit-Dag and Repetek areas. The rock outcrops and accumulation of gravels, pebbles and stone occur in the Usturt plateau region and Zaunzug Karakum which inhibit the growth of the vegetation (Fig. 6.1).

In the western part of the desert, slight to moderate and severe to very severe salinity has developed in the coastal plains along the

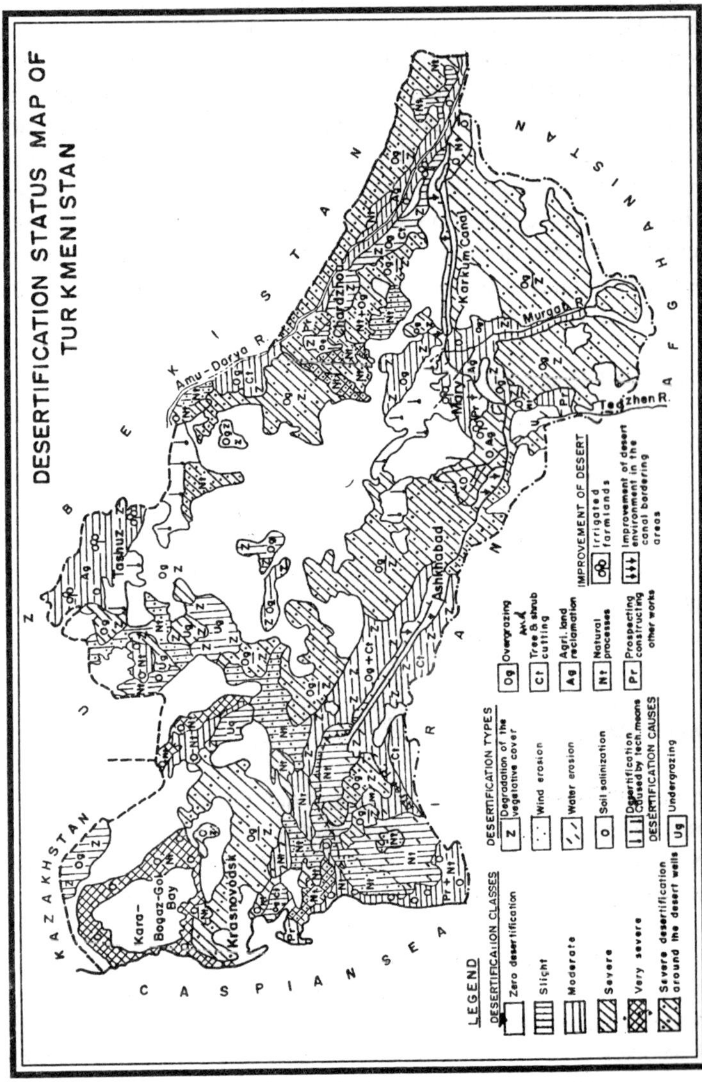

Fig. 6.1: Desertification Status Map.

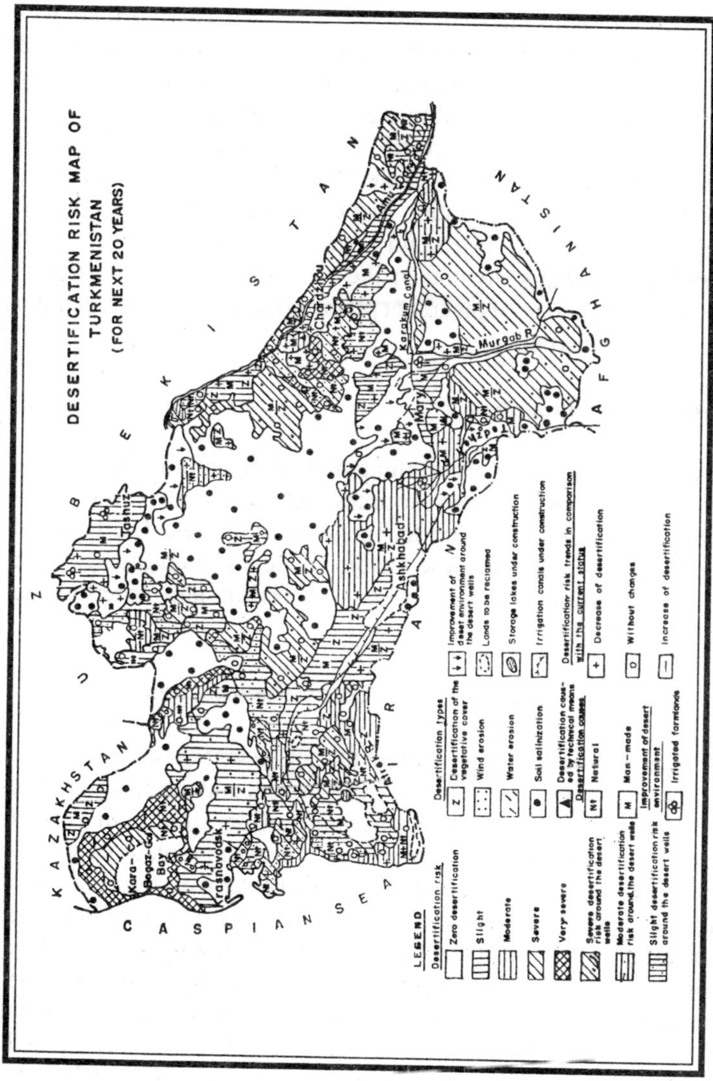

Fig. 6.2: Desertification Risk Map.

Caspian area and in the peripheries of the Kara-Bogaz-Gol Bay and Sarykamysh Sea. In these areas due to regression and transgression of sea, the salinity problem has increased manifold (Fig. 6.1).

The desertification risk map (Fig. 6.2) based on the current desertification status map indicate that in the next 20 years there will be no improvement in the areas desertified by natural salinity and wind erosion processes. However, the naturally desertified areas may slightly increase in the near future due to seepage.

Takyrs located in the north of Kopet-Dag mountains are characterised by clay deposits, sealed surfaces and salt encrustations and do not allow the vegetation and crop to grow. As a result, the biological productivity in certain areas has decreased resulting in desertification. Water erosion in the form of rills and gullies has desertified some areas of piedmont plains in the western and southern parts of Turkmenistan.

Anthropogenic Processes

In Turkmenistan, apart from climatogenic or natural factors, the desertification of natural resources has also been induced by anthropogenic, zoogenic and technogenic processes. The types of desertification caused by these factors are degradation of vegetative cover, wind erosion, water erosion, salinization and waterlogging (Kharin *et al.*, 1985).

Degradation of Vegetative Cover Type of Desertification

This type of desertification around the settlements, watering points and oases has been caused by the overgrazing, undergrazing, cutting and lopping of trees and shrubs and eating and trampling of vegetation. The desertification around watering points is of a very severe, severe and moderate nature in the western, eastern, southern and south-eastern parts of Turkmenistan (Fig. 6.1). The desertification in the first one-km-wide grazing belt, directly surrounding the watering points, is very severe due to the complete degradation of vegetation, intensive sand deflation resulting into the sand movement and formation of barchan dunes. In the second grazing belt of 1.5 km. width where grazing load is high and the vegetation cover is 20-30 per cent, many shrubs and grasses have disappeared and the sand deflation and deposition activities increased, the desertification is of severe nature. In the third grazing belt of 2.5 – 3.5 km. width, due to more distance from the watering point, moderate grazing and 70-80 per cent vegetation cover, the desertification is moderate (Fig. 6.3).

It also occurs in those areas of Turkmenistan which remained ungrazed for a long time and surface crust has developed due to the formation of lichen-moss-algae. It has resulted in the deterioration of plant habitats and regeneration and finally led to a decrease in pasture productivity. The cutting and lopping of vegetation from the different types of landforms, particularly aeolian landforms and its eating and trampling by rodents, have resulted in degradation of the vegetative cover type of desertification. This type of desertification in future will decrease around wells and oases due to proper soil and water conservation measures (Fig. 6.2).

Wind Erosion Type of Desertification

This desertification is manifested by the deflation hallows and accumulation of aeolian sand in the form of ripples, sand sheets, hummocks, reactivated dunes and barchans and barchanoids (Fig. 6.3). This type of desertification occurs around the oases and in sandy alluvial and coastal plains located in the western, central, eastern and southern parts of Turkmenistan and along the Karakum canal (Fig. 6.1). In the sandy desert due to the construction of the Karakum canal, roads, new settlements and extraction of oil and gas, the large areas have been affected by the slight to moderate and very severe wind erosion type of desertification. The moving sand and barchans caused by technogenic factors occur in a 0.5-km-wide belt along the Karakum canal. The deflation process has also been accelerated by the zoogenic factor in the stabilised dunes due to burrowing and eating and trampling of vegetation. The average number of burrows on one hectare of land in the Murgab oasis is 850-900, in the Amu Darya oasis 740-800 and in the Tedzhen oasis 490-550 (Babaev, 1973).

In future due to irrigation of large pasture areas, grazing regulations, afforestation and other suitable conservation measures, the deflation processes are supposed to decline resulting in a decrease in the sand drifting activities. But due to an increasing human impact along the Karakum canal, aeolian activities will increase and affect the large acreage of agricultural land (Fig. 6.2).

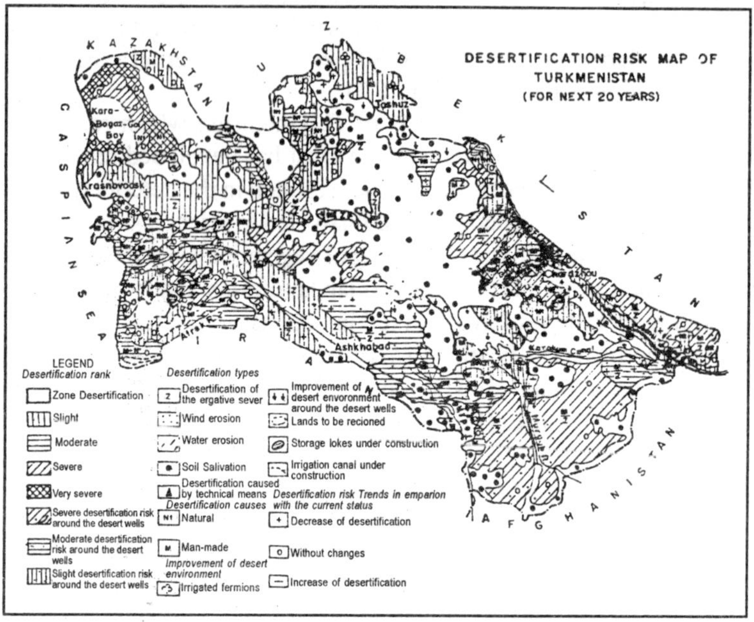

Fig. 6.3: Desertification Status Map of Turkmenistan.

Water Erosion Type of Desertification

The rills, gullies, ravines, gravels, stones and rock outcrops are the manifestations of this type of desertification. It is prevalent in takyrs, takyr-like areas and on mountains and piedmont plains and plateaus in western, northern and southern parts of the Karakum desert (Fig. 6.1). It is observed that on takyrs the average perennial water turbidity is about 3 kg/m^3. In takyr-like areas where the slope is greater than takyrs, water turbidity reaches 5 or 13 kg/m^3 and in some years 17-20 kg/m^3 (Leshchinsky, 1974). In mountains and piedmont areas due to heavy floods and small rivers, water turbidity exceeds 50-100 kg/m^3 (Krista, 1976). The Maly Balkhan and Kopet-Dag mountains, gullies 5-10 m. deep and 10-30 m. wide have developed resulting in the formation of typical badlands. About 120 gullies occur along one kilometre and the erosional network sometimes reaches 60-85 km^2. The stones, gravels and rock outcrops are also covering the surfaces of these landforms (Fig. 6.4).

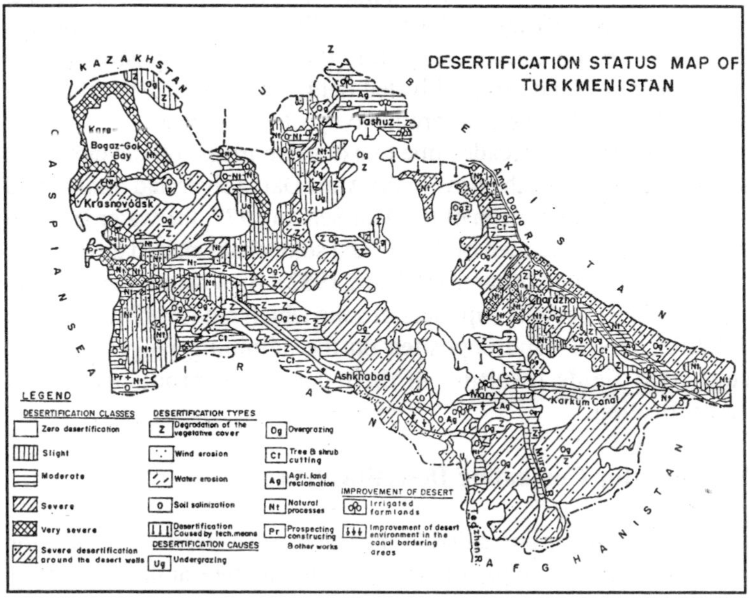

Fig. 6.4: Desertification Risk Map of Turkmenistan (For next 20 Years).

Salinization and Waterlogging Type of Desertification

The manifestations of this type of desertification are saline/alkaline and waterlogged areas, salt encrustations and surface crust developed due to excess irrigation by flooding, poor drainage, canal seepage and rise in groundwater table. This type of desertification largely occurs in four cases, namely, Chardzhou, Tashauz, Tedzhen and Murgab (Fig. 6.3). It is also developing in newly irrigated areas of the Karakum canal and Kopet-Dag piedmont plains (Fig. 6.1). In Tedzhen oasis, the moderate, strongly and extremely saline soils cover 60 per cent of farm areas. In the Chardzhou and Tashauz regions, the soils are mainly developed under hydromorphic conditions with mineralization reaches up to 20-30 g/l. In the Mary region, the sols are strongly salinized and the average mineralization of the subsoil is about 11 g/l (Kharin *et al.*, 1985). In future, by the year 2000, this type of desertification due to irrigation farming will develop in large areas of the Karakum canal (Fig. 6.2).

In the year 1980, the flow of the Caspian Sea into the Kara-Bogaz-Gol Bay was stopped by complete damming of the western shore of this bay. As a result, the surface of the bay dried up, salt encrustations developed and its large acreage turned into solonchaks resulting in desertification. Due to the deflation process, large amounts of salts are being carried away from this bay and deposited on the adjoining rangelands and cultivated fields resulting in desertification and depletion of biological productivity of these areas.

In the last few decades due to extensive irrigation development and flow regulation, the flow in the Amu-Darya river had diminished resulting in the extinction of the Aral Sea. In 1977, the water level on the eastern shore receded by 50 km. and solonchaks started developing on 1,000 km^2 dried off area (Babaev, 1982). As a result, these dried-off areas have become suppliers of salts which are transferred by the deflation process and deposited on the adjoining rangelands and cultivated fields resulting in their decreased productivity and desertification.

Measures for Combating Desertification

The following various suitable measures have been suggested for combating different types of desertification in the Karakum desert of Turkmenistan (Kharin *et al.*, 1985).

Combating of Degradation of Vegetative Cover Type of Desertification

This type of desertification around unevenly distributed watering points could be checked by even distribution of watering points over the grazing areas by digging out more wells and use of surface water by harvesting in tanks and reservoirs.

In order to avoid the development of moss-lichen on protected grazing lands and degradation of vegetation, such lands should not be left ungrazed for more than six years.

Rotational grazing and improvement of pastures will allow to periodically replenish the seed stock of valuable fodder plants in soil for subsequent reproduction and 3-6 three to six-fold increase in pasture productivity.

Combating Wind Erosion Type of Desertification

The mechanical protective structures such as standing row and checker-board pattern have been found useful to check wind erosion and subsequent deposition of dune sands. Seeds of Haloxylon persicum (3 kg.), Salsola paletzkiana (4 kg.), Calligonum species (8 kg.) and cutting and seedlings 3,000 each are needed for afforestation of 1 hectare area of moving dune.

Mechanisation of sand stabilisation and afforestation operations have proved to be successful through chemicals ensuring the formations of stable anti-deflation crusts on sand surfaces. The promising chemicals are nerosine, waste oil products, sulphate-alcohol distillery refuse, gossipol oil, etc.

The by-product of oil locally known as Majud used at the rate of 4 tons/ha. for sand-binding has been proved more effective and cheaper than other chemicals. The total cost of the combined chemical and biological measures for the stabilisation of moving sand comes to 200 to 250 roubles/ha. whereas the total cost by using mechanical and biological measures is about 2,000 to 25,000 roubles/ha.

Combating of Water Erosion Type of Desertification

This type of desertification could be controlled by regulation of stream flow, slope run-off and soil protection on slopes by contour bunding and contour furrows. The other important measures are pasture management and improvement, protection of trees and shrubs and afforestation.

In dry farming which is only possible in mountain regions, the soil tillage should be across the slopes (horizontally) and proper crop rotation should be used for soil erosion control. Cultivation of annual crops with deep ploughing (25-30 cm.) in autumn and furrow-making should be done on slopes below 5° to 10°. The cultivation of perennial forage plants combined with protective belts of perennial forage plants is advised on slopes steeper than 10°.

To prevent irrigation erosion on the slopes of 2° to 3° and the furrow length 150 m., irrigation should be started with the pressure gradient of 0.07 litre per second. On the slope of 1° to 4° on the furrow length within 100 m., the pressure gradient should be 0.15 to 0.01 litre per second and that on slopes of 3° to 6°, it should be 0.01 to 1.15 litre per second.

Combating of Salinization and Waterlogging Type of Desertification

To avoid this type of desertification, the flood method of irrigation should be replaced by drip and sprinkler irrigation depending upon the terrain conditions. This will check the rise of ground water table and will keep it below the critical limit of 2 to 2.5 m. The crop rotation should be strictly planned according to soil salinity. It should be 3:6 to 3:7 (3 fields under alfalfa and 6 and 7 fields under cotton).

The collector-drainage network is one of the efficient means for desalinization of the irrigated land and also for those to be irrigated in the near future. The agro-technical practices such as land levelling, leaching, flooding and inter-row oil tillage eliminate the negative impact of secondary salinization.

Human interference with natural drainage systems like complete damming and diversion of stream flow, etc., should be avoided. The salinization and waterlogging conditions developed in the Kara-Bogaz Gol Bay and Aral Sea during the last few years could be improved by demolishing embankment along the Bay and allowing the waters of the Amu-Darya into the Aral Sea.

Conclusion

Remote-sensing techniques in conjunction with ground truth have been proved useful for the identification and mapping of different types of desertification caused by the natural and anthropogenic processes in the Karakum desert of Turkmenistan. The anthropogenic processes are more dominant and widespread affecting 80 to 90 per cent area of the different ecosystems in the desert. Whereas, the natural desertification processes are only responsible for the degradation of 5 to 10 per cent area of the Karakum desert. The mechanical, biological and chemical methods could be successfully employed for combating desertification processes.

Acknowledgements

The author is deeply indebted to the Director, Head of Resource Survey and Monitoring Division, Central Arid Zone Research Institute, Jodhpur, and the Indian Council of Agricultural Research and Education, Department of Agricultural Research and Education, Department of Science and Technology, New Delhi, and the former USSR Academy of Sciences, Moscow, and Turkmen SSR Academy of Sciences, Ashkhabad, for providing me an opportunity to visit Moscow and Turkmenistan for studying arid zone ecology and control of desertification under the Indo-Soviet Integrated Long Term Programme (ILTP). I am extremely thankful to the Director, Deputy Director, Scientists and interpreters of the Institute of Desert, Ashkhabad, Turkmenistan, for providing me the facilities to visit laboratories and field experiments and also for sparing their valuable time for scientific discussions on the different desertification problems and their control and also to permit me to use their published reports, monographs, maps, etc. for writing this paper.

REFERENCES

Anonymous, (1976), Problems in the Development and Conservation of Desert and Semi-Desert Lands, *Pre-congress Symposium K 26*, Ashkhabad, USSR, p. 216.

Anonymous, (1977), Country Report: India. United Conference on Desertification, UNEP, Nairobi, p. 71.

Babaev, A.G. (1973), *Oases Sands in Turkmenistan and their Development Trends*, Ylym, Ashkhabad, p. 353.

Babaev, A.G. (ed.), (1982), *Combating Desertification in the USSR; Problems and Experience*, USSR/UNEP Publication and Information Support Project, Moscow, p. 120.

Dregne, H.E. (1991), "Global status of desertification", *Ann. Arid Zone* 30: 179-185.

FAO, (1984), *Provisional Methodology for Assessment and Mapping of Desertification*, Food and Agriculture Organisation of the United Nations, Rome, p. 84.

Ghose, B., Singh, S. and Kar. A. (1977), "Desertification around the Thar: A geomorphological interpretation", *Ann. Arid Zone* 16(3): 290-301.

Jaiswal, P.L. (ed.) (1977). *Desertification and its Control*, ICAR Publication, New Delhi, p. 358.

Kharin, N.G., Nechaeva, N.T., Nikolae, M.V. Babaev, T. Dorbin, L.G. Babaev, A., Orlovsky, N.S., Redzhepbaev, K., Krista, B.T., Nurgeldyev, O.N., Batyrov, A. and Svintsov, I.P. (1985), *Methodological Principles of Desertification Processes, Assessment and Mapping Arid Lands of Turkmenistan Taken as Example.* CED. A.G. Babaev; Desert Institute Turkmen, SSR Academy of Sciences, Ashkhabad, p. 116.

Krista, B.T. (1987), "Outlook for use of local surface water", *Jour. Problems of Desert Development*, (5): 30-37.

Krista, B.T. (1976), *Hydrological Features of Western Areas in Central Asia*, Ylym Publ., Ashkhabad.

Leshchinsky, G.T. (1974), *Resources of Ephemeral Surface Runoff in Desert of Central Asia and Kazakhstan*, Ylym. Publ., Ashkhabad.

Paylore, P. and Haney, Jr. R.A. (1976), *Desertification: Process, Problems, Perspectives*, Office of the Arid Lands Studies, University of Arizona, U.S.A., p. 125.

Rozonov, B.G. (1992), "Present-day human-induced desertification", *Abstracts, Int. Symp. on Evolution of Deserts*, Ahmedabad, pp. 166-167.

Singh, S., Ghose, B. and Kar, Amal, (1978), "Geomorphic changes as evidence of Palaeoclimate and desertification in Rajasthan desert, India (Luni Development Block – A case study), *Man and Environment*, 2: 1-13.

Singh, S., Kar. A, Joshi, D.C., Ram, B., Kumar, S., Vats, P.C., Singh, N., Raina, P., Kolarkar, A.S. and Dhir, R.P. (1992), Desertification mapping in western Rajasthan. *Ann. Aridzone* 33(3).

7

Impact of Crop-Husbandry Production on Desertification in Kenya

CLEOPHAS LADO

Introductions

The broad objective of this paper is to assess the impact of crop-husbandry production on desertification in Kenya's arid and semi-arid environments. The classification of existing crop area by type, including tree and agricultural crops is shown.

Second, the impacts of intensive crop husbandry and existing farming systems on desertification are discussed.

Third, the identification of the potential use of specific crops for desertification control in relation to the possible agro-forestry systems is noted. Finally, the paper proposes some appropriate monitoring strategies required for the improvement and sustainable crop production systems and environmental management in Kenya.

Preamble

This paper addresses itself to the impact of crop-husbandry production on desertification in Kenya. The fundamental and persistent desertification "threat" in the Arid and Semi-Arid Lands (ASAL) of Kenya obviously forms one of the foci of the National Environment Secretariat (NES). Hence, a central concept of this paper is the notion of desertification control in the ASAL.

Arid and semi-arid environments seem inseparable since both environments differ only slightly in terms of their degree of aridity (Obara, 1980). Aridity refers to the effectiveness of precipitation, which is the total precipitation actually available for use by plants and animals. When precipitation is less than annual potential evapotranspiration, vegetation cover tends to become sparse and the area may be described as arid with or without the prefix "semi".

Dresch (1975) observed that semi-arid zones occur between arid and humid zones. In Kenya arid and semi-arid areas are broadly classified as those areas receiving less than 760 mm. annual rainfall (Brown, 1963a; Ominde, 1971), the minimum precipitation under which rainfed agriculture is normally expected to be moderately practicable, although Swynnerton (1954) makes this figure less than 635 mm. per annum. The area most prone to desertification is estimated to cover about 83 per cent of the total land surface of Kenya. This area has been classified as semi-arid, arid and very arid on the basis of the index of available water, ranging from 31-45, 12-30 and 5-11 per cent respectively (Woodhead, 1970). These agro-ecological zones know no international boundaries since they extend to Northern Uganda, Southern Sudan, Central Tanzania, Somalia and Eastern Ethiopia. This implies that any realistic desertification control in the ASAL requires international cooperation and participation.

Arid and semi-arid zones have been significant in the context of range management and development (Ominde, 1971; Heady, 1972; Pratt and Gwynne, 1977). Semi-arid zone alone comprises about 15 per cent of Kenya, which is an area of marginal agro-ecological potential and potentially productive rangeland, although it is being affected by pockets of encroaching woody species (Republic of Kenya, 1984). The arid zone accounts for approximately 46 per cent of the total area of the country, which is generally an area of potential rangeland containing "islands" suitable for agriculture. Wildlife and tourism could probably constitute a better alternative to pastoralism in the land use system in this low agro-ecological potential zone. Conversely, the very arid land covers about 22 per cent of the country, which is rangeland of low potential consisting of very dry form of bushed grassland.

Although the above three agro-ecological zones are generally too dry to sustain dependable agriculture except where terrain and irrigation permit, the relatively high population growth and the decline in the agricultural development cores (high agro-ecological potential zones in the Lake Victoria Basin, Highlands and Coastal region) of Kenya now presuppose that utilisation of water deficit regions for crop production must be reinforced. It is the Government's plan to focus more attention to the development of ASAL (Republic of Kenya 1974, 1979). But in order to provide a sound scientific basis for rehabilitation and sustainable development of the arid and semi-arid ecosystems, it is a necessary prerequisite to combat desertification. Therefore, the main objectives of this paper are:

(a) To highlight the principles of desertification;
(b) To classify existing crop area by type, including both tree and agricultural crops in ASAL;
(c) To assess the impacts of intensive crop husbandry and existing farming systems on desertification giving specific examples;
(d) To identify the potential use of specific crops in desertification control (e.g. as relates to possible agro-forestry systems); and
(e) To propose the appropriate monitoring mechanisms required for improved environmental management in crop production systems.

Principles of Desertification

Concepts of Desertification

What is the concept of desertification? A clear conception of the desertification process, its pre-conditions, the controlling factors and the consequences is essential in order to obtain a satisfactory answer to this question. Most of the decision-makers and even development experts who draw up the schemes for controlling desertification use the term without having any clear idea of what it means. This uncertainty is often expressed in the fact that desertification is confused with drought, shortage of water, dune migration, or simply with the existing deserts (Ibrahim, 1983). This scholar observed that the confusion about the process of desertification has its origin in the fact that ecologists have so far only sporadically investigated the precise nature of this process. He noted that a similar situation exists with regard to the anthropogenic aspect of desertification. There are two reasons for this (Ibrahim, 1983):

(a) The observation period has to extend over several decades if reliable findings are to be made.

(b) Neither the ecological nor the socio-economic aspect of the desertification process can be investigated separately and isolated from the other natural and socio-economic processes within the area.

Out of purely practical considerations, Ibrahim defined desertification as the process of creeping destruction of the ability of arid and semi-arid ecosystems to generate themselves, caused by land use methods which are not adapted to the natural conditions. Because of this problem, soil, vegetation, water balance and hence also the micro-climate are so seriously affected by desertification that desert conditions become widespread and this ecological degradation triggers off series of socio-economic consequences (Fig. 7.1).

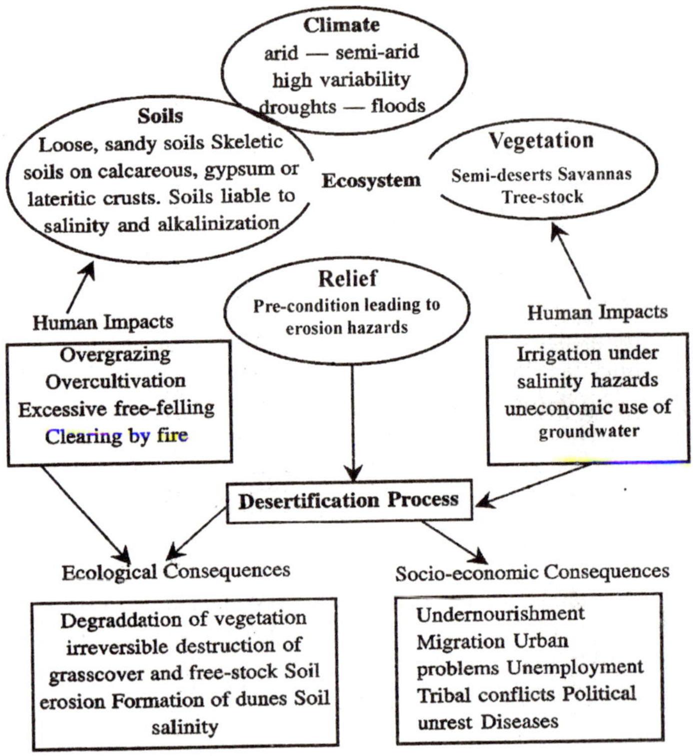

Fig. 7.1: Model of the Desertification Process.

Figure 7.1 shows a conceptual model of the kind of desertification process presumably experienced in the ASAL. This model shows the forms of human intervention (e.g. overgrazing, overcultivation and irrigation). One form of human intervention is overgrazing and the resultant destruction of the grass cover. However, both the nomads and sedentarized population can be responsible for overgrazing. This model of desertification does not show the causal grading of different human interventions in the ecosystem, since the complex set of causes can be seen to have different compositions in the different areas. For example, overgrazing in the rangelands of Kenya is a more important cause of the process of desertification than in the humid and sub-humid agro-ecological zones, where intensive cultivation of cereals is probably the most important factor.

An extensive review of the various issues related to the concept of desertification has been given elsewhere (Verstraete, 1983). The concept of desertification was introduced in 1949 to designate a number of ecological degradation process in tropical Africa, especially the progressive transformation of tropical forests into savannahs or even drier ecosystems (Verstraete, 1983; Ojany, 1986). Since then, the word has received a number of other meanings, some of them contradictory, most of them ambiguous. However, it is generally agreed that desertification is a dynamic process, a climatic manifestation, socio-cultural issue and the result of economic or political constraints (Verstraete, 1983).

Desertification is the spatial extension of desert-like conditions as a result of man's impact on the ecosystems of semi-arid regions (Mensching and Ibrahim, 1977). These scholars observed that desertification occurs mainly in desert-boundary regions and involves a complex of physical geographical processes having negative consequences of landuse and, ultimately, leading to a strong disturbance of the natural ecological equilibrium. Desertification is differently used and interpreted by different scientists (Rapp, 1974; Le Houerou, 1975). Both scholars prefer to call this ecological process desertization. However, the term "desertification" is of wider international usage.

According to Chanda (1980) desertification is the intensification of the extension of desert-like conditions to areas outside the present deserts. He pointed out that desertification is the process of impoverishment of arid, semi-arid and sub-humid ecosystems leading to reduced productivity of plants, and to a reduction in biomass, animal population density or crop yields. This implies that desertification cannot occur in extremely arid or hyperarid regions devoid of life because there is nothing to desertify in such areas. The areas currently in danger of desert encroachment are those bordering present hyperarid deserts.

Desertification is not a problem restricted to deserts, but to all areas or ecosystems where mismanagement of the environment due to human action prevails, leading to the impoverishment of the land and its natural resources (UNEP, 1979). UNEP conceived desertification as a process of ecological degradation in arid, semi-arid and sub-humid lands by which the productivity of land is lost or substantially diminished: grazing lands cease to produce palatable pasture, dryland agriculture fails and irrigated fields are abandoned due to salinization, waterlogging and other forms of soil deterioration. Although the same processes leading to desertification are operational in almost all areas of Kenya and many areas show latent stages of the process, it is in the ASAL where the process has significantly developed. The present paper is concerned with the impact of crop husbandry on desertification in ASAL.

Limits of Desertification as a Scientific Subject of Study

Besides the discrepancy in the concept of desertification, controversy also prevails as to the limits of the phenomenon as a scientific subject of investigation:

(a) Differing accent has been placed by different scientific disciplines tackling the same problem. Generally, one can distinguish two principal streams of scientific investigation: the one analyses the natural processes, their climatic causes, characteristic forms and their effects on the ecological system; the other deals with socio-economic aspects and their political implications. As a geographical subject of study, however, desertification should concentrate on the field of interaction between the natural conditions in desert marginal zones and man's destructive activities. Participation in investigating and solving this interdisciplinary problem is an important task for geographers in Kenya and elsewhere in the world.

(b) The geographical boundaries of the zone within which the processes of desertification occurs are not exactly definable. The greatest number of changes, causing the heaviest damage, are concentrated in the arid and semi-arid zones. Analogous changes elsewhere, for example in the humid tropics, are irrelevant to the present paper. One should not list every anthropogenic impact on ecosystems under desertification. This paper concentrates on the impact of crop husbandry on desertification in the ASAL in Kenya.

(c) Because climatic fluctuations and dangers have a supra-regional character and extend over long periods of time, many scientists are tempted to involve themselves with palaeo-climatic changes. The investigation, therefore, assumes an inappropriate time-frame, such that vague conceptions of the significance of the processes of drastic climatic change occurred many times during the course of geological history; noticeably during the *Pleistocene* and *Holocene*, when the present day deserts became established 3000 to 6000 years ago. Although we should limit ourselves to the last 6000 years, special attention should be paid to the last 100-200 years, because man's impact on natural ecosystems is strongest and the processes of desertification can be more easily observed and controlled.

(d) The term "desert encroachment" is often used instead of "desertification". The latter term has, however, a much broader sense. Besides, the term "desert encroachment" lays the blame on the desert, i.e. on nature, while man is to be blamed for greater part of desertification. It is man who is said to be encroaching on the fragile ecosystems of the arid and semi-arid environments in Kenya changing them into desert-like conditions. The present paper is partly concerned with the processes that lead to such "man-made deserts" and largely concerned with their control.

Extent of the Phenomenon of Desertification

The transitional zone between deserts and dry savannahs, is an ecologically vulnerable region in which man's destructive activities are most far-reaching. Naturally, the expansion and intensification of desert-like conditions do not occur along a continuous front, systematically pushing back the cultivated, pastoral and wildlife areas.

It is more a spot-like extension within the transitional zone itself, for landuse has there to be concentrated in certain favourable areas. In extremely dry periods running over several years, these degraded areas may expand considerably and grow together into a vast desertified region.

In addition to climatically determined zonal extension of the desertified areas, anthropogenic causes play a decisive part. Generally, areas of long periods of settlement tend to have a higher degree of degradation than relatively recent settlements. The process of desertification is also enhanced by the human population pressure and the intensity of land use. Practices, such as clearing by burning, extensive ploughing of grasslands, overgrazing and overcultivation accelerate the process of soil degradation.

It is difficult to draw a sharp line between climatic and anthropogenic factors. Grove (1974) makes the differentiation clear-cut when he says: "Drought is part and parcel of the climatic pattern. Desertification is the work of man". Desertification is a dynamic process which spreads imperceptibly, and if not checked before critical threshold is reached, can be irreversible (Ojany, 1986). But this is not necessarily true, it essentially depends on: the entropy of the ecosystem, their ability to heal, and their resilience (Chanda, 1980). This scholar states that resilience, in turn, depends on climatic, soil and biological conditions, particularly the dynamic status of plant communities, the availability of diaspores from pioneer species, etc. Desertic ecosystems are essentially water-controlled ecosystems with frequent, discrete and largely unpredictable water inputs (Noy-Meir, 1976). And because water or precipitation, which we may consider as the driving variable comes in 'pulses' of short duration, Noy-Meir observed that the ecosystems must of necessity respond in 'pulses' too to make maximum use of this resource. Thus, Chanda argues that when an effective rainfall event occurs, biological activity is rapidly stimulated, subsequently leading to increased biomass. However, we cannot pursue further the resilience theory because it is irrelevant to the desertification control in the ASAL where anthropogenic causes play a major part.

Indicators of Desertification

Desertification monitoring, assessment and intervention cannot be carried out without a knowledge of the indicators (symptoms) of the process. How can we diagnose that arid and semi-arid ecosystems in Kenya are undergoing desertification? This can be noticed through: (a) reduced productivity of the dry matter (or living phytomass) in the environment, and decline in food production; (b) the reduction in the genetic diversity of the large plants and animals; (c) deforestation, catchment loss, accelerated erosion and silting; (d) creation of desert-like conditions; and (e) increased hazards for human occupancy, thus leading to migration to more favourable areas.

The above changes may lead to destruction of good land by the creation of destructive erosional channels and the production of rugged terrain by water and wind. This is the outcome of a process of vegetational and soil deterioration. It has been rightly stressed that in marginal lands the process of desertification affects the water balance and ensuing ecological stress triggers off a series of deleterious (harmful) socio-economic consequences of which drought and famine are the most obvious (UNCD 1977, Ibrahim 1982, Eckholm, 1984). But what is the difference between desertification, drought and famine?

Drought refers to a temporary drop in expected amount of precipitation (particularly rainfall) for an area. Usually, it results in decline in water and food productivity in the area for both man and animals. In the ASAL, it is associated with high temperatures and desicating winds. Without provision for appropriate adjustments (or interventions) to drought, famine occurs. Famine is starvation due to shortage of water, grazing land and food for human consumption. Its impact can be far-reaching depending on the magnitude and length of the drought. However, the shortages of water and food can also be attributed to: (a) the decline in land productivity; (b) inadequate food which is a result of production not matching increased human and livestock populations; (c) lack of new areas for agricultural expansion; and (d) biological limitations, particularly weeds, pests and diseases.

The foregoing difficulties demonstrate lack of administrative adjustments to drought hazards in the ASAL. Overgrazing as a form of human intervention in the ASAL comes when land is grazed beyond its carrying capacity. Bad agricultural practices are significant, especially when man extends cultivation into the ASAL. Such land use which is incompatible with the arid and semi-arid ecosystems in Kenya can lead to an enhancement of fluvial and aeolian morphodynamic processes (water and wind erosion) which serve as important indicators of desertification.

Status of Desertification

About 83 per cent of Kenya can be classified as arid and semi-arid (i.e. low agro-ecological potential zones). In this region about 20 per cent of the country's inhabitants reside. The remaining are compressed into the high agro-ecological potential zones of the Lake Victoria Basin, Highlands and Coastal regions. There is a tendency to extend this physical division into a typology of environmental degradation so that losses in productivity and the natural resource base in the low agro-ecological potential zones are attributed to desertification. In the remaining zones the more traditional categories of soil erosion and deforestation are usually identified separately. Although it is true to say that the end state of degradation in these two contrasting agro-ecological zones may appear different, to assume that the causal process is different is quite misleading. So the physical preoccupation is confounded by an arbitrary spatial division of the process (Baker, 1986).

The process of desertification is becoming serious in many agro-ecological zones of the country since it is a symptom of a much wider problem of environmental mismanagement resulting into degradation. It is said that the evidence of environmental degradation is widespread and often alarming in Kenya (Baker, 1986). Current estimates show that over 75 per cent of the country indicates moderate to high degree of desertification hazard (Republic of Kenya, 1984). However, such estimates must be cautiously treated because of inadequate monitoring and assessment of desertification, especially in the ASAL.

Causes of Desertification

Out of the total surface land area of Kenya of about 569,250 km^2, approximately 483,830 km^2 are already experiencing some form of desertification and out of this proportion about 110,000 km^2 are said to be severely degraded, while 53,500 km^2 indicate latent to moderate symptoms of desertification. However, there is lack of systematic assessment and monitoring of the causes of desertification in the ASAL. This section attempts to enumerate some of the causes of desertification in the ASAL with a view to understanding the dynamic and interrelated nature of the process. These causes include:

(a) excessive human population pressure on the arid and semi-arid ecosystems;

(b) incompatible land uses, such as overgrazing, overcultivation, and shifting cultivation;
(c) deforestation for fuelwood, charcoal (woodfuel), cultivation, timber, fencing materials, settlement and lopping of trees for fodder;
(d) extension of agriculture and uncontrolled settlements into ASAL;
(e) lack of water from over-harvesting of existing sources in ASAL;
(f) periodic droughts and variability of rainfall in the ASAL:
(g) improper revegatation and inadequate forest laws;
(h) lack of adequate environmental perception of desertification hazards;
(i) lack of education on environmental hazards, especially soil erosion, salination, drought and desertification;
(j) lack of protection of water catchments, river banks and steep and unstable slopes from cultivation and clearance;
(k) inacceptable standards of civil engineering and other building works in most work sites;
(l) unnecessary burning of vegetation and other bush-fire which destroys forage and vegetation;
(m) lack of comprehensive rational and utilisation policies including human settlement;
(n) the processes of salination and alkalinization in the ASAL;
(o) lack of maintenance of many large or small engineering works, e.g. water dams and irrigation schemes;
(p) restriction of the movement of nomadic pastoralists and the resultant reduction of the area formerly occupied by them as a result of siting and realignment of political and administrative boundaries; tribal antagonism, development of forest reserves, national parks and game reserves; and the establishment of commercial ranches; and
(q) increasing sedentarization of nomadic people: nomads and their livestock have recently tended to concentrate around a few springs and wells and especially boreholes which have been installed. Overgrazing and denudation and the resultant outward movement of people in search for wood accelerates the rate of degradation of vegetation cover in ASAL.

It is clear from the above list that a multitude of causal explanations (if not speculations) have been advanced for desertification in the ASAL. However, the causes of desertification are not yet definitely known, being a complex process (Fig. 7.1). Its factorial interaction cannot be any less complex. What is required is constant systematic monitoring and assessment of these causes with a view to identifying when and where man's intervention on desertification in the ASAL may be appropriate.

Classification of Tree and Agricultural Crops in ASAL

Desertification monitoring, assessment and intervention cannot be performed effectively without the knowledge of agricultural typology of the ASAL. This section attempts a classification of both tree and agricultural crops grown in the arid and semi-arid agro-ecological zones of Kenya (Table 7.1).

In Table 7.1 agricultural typology shows that there is emphasis on agricultural diversification in the transitional (sub-humid to semi-arid), arid and semi-arid agro-ecological zones. This emphasis reflects risks on environmental constraints (drought, soil and biotic factors), cultural preferences for certain food crops, and government policy to increase food and cash crop production in the ASAL. However, there is much to be learnt about the characteristics of individual tree and agricultural crops (Table 7.1) in relationship to agro-forestry and desertification control in the ASAL. A growing problem is how to establish an adequate system of monitoring and assessing their role in desertification control. Although the area under rainfed crops in ASAL is much smaller than that used by nomadic pastoralists, the population supported by farming is far greater. With recent policy to attain self-sufficiency in food production (Republic of Kenya 1981, 1986), the proportion of the land used for arable farming in increasing. Generally speaking, arid and semi-arid lands with an annual average rainfall exceeding 350 mm. are, depending on soil conditions, considered suitable for the production of annual agricultural crops. Nevertheless, the principal characteristics of the agriculture in the ASAL are that crop productivity per unit area of land is low and soil fertility is restored by fallows. The drier the area the lower the crop yield and the longer the fallow period, depending on land availability. But this situation of fallow period is declining in some areas despite their degree of aridity.

Table 7.1: Agricultural diversification in the transitional (sub-humid, semi-arid), arid and semi-arid agro-ecological zones.

Group	*Common Name*	*Botanical Name*	*Transi -tional Zone*	*Semi -Arid Zone*	*Arid Zone*
(1)	(2)	(3)	(4)	(5)	(6)
Cereals:					
	Bulrush millet	*Pennisetum typhoideum*	*	*	*
	Durum wheat	*Triticum durum*		*	*
	Finger millet	*Eleusine coracana*	*	*	*
	Fonio millet	*Digitaria exillis*	*	*	
	Foxtail millet	*Setaria italica*	*	*	*
	Maize (katumani maize)	*Zea mays*	*	*	*
	Napier grass	*Pennisetum purpureum*	*	*	
	Proso millet	*Panicum miliaceum*	*	*	*
	Rice	*Oryza sativa*	*	*	*
	Sorghum	*Sorghum vulgare*	*	*	*
	Tef	*Eragrostis tef*	*	*	
Legumes:					
	Bambarra groundnuts	*Voandzeia Subterranea*	*	*	
	Black gram	*Vigna mungo*	*	*	*
	Bonavist bean	*Dolichos lablab*	*	*	*
	Butter bean	*Phaseolus lunatus*	*	*	*
	Chick pea	*Cicer arietinum*	*	*	*
	Cow pea	*Vigna unguiculata*	*	*	*
	Gao tree	*Acacia albida*	*	*	*
	Garden pea	*Pisum sativum**	*	*	
	Green gram	*Vigna aureus*	*	*	*
	Groundnuts	*Arachis hypogaea*	*	*	
	Haricot bean	*Phaseolus vulgaris*	*	*	*
	Horse gram	*Dolichos uniflorus*	*	*	*
	Horse tamarinds	*Leucaena leucocephala*	*	*	*
	Moth bean	*Vigna aconitifolia*	*	*	*
	Pigeon pea	*Cajanus cajan*	*	*	*
	Runner bean	*Phaseolus coccineus*	*	*	*
	Soya bean	*Glycine max*	*	*	
	Tepary bean	*Phaseolus acutifolius*	*	*	*
Tubers:					
	Buffalo gourd	*Cucurbita fastidissima*	*	*	*
	Carrot	*Daucus carata*		*	*
	Cassava	*Manihot esculenta*	*	*	*

(*Contd.*)

(1)	(2)	(3)	(4)	(5)	(6)
	Irish potato	*Solanum tuberosum*	*	*	
	Murama bean	*Tylosema esculentum*	*	*	*
	Sweet potato	*Ipomoea batatas*	*	*	*
	Vigna	*Vigna labatifolia*	*	*	*
	Yam	*Dioscorea esculenta*	*	*	
Vegetables:					
	Cabbage	*Brassica oleracea*	*	*	
	Cauliflower	*Brassica botrytis*	*	*	
	Collard green	*Brassica acephala*	*	*	
	Cucumber	*Cucumis sativus*	*	*	
	Egg-plant	*Solanum melongena*	*	*	
	Lettuce	*Lactuca sativa*	*	*	
	Onion	*Allium cepa*	*	*	*
	Pumpkin	*Cucurbita maxima*	*	*	
	Spinach	*Spinach oleracea*	*	*	
Fruits:					
	Apple	Malus pumila	*	*	
	Apricot	*Prunus armeriaca*	*	*	
	Avocado	*Persea americana*	*	*	
	Banana (sweet)	*Musa sapientum*	*	*	*
	Cashewnuts	*Anacardium occidentale*	*	*	
	Coconuts	*Cocos nucifera*	*	*	
	Grape fruit	*Citrus paradisi*	*	*	
	Guava	*Psidium guajava*	*	*	
	Jojoba	*Simmondsia chinensis*	*	*	*
	Lemon	*Citrus limon*	*	*	
	Lime	*Citrus aurantifolia*	*	*	
	Loquat	*Eribotrya japonica*	*	*	
	Macademia nuts	*Macadamia spp*	*	*	
	Mango	*Mangifera indica*	*	*	
	Orange (sweet)	*Citus sinensis*	*	*	
	Pawpaw	*Carica papaya*	*	*	
	Peaches	*Prunus Persica*	*	*	
	Pears	*Pyrus communis*	*	*	
	Pineapple	*Ananas comosus*	*	*	
	Plantain	*Musa paradisiacal*	*	*	
	Pummelo	*Citrus grandis*	*	*	
	Purple passion fruit	*Passiflora edulis*	*		
	Sour orange	Citrus aurantium	*	*	

(*contd.*)

(1)	(2)	(3)	(4)	(5)	(6)
	Sweet melon	*Cucumis melo*	*	*	
	Tangerine	*Citrus reticulata*	*	*	
	Tomato	*Lycopersicon esculentum*	*	*	
	Water melon	*Citrullus lanatus*	*	*	
	Yeheb nuts	*Cordeauxia edulis*	*	*	*
	Yellow passion fruit	*Passiflora flavicarpa*	*		
Fibres:					
	Cotton	*Gossypium hirsutum*	*	*	*
	Flax	*Linum usitatissimum*	*	*	
	Kapok	*Ceiba pentandra*	*	*	
	Kenaf	*Hibiscus cannabinus*	*	*	
	Sisal	*Agave sisalana*	*	*	*
Medicinal Crops:					
	Miraa	*Catha edulis*	*	*	*
	Tobacco	*Nicotiana tabacum*	*	*	*
Spice Crops:					
	Chilli	*Capsicum annum*	*	*	
	Coriander	*Coriandrum sativum*	*	*	
	Pepper	*Piper nigrum*	*	*	
Sugar Plants:					
	Sugar-cane	*Saccharum officinarum*	*	*	
Oil Crops:					
	Castor	*Ricinus communis*	*	*	
	Linseed	*Linum usitatissimum*	*	*	*
	Rai (oilseed)	*Brassica juncea*	*	*	*
	Simsim	*Sesamum indicum*	*	*	
	Sunflower	*Helianthus annuus*	*	*	

* Applicable

Subsistence crops, mainly graminaceous or leguminous grains, such as millets, sorghums, maize, beans, cow-pea and vegetables are often grown under more marginal conditions than cash crops like castor, groundnuts, citrus, coconuts, cashewnuts, rice, cotton, miraa and sisal. It is, therefore, in the more arid parts of the country that tree-based cash crops are required. More recently children have identified uses of many trees and shrubs in rural Kenya (Smith, D.L. and Obel 1980). Some of these trees, such as Acacia albida, Euphorbia candelabrum, Acacia tortilis and Adansonia digitata are useful in agro-forestry and desertification control in ASAL.

Impact of Intensive Crop-Husbandry Production and Existing Farming Systems on Desertification

Because of the harsh climate only small areas are available for arable agriculture in the ASAL. The cultivated areas include food plains and irrigation schemes. But the main crops are millets, sorghums, simsim, cassava, sweet potatoes, maize, rice, beans, peas, water melon, chillies, onions, tomatoes, groundnuts, castor, cotton, sisal and vegetables (e.g. cabbages, sukuma wiki or collard green). Promotion of agriculture in the ASAL has been necessary because of high population pressure in the high potential agro-ecological zone.

Irrigated Agriculture and its Consequences

It is Kenya's policy to increase the area under irrigation in the ASAL. Although irrigation development is intended among other things to improve the ecology of the drylands, problems associated with it, such as deforestation, mechanization (or tractorization), clearance of vegetation, etc., have contributed to desertification in the ASAL. Deforestation in the neighbourhood of irrigation scheme has been the result of woodfuel demand in the absence of cheap alternative energy sources (Republic of Kenya, 1984).

Cultivation on the basis of irrigation both in the arid and semi-arid zones, is known to lead to salinity of the soil, which diminishes its productivity (Mensching and Ibrahim, 1977). Since the 1950s, irrigation in the ASAL has been greatly intensified through the building of dams and deep borings, which have basically altered traditional systems of irrigation, especially the Perkerra Irrigation Scheme in Baringo. Those irrigation schemes are less adopted to the hyrological conditions of the ASAL. Because of high evapotranspiration rates and highly irregular water supply, salination has occurred to a limited extent in Perkerra and Taveta Irrigation Schemes (Stein and Schulze, 1978). Similar salination conditions have been reported in Bura Irrigation Scheme.

While the process of salinization and alkalinisation are not serious problem in large-scale irrigation schemes, their widespread occurrence, especially in north-eastern Kenya, due mainly to natural cause of past time saline ground water table is a problem directly related to desertification in the ASAL. Such saline conditions gave rise to Chalbi Desert in Marsabit District.

The above problems associated with agriculture in the ASAL have direct interrelationships to mismanagement or irrigation projects (as has been the case of Bura Irrigation Scheme), lack of funds, inadequate technological and scientific "knowhow". Although feasibility studies on soil and their drainage capabilities, water quality and crop management are carried out before undertaking most of the irrigation projects, improperly managed irrigated projects such as Perkerra and Bura are a possible or potential cause of desertification in Kenya. However, it is our conviction that with adequate technological and scientific "knowhow" combined with proper management, irrigation undertakings in the ASAL should not pose a major threat of desertification. But frequent monitoring and assessment of the process in all the irrigation schemes in the ASAL is a necessary prerequisite because of the usual vulnerability of irrigation to waterlogging, salinity and alkalinity. Elsewhere intervention on salinisation has been based on the following methods (NAS, 1974; Elgabaly, 1977; Schleiff, 1977; Obara, 1984: (a) Leaching excess salts from the soil profile with freshwater; (b) The addition of ameliorative additives to the soil to assist the washing out of salts (e.g. gypsum and to suppress capillary rise and the evaporation of the soil solution) (e.g. soil cover, mulch); (c) Husbandry practices to improve the depth of the rooting zone (e.g. deep ploughing or deep loosening of the soil) and to diminish the salinity of the soil solution in the zone (e.g. irrigation technique); and (d) Measures related to plant culture : the choice of species and varieties which are salt tolerant, the crop rotation and fertiliser application.

Overcultivation

The expansion of arable agriculture into arid and semi-arid lands without proper consideration of environmental impact is another cause of desertification. This is occurring because of the effects of arable agriculture on pastoralism, especially in : (a) completely arid districts: Mandera, Wajir, Garissa, Marsabit, Isiolo and Turkana; (b) arid and semi-arid districts: Tana River, Taita-Taveta, Kitui, Kajiado, Samburu, Laikipia, Embu, Meru, Baringo, West Pokot, Machakos, Kilifi and Kwale; and (c) Semi-arid districts: Lamu, Narok, Nakuru, and Elgeyo Marakwet.

In the above districts, the nomadic pastoralists are forced to move further into more arid and vulnerable ecosystems (e.g. Kajiado,

Baringo and Marsabit Districts). The driving back of the nomads from the rangelands leads to degradation of sparse vegetation cover in the ASAL, on the one hand, and to a disturbance of the ecological balance in the cultivated areas, on the other. Serious consequences result from the ploughing of dry soils in particular. The cultivated topsoil is regularly loosened by the plough and is partly cleared of rock debris, which can lead to deflation of huge amounts of fine soil cover over a short span of time. Ojany (1986) reported that the unsupervised mechanisation or tractorisation in large-scale rainfed agriculture accelerates aeolian (wind) and fluvial (water) morphodynamic processes which serve as good indicators of desertification in many parts of Kenya. In the ASAL, cereal lands as well as tree plantations are ploughed. Such areas have mostly shallow topsoil which is exposed to strong deflation after being loosened by the plough. Over a period of time the topsoil is blown away and the bedrock or the calcareous pan is laid bare, e.g. in Marsabit and Turkana districts.

Observation in the ASAL has revealed that the cultivation of areas which should be used as grazing lands leads ultimately to the formation of gullies and badlands, particularly in Machakos, Kitui, Kajiado, Baringo, Samburu, Turkana, Meru and Marsabit (Brown, 1963a). The badlands render any further cultivation impossible in these districts. It has been postulated that one can correlate periods of overcultivation with phases of intensified desertification in the ASAL.

Shifting Cultivation

Shifting civilization embraces a large variety of traditional forms of agriculture. Generally speaking, the systems so described are characterised by the felling and burning of woody vegetation, followed by one to several years of cultivation, and consequently by a period of bush fallow. After fallow, the cycle is repeated (King and Chandler, 1978). Shifting cultivation is also known as "slash and burn agriculture".

It is said that shifting cultivation is a form of agricultural land-use well adopted to the environment under certain socio-economic conditions. Where there is relatively low population pressure, the system permits the resuscitation of the soil before the cultivators return to a fallow area. However, shifting cultivation in the ASAL has been carried out under a number of climatic and ecological conditions varying spatially. The period of cultivation, the size of area, duration of fallow, the methods of planting and the crops involved determine the system's viability.

Throughout the ASAL, the practice of shifting cultivation is changing rapidly although the change is not being monitored and assessed. However, the changes in the ASAL have been induced by demographic factors, such as increased population growth rates and population migration. The introduction of cash crops into communities has also had a significant impact on shifting cultivation, especially in Machakos and Kitui districts. Moreover, the policies of the government have led to a reduction of the area of land available for shifting agriculture in the ASAL. For example, the large-scale reservation of land for forests and wildlife national parks, though desirable, has reduced the traditional fallow periods of shifting cultivation.

Although viable new intensive farming systems (e.g. irrigated agriculture) have emerged in the ASAL, there has been very little research and, consequently, little innovation, on the rationalisation and improvement of the farming systems of shifting cultivators. It is said that slash and burn cultivation and snatch cropping inevitably leads to land degradation in ASAL since it sets in motion processes of degradation, such as wind and water erosion, soil compaction and sealing (NES, 1988). What has happened is that there has been a shortening of the fallow period of the cultivation cycle, which has inevitably led to soil degradation often to the point of making the land incapable of supporting further vegetation growth and crop production.

Intensive Dryland Farming Practices

Many intensive dryland farming practices, while increasing crop yields in the short-term, have made the soils more vulnerable to erosion in the ASAL. Because of the low moisture holding capacity of the soils and the leaching of nutrients, abandoned farmlands in the ASAL do not recover quickly and sometimes not at all. This is a serious form of desertification since the limited areas of maximum agricultural potential, e.g. in Kitui, Machakos and Marsabit, are degraded to a state often beyond that caused by overgrazing. In areas (such as Kajiado and Machakos) where there is a close linkage between dryland cultivation and animal-based systems, through the grazing of stubble, use of manures, etc., the collapse of the crop-based system tends to lead to failure of pastoralism.

Cultivation of Cereals Beyond the Agro-ecologically Set Zones

Desertification has occurred as a result of ignoring the agro-ecologically set zones for maize, millet and sorghum cultivation. The zone of cultivation which ideally suits the agro-ecological conditions has been defined (Jaetzold and Schmidt, 1982). Although this is difficult to trace in the field for practical purposes, one can say that it is within the arid and per-arid (very arid) agro-ecological zones where the effects of desertification through cereal cultivation are quite perceptible (such areas include Turkana, Marsabit, Mandera, Wajir, Kajiado, Machakos and Kitui). These areas of low precipitation have less latitude for errors in land use methods than areas of better precipitation and the effects of badland use are consequently seen more quickly and more severely (Brown, 1963a). Unless the present systems of land usage can be extensively modified, about 20 per cent of the Kenya's population living in the ASAL are in danger of desertification. Because of the irregular rainfall inadequate for cereal cropping, aggravated by human population pressure, erosion, denudation and desertification are expected to increase in the ASAL unless improved crop-husbandry production is undertaken to prevent them from occurring.

On the sandy soil of the ASAL, the 500 mm. isohyet seems to be an appropriate border for cereal cultivation while on the clayey soil the 600 mm. isohyet is more reasonable for millet cultivation (Mensching and Ibrahim, 1977). Sandy soil is more favourable because of the high rate of seepage, which enables the crop plants to make full use of rain water.

Field studies elsewhere have confirmed that transgression over the potential, i.e. the climatically controlled border of millet cultivation, is a major factor in the process of desertification in the ASAL (Mensching and Ibrahim, 1977). Although no equivalent observation has been made in the ASAL of Kenya, it is of paramount importance to monitor and assess the impact of cultivation of cereals: maize, millets and sorghums, beyond the agro-ecologically set zones on desertification. To prepare the field for planting of cereals, the agriculturalists cut down all the trees and clears the land completely of all weeds and grasses to secure every drop of water for the cereal plants in the ASAL. Repeating this, year after year leads to the ultimate destruction of the natural vegetation cover, which enhances the deflation of the topsoil. This results in the degradation of soil and the decline in its productivity. The rotation system of land use which formerly allowed the regeneration of the soil has almost disappeared owing to an immense increase of human population in the last two decades. And though the land has thus become exhausted, cultivation of maize, millets and sorghums is expanding, for the less productive the land becomes the more land is brought under cultivation to attain an equal amount of crop.

Intensive Subsistence Cultivation

Until recently Machakos, Kitui, Meru, Embu, Baringo, Kajiado, Turkana and Marsabit districts were basically dependent on subsistence cultivation, but also keeping livestock simultaneously. Because of the increasing population pressure, these districts have been forced to adopt intensive subsistence cropping with emphasis on the following crop husbandry which obviously helps in checking desertification in the ASAL:

(a) Effective control of erosion on cultivated land using narrow base or bench terracing, trash lines, or lines of a planted grass, such as *Panicum makarikaneusis*, or a herb, such as *coleus spp*. The use of sisal planted in lines or strip along the contour is also encouraged both as a cash crop and as a conservation measure, particularly in Machakos and Kitui districts.

(b) Early planting to make the best use of rain which falls; many crops are dry-planted to germinate with the first rains.

(c) The use of drought escaping crops such as sorghum, bulrush millet or the early maturing maize varieties such as *Katumani maize* and *dryland composite maize*, have been encouraged. Preference is also given to cowpeas, grams, finger millet, and early maturing bean varieties in the above districts.

(d) The planting and maintenance of substantial areas of cassava by every family as a reserve in times of real food shortage.

(e) The use of phosphate or animal manure or both to accelerate crop growth and to produce heavier yields in any season of adequate rainfall.

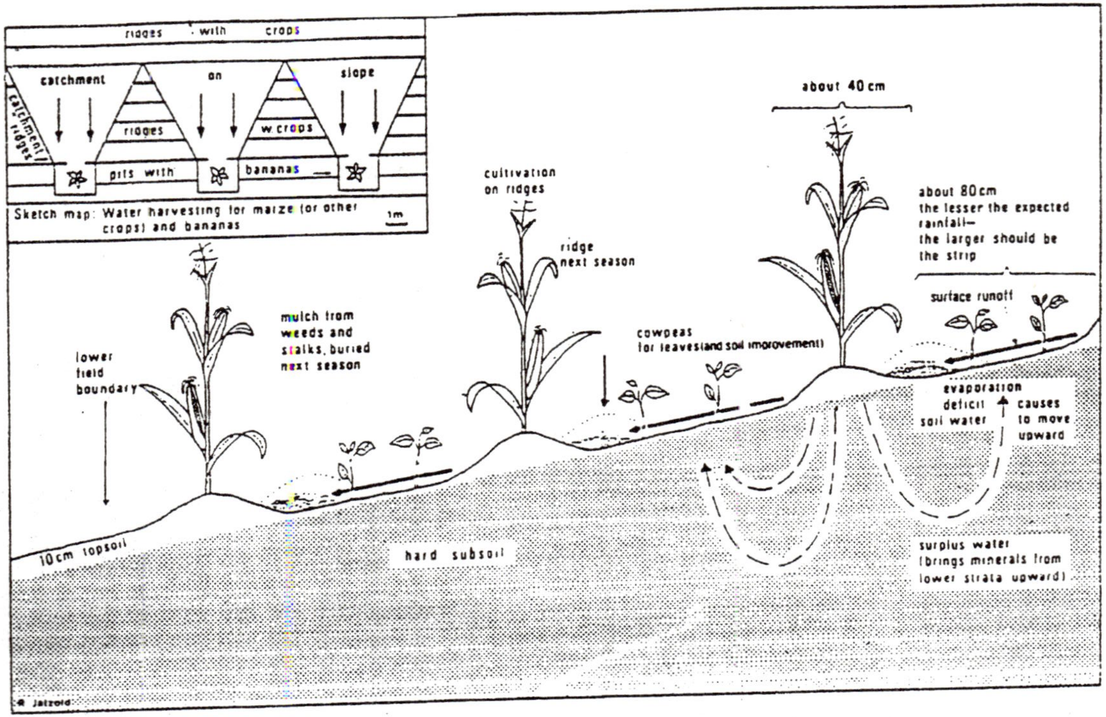

Fig. 7.2: Runoff-catching agriculture for areas with 300-400 mm. average precipitation per rainy season.

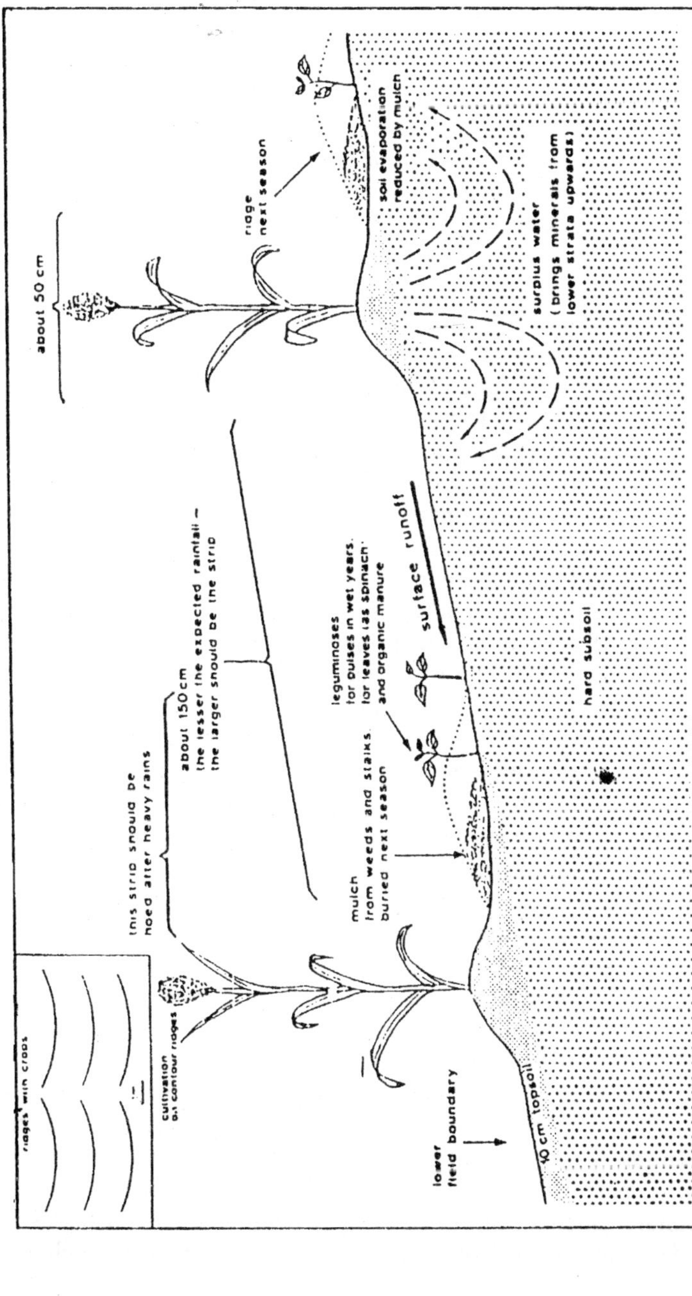

Fig. 7.3: Runoff-catching agriculture for areas with 200-300 mm average precipitation per rainy season.

Runoff Agriculture

Once rainwater runoff has been harvested from slopes, it can be used for crop production in the ASAL. The runoff water from the catchments is collected in constructed channels on the hillslopes. Channels direct the water to cultivated fields which are often terraced. Some farmers dam the small channels between the catchment and the fields with rocks by removing strategic rocks from the channel walls which guide the water to different fields. The technique described above is known as runoff agriculture. Runoff agriculture has been developed to permit crop production on arid and semi-arid lands receiving as little as 100 mm. per annum (NAS, 1974). The techniques of runoff agriculture are currently applicable in Turkana, Machakos and Kitui districts. Runoff farms using modern technology and crop varieties selected for local conditions (Figures 7.2 and 7.3) could benefit other arid and semi-arid agro-ecological zones in Kenya. Maize, bananas, sorghums, millets, citrus fruits, forages and leguminoses (e.g. cowpeas and beans) all are potentially important crops for runoff agriculture in ASAL; most are new grown in runoff agriculture in Turkana, Machakos and Kitui.

In runoff agriculture the principles and practices depend on rainwater harvesting. The basic need is a rainwater catchment that provides enough water to mature the crops. Obviously, the crops own water requirements and general water conservation techniques are crucial to a successful harvesting in the ASAL. Poor crop yields in drought years are often offset by production in good years.

The type of farming practiced in the ASAL must make the best use of the water. Generally speaking, perennial crops with deep root systems adapt better to runoff agriculture, because they can use runoff water stored deep in the soil, safe from evaporation. Some deep rooted, drought-resistant fruit trees, such as mangoes, lemons and oranges can be very successful. Shorter lived crops can also be grown; grains, such as millets, sorghums, *katumani* and *dryland composite maize* varieties, that mature rapidly and require only one rainfall hold particular promise in the runoff agriculture in the ASAL. Plants that become dormant during dry period and begin growing when water becomes available are particularly suited to runoff agriculture in the ASAL. Therefore, some of the crops recommended for runoff agriculture which would aid in alleviating desertification in the ASAL include: peaches, apricots, grapes, wheat, peas, maize, sorghums, millets, citrus fruits, beans, peas and sunflower. What are the overall advantages of runoff agriculture in the ASAL? There are several advantages of runoff agriculture, namely:

(a) Runoff water can allow crop plants to grow in otherwise too arid habitats (e.g. perarid agro-ecological zones in Kenya);
(b) Runoff agriculture can be used to make new agricultural lands where water is otherwise inadequate to support agriculture. And yields from already cultivated areas can be increased without installing costly irrigation projects to bring in water from neighbouring agro-ecological zones. It has particular promise for the ASAL; runoff agriculture can lower the risk of crop failure;
(c) Runoff used to grow forage can alleviate grazing pressures on nearby rangelands. Overgrazed areas can be revegetated and the carrying capacity of grazing land considerably increased;
(d) Runoff agriculture can extend the season during which forage is succulent and nourishing, providing green forage when it is especially required; and
(e) Water-spreading can provide erosion control, for it deflects the torrent of water and dissipates its energy.

The above advantages of runoff agriculture have not been sufficiently monitored and assessed in the ASAL. These advantages should be properly assessed in the context of desertification control in the ASAL.

Ley Farming Systems

Ley is used wherever several years of arable cropping are followed by several years of grass and legumes utilised for livestock production. However, a distinction should be made between unregulated and regulated ley farming systems in Kenya. Unregulated ley systems in the country are characterised by natural fallow vegetation of various grass species, a certain amount of bush growth on the pasture, communal grazing and a lack of pasture management, all of which make such systems often more akin to short-term fallow systems than to regulated ley systems (Ruthenberg, 1976). While the borderline between unregulated and regulated ley system is difficult to draw, we are more interested in the intensive ley systems (or regulated ley systems) found in Kenya, particularly in West Pokot, Machakos, Samburu, Narok, Kericho and Elgeyo Marakwet Districts. Individual grazing, fencing pasture management, and rotational use of the grassland are the usual characteristics of regulated ley systems.

The seeding of planting of grass leys, the production of hay or silage and the application of mineral fertilisers are indicators of intensive ley farming systems. Ley stems show a more or less alternation between arable farming and leys. The sum of the arable years and the ley years indicates the rotation cycle. Noteworthy are some modernised smallholder ley systems with intensive dairy production in Kenya (MacArthur, 1964). In Kenya Highlands, settlement of ex-European farms and estates led to the establishment of numerous smallholdings with ley farming and intensive milk production as major planning objectives, which achieved a certain degree of success.

Most established ley systems are found on large, mechanised farms, primarily because the use of the tractor is possible only with a complete clearance of bush and stumps, i.e. it requires a fallow vegetation of grass. Furthermore, it is said that the profitability of the leys depends on modern forms of animal husbandry, which are much more easily introduced in large than small farms. In Kenya highlands, wheat, barley, pyrethrum; sheep, and beef and dairey cattle are the main branches of activity in large modern ley farms. Leys are required because mechanised wheat, the most profitable arable crop, cannot be grown as a monoculture as could be with maize.

The farming pattern developed at Mau Narok shows a spill over effect in neighbouring Maasailand where semi-nomadic grazing is giving way to a ley system. The sequence in land developed tends to proceed as follows:

(a) Contractors are allowed to clear Maasailand and to plant wheat, but some share of the crop goes to those Maasai who claim the land;
(b) After 5-8 years of wheat monoculture the land goes to grass again and offers as much improved pasture;
(c) Some part of the return from wheat is invested in wells and fencing for more effective livestock production;

(d) Some Maasai enter into mechanised wheat production and the resulting pattern of land use is likely to be an alternation of wheat years and grass utilised for beef.

In other areas potatoes or maize-ley rotation is carried out. Ley farming has become a common system in the rangelands of Machakos, Samburu, Kajiado, etc. where reseeding to get more palatable grass has become a common practice. Ley systems are commendable in desertification control but there is need for their monitoring and proper assessment.

Terracing

Terraces can be constructed as ditches across the slope, thus decreasing the length of a slope, as ditches divide the slope into shorter stretches. This type of terrace, however, does not influence rain erosion, and rill erosion is only partly affected (Wenner, 1979). Therefore, bench terraces are preferable, as these also change the degree of slope. They retain eroded soil, moisture, and nutrients. Bench terraces can develop from grass strips acting as wash stops. Grass should be planted on the ridge to protect it. The channel will be filled up mainly through cultivation and by deposits but part of it should be maintained as a drain for water passing through the grass strip during heavy rain storms. To hasten the formation of a bench terrace you can make a ridge by digging a channel (0.61 to 0.91 m. deep) and throwing the soil uphill, using the so-called "Fanya Juu" (Wenner, 1979; Jaetzold and Schmidt, 1982). "Fanya Juu" terracing is commonly used in Machakos and Kitui districts, because the risk of drought makes grass strips without ridges unreliable for conservation, especially on steep slopes.

On gentle slopes between 2 and 12 per cent terracing is not obligatory according to the present Agricultural Act, but terracing is usually desirable on slopes exceeding 5 per cent (Jaetzold and Schmidt, 1982). In semi-arid areas and areas with erodible soils, even slopes between 2-5 per cent, usually need to be terraced. On slopes exceeding 12 per cent, but not exceeding 55 per cent, terraces (preferably developed bench terraces) should be used if the depth of the soil is more than about 0.70 m. For very steep slopes modified bench terraces are recommended, i.e., narrow ridges cut into the slope, suitable for fruit trees, fodder trees, forest trees and coffee. Terracing has been carried extensively in some districts, e.g. Machakos. It is a good method of soil conservation which reduces environmental degradation and desertification in the ASAL.

Contour Strip Cropping

Strip cropping means wide strips with alternating crops under rotation. It is preferably used when slopes are less than 20 per cent and where the soil is permeable. Suitable crop rotations may be employed in the practice of strip cropping in which relatively narrow strips of different crops are planted on the contour, strips of erosion susceptible crops being separated by strips of close-growing protecting crops, such as groundnuts, sweet potatoes, peas, beans and grass. The densely planted strips is effective in reducing runoff and erosion and desertification in the ASAL, e.g. in Machakos district. Maize and other cereal crops (e.g. millets and sorghums) should not if possible, be cultivated on a monoculture basis in the ASAL. A three-year rotation of maize and grass is recommended for maintaining a soil structure which will reduce the rate of erosion (Wenner, 1979) and, consequently, desertification in the ASAL. However, in semi-arid regions with somewhat more precipitation we find a tendency for maize monoculture which is a risky undertaking likely to cause nutrient depletion, severe soil erosion and desertification. Although multiple cropping is impossible in the ASAL without irrigation, mixed cropping or intercropping of maize, millet and pigeon pea has been a commendable practice in Machakos and other dry districts.

During the establishment of the plantation, when the trees themselves cannot protect the soil in the ASAL, it is usual to plant cover crops in the interrows, a method called intercropping. The creeping legumes (e.g. cowpeas, groundnuts and beans) commonly used as cover crops in the ASAL provide a dense protective cover for the soil to be eroded by water and wind alike.

Potential Use of Specific Crops in Desertification Control in ASAL

Although desertification as a process has similar criteria everywhere, its causes, accompanying phenomena and consequences vary considerably according to the different ecological and socio-economic conditions of the different regions. Because of this difficulty, measures taken to control desertification must be adopted to local conditions.

This means the agricultural crop components in any existing agro-forestry system, or in anyly new postulated one, should be largely restricted to species which satisfy existing consumer and market preferences in the ASAL. Introducing completely new food or cash crops is, therefore, a lengthy business, which may require the farmers' participation in planning and decision-making. Past experience shows that localised interventions, limited to certain sectors have frequently given only temporary solutions to desertification control and in most cases they have even created new and worse problems (Rapp, 1974).

The main thrust of the Plan of Action is the application of existing knowledge in science and technology to the development of specific crops in desertification control in the ASAL. The potential use of specific crops in desertification control in the arid and semi-arid ecozones is succinctly summarised below:

Potential use of Specific Crops in Irrigated Agriculture in the ASAL

The potential crops recommended for irrigated agriculture in the ASAL include: rice, cotton, maize (*Katumani* and *Dryland composite maize*), sorghums, millets, beans, groundnuts, simsim, wheat, barley, cotton, melons, grapes, apricots, citruses, tomatoes, spinach, cowpeas, safflor, coconuts, triticale and vegetables. Under saline conditions the adaptation of the crop plants to grow on saline soils, to achieve reliable harvests from the economic point of view is essential. Salt-tolerant crop plants, such as spinach, cotton, safflor, coconuts, and Triticale (*Triticum xsecale*), have potential use in the ASAL. The irrigated crops would help in desertification control in the ASAL by settling the nomads and reducing dependence on livestock. This would alleviate overgrazing which is one of the major causes of desertification in the ASAL.

Dryland/desert Crop Development Project

This project aims at promoting dryland adapted crops and plants or crops indigenous to the arid and semi-arid ecozones through demonstration, improvement and trials, seed multiplication, commercialisation of production, processing and marketing (Republic of Kenya 1984). The traditional dryland recommended crop plants embrace: sorghums, millets, miraa *Eragrostis tef*, sisal cowpeas, grams, peas and maize, while indigenous desert crops: *Prosopis*, jojoba (*simmondsia chinesis*), *Glyricidia sepium*, *Tephrosia Candida*, Guar, Guayule, *Leucaena*, Jerusalem Thorn, and *Euphorbia candelebrum*. Besides these crops ane may add cassava, sweet potatoes, sunflower, cashewnuts, sisal, and Yehab nuts.

Potential use of Trees and Crops in Agro-forestry

Agro-forestry is a relatively new discipline which combines elements of agriculture, i.e., crops or animals, with element of forestry (the science of planting and caring for trees) in sustainable production systems. Huxley (1983) pointed out that agro-forestry is concerned with both production and conservation. Agro-forestry involves the cultivation on the same land unit of herbaceous plants and trees. However, multipurpose trees, such as *Acacia albida*, *Acacia Senegal*, *Prosopis juliflora*, *Leucaena leucocephala* and *Eucalyptus camaldulensis*. But what are multipurpose trees?

Multipurpose trees are woody plant species which in addition to the normally expected products and services, such as wood, micro-climatic influences, soil improvements, addition of organic matter, provide significant additional products and services, such as nitrogen fixation, fodder or forage, edible products for humans, gums, resins, fibres and medicinal products (Burley and Von Carlowitz, 1984). However, regular crop plants commonly cultivated essentially for a single purpose, e.g. tea, coffee, cirtrus, etc., are excluded in agro-forestry. Borderline cases, e.g. cashewnuts and coconuts are judged for inclusion or exclusion on an arbitrary basis.

A large number of multipurpose trees and shrubs in the ASAL are often retained or incorporated on farms in subsistence farming systems, e.g. *Adansonia digitata* (Baobab tree). Woody perennials are maintained in contour strips across the slopes and around the fields to act as windbreaks. The contribution of these trees is the product of fodder and firewood and their protective functions in reducing the erosion hazards and desertification thereby making crop production possible in those steep slopes like those in Machakos district, where profitable cropping would otherwise be extremely difficult.

(a) *Trees recommended for intercropping:* There is little literature on the intercropping of trees and food crops in the arid and semi-arid ecozones except for the important *Acacia* species:

Acacia albida and *Acacia senegal* (Felker, 1978; El Houri Ahmed, 1979). However, trees suitable for intercropping with food crops where food crop phase is not merely an establishment phase for the trees should possess one or more of the following attributes (Redhead et al., 1983):

— Fast growth: producing fuel and house building poles within a period of not more than 8 years.
— The characteristic of shedding leaves during the wet (crop growing) season: e.g. *Acacia albida* (intercropped with maize and field beans).
— A straight form and dominant leader growth, even when growth at wide espacement: e.g. *Eucalyptus camaldulensis* (intercropped with maize and field beans), and *Eucalyptus melliodora* (intercropped with maize, sorghum, or field beans).
— A high value crop as well as fuel: e.g. *Acacia nilotica* (tannin), *Acacia senegal* (gum Arabic).
— The capability of being repeatedly lopped for fodder: e.g. *Leucaena leucocephala* and *Acacia albida* (intercropped with maize and field beans).
— The ability to fix nitrogen: e.g. *Acacia albida*, *Acacia Senegal* and *Leucaena leucocephala*, *Acacia albida* is further associated with sorghums, millets, groundnuts etc.

(b) *Acacia albida:* To grow unirrigated crops under *Acacia albida* is a useful idea because this legume is a good protection for cultivated fields in the ASAL. This is the arid and semi-arid zones tree that has received considerable attention in desertification control (King and Chandler, 1978). This tree has a long tap root which does not compete with surface crops for nutrients and moisture, but which acts as a nutrient pump bringing nutrients back to the upper soil root zone of most agricultural crops and grasses.

As it is a legume, it is also capable of fixing atmospheric nitrogen. It is a deciduous tree which sheds its leaves at the beginning of the rains when they decompose rapidly to release plant nutrients at the same time they are most needed. It simultaneously avoids the shading of agricultural crops (such as maize, sorghums, millets and beans) planted beneath it. During the dry season the large crown of the tree shades the soil and animals, and provides a windbreak which reduces dessication. The branches may be lopped for emergency cattle fodder and large amounts of nutritive seed pods are produced annually. Finally, the wood may be used as fuel.

(c) *Leucaena Leucocephala:* Intercropping observations elsewhere in the ASAL have demonstrated the benefits of interplanting *Leucaena* with annual food crops (particularly *phaseolus vulgaris* to reduce the competition of weeds) such as maize, rice, climbing beans, roots and tuber crops (e.g. Cassava, potatoes, yams, etc.). *Leucaena* prunings applied to the adjoining rows of maize contribute to increased yields when nitrogen fertilisers are unavailable. The potential of *Leucaena leucocephala* has been demonstrated elsewhere (Rachie, 1983). Improved varieties of *Leucaena* have proved exceptionally well-adapted. It has been demonstrated that: Plant growth is rapid; the foliage provides an excellent source of forage; various planting configurations are possible; it is relatively compatible with annuals in associated plantings; growth can be controlled or modified by hand or machine pruning and it seeds profusely. Therefore, *Leucaena leucocephala* has a great potential in desertification control in the ASAL.

(d) *Perennial pigeon peas:* These peas (*cajanus cajan*) used both green and dried, and as edible browse, are really short-term perennials, although some bushes (small trees) have been reported to live for 'decades' unless injured or attacked by insects and diseases. The stems are useful as a limited fuelwood source. Deep-rooted perennial plant usually intercropped with other crops, such as maize, finger millet or beans.

(e) *Prosopis Juliflora:* Several *Prosopis* species produce edible pods and seeds which are still used for food in some areas. Some species are also useful as a dry season browse (it is palatable). Although their importance has diminished in recent times, they are useful in desertification control in the ASAL.

(f) *Field beans (Phaseolus vulgaris):* The common bean is variously known by many different names, especially kidney beans, haricot beans or French beans and several hundred varieties are known, both dwarf and climbing in habit. They can be eaten green in the pod but are generally harvested as the dry bean, in which condition they can be stored as a grain legume. They thrive best at altitudes from 600 to 2400 m. The crop is grown in pure stand by largescale farmers, but it is generally intercropped with maize in smallholdings e.g. in Machakos district.

(g) *Cowpea (Vigina unguiculata):* This is a dual purpose annual grown as much for its leaves, as a green vegetable, as for the dried beans. The plant is drought resistant and is widely grown throughout the ASAL up to 1500 m., usually interplanted with maize, sorghum, or yams.

(h) *Groundnut (Arachis hypogea):* The groundnut is better known as an oil seed than as a grain legume. It is grown in many parts of the ASAL. Groundnut oil is a high quality cooking oil, also widely used in making margarine. The seeds are highly nutritious, contain about 30 per cent protein and are rich in vitamins B and E. They can be processed into flour and used in soups and stews, or eaten fresh. The leaves and stalks can be used as green fodder or dried for hay. Groundnuts prefer a light, sandy soil, a rainfall of 600 to 1200 mm. The crop is ready for harvest in 14 to 20 weeks.

(i) *Soya bean (Glycine max):* Soya bean is one of the most important food crops. It is important both as an oil seed and a grain legume. Although the crop is not widely grown in the ASAL, it has enormous potential for desertification control. Active research is being undertaken to promote its use as a high protein grain legume.

(j) *Other useful plant species in desertification control:* Other grain legumes which deserve attention are bambara groundnuts (*voandzeia subterranea*), bonavist beans (*lablab niger*), chick pea (*Cicer arietinum*), green and black grams (*Vigna aureus*, V. mungo), lima bean (*Phaseolus lunatus*), sword bean (*Canvalia ensiformis*), winged bean (*Psophocarpus tetragonalobus*), the yam bean (*sphenostylis stenocarpa*).

Monitoring Mechanisms for Improved Environmental Management in Crop Production Systems

The strong impact of man on the fragile arid and semi-arid ecosystems in Kenya significantly contribute to badlands and causes of desertification. The most recent formation of badlands started with the sedenterization of the nomadic pastoralists during the colonial administration (Brown, 1963a; Baker, 1986). A plan to intensify the development of African agriculture (Swynnerton, 1954) and a national cash crops policy (Brown, 1963b) probably enhanced desertification process in the ASAL. However, the impact of such plans and policies on desertification have not been closely monitored with a view to improving environmental management in crop production systems in the ASAL. It is recommended that the impact of agricultural intensification and cash crop production on desertification should be monitored frequently. More specifically:

(a) Over-intensive farming and the growing of ecologically unsound crops in the ASAL should be frequently monitored with the help of aerial photography in order to assess desertification process. However, aerial photographs provide a reliable and easily accessible basis of monitoring crop production systems only when they are regularly taken and correctly interpreted. Aerial photographs are useful for:

  — mapping uncultivated and cultivated fields including fallow land with shrubs in the ASAL.
  — monitoring changes on shifting agriculture or bush fallowing caused by human population pressure and introduction of cash crops in the ASAL.

(b) Monitoring of yield differences in alley cropping trials in the ASAL should be given priority because this method would boost agricultural production and simultaneously alleviate desertification.

(c) Although runoff agriculture has been developed to permit crop production on arid and semi-arid lands with 100-400 mm. average precipitation per growing season, constant monitoring of the crops water requirements and water conservation techniques is crucial to a successful harvesting. This would further assist in selecting potentially important crops for runoff agriculture and desertification control in the ASAL.

(d) Vulnerability of irrigation to waterlogging, salinity and alkalinity in the ASAL should be constantly monitored since such undesirable environmental conditions could lead to low soil productivity and desertification process.

(e) Withdrawing cultivation from agronomically perarid ecozones requires intensification of agricultural production in edaphologically suitable areas in the ASAL. But monitoring such changes would depend on:

— improving the agricultural methods based on better seeds and improved field machinery.
— making full use of the existing landuse possibilities of the ASAL region based on introduction of a more favourable crop rotation, ley farming, suitable fertilisers and improving farmers animal husbandry.
— improving the use of ground water and runoff agriculture.
— intensifying technical service advice, providing credits, establishing market services and acquiring capital goods.

(f) Monitoring the development of agro-forestry landuse systems (field crops combined with soil-conserving tree species, such as Acacia albida, *Acacia Senegal* and *Leucaena leucocephala*) in the ASAL.

(g) It may be said that with proper management of land; subsistence and a moderate cash income could be provided for each family in the ASAL provided that the Government on its part maintains services at an adequate level and effectively induces the people to adopt the agronomic practices proven and advised by the extension workers (bwana shambas). However, the full potential of the ASAL will never be realised as long as desertification is not monitored, assessed and controlled. It is recommended that:

— the uncontrolled land utilisation methods which destroy the natural vegetation cover, drastically reduce soil fertility and cause soil erosion and desertification hazards should be closely monitored.
— the development of desertification in the arid and semi-arid ecozones should be monitored and compared in the short-, medium-, and long-terms using the following

parameters: precipitation, extension of land use, and soil productivity. Such approach would establish how long it would take for desertification to occur. It is doubtful, whether climatic changes could be perceived over a short period to cause desertification, while increased soil productivity and an expansion of cultivation in the ASAL could be attributed to land reclamation, intensive irrigation and fertilise applications. In the arid ecozone, such improvements could be temporary and in the long run could lead to enhancement of desertification as indicated by the literature cited.

(h) Two parameters are probably indespensable for the monitoring and evaluation of the complex of desertification in the arid and semi-arid ecozones: degradation of the vegetation to a sparser form and the impoverishment of the soil so that the natural potential becomes lowered. On the basis of these parameters the following methods are proposed for the appropriate monitoring mechanisms and assessment of desertification in the ASAL:

— Comparison of aerial photographs and topographical maps of various dates.
— Filtering of remote sensing satellite imagery for a general evaluation of crops, vegetation, soil moisture and state of the hydrological systems and micro-relief. This should be done in collaboration with establishing ground truth.
— Investigation of the morphodynamic processes which are occurring currently in order to perform a field analysis of rates of fluvial and aeolian erosion in the ASAL.
— Evaluation of historical and archaeological data, as well as a critical assessment of statements by local people concerning changes in the last decades, especially in areas of intensive arable agriculture.
— Study traditional and modernised land use systems, particularly shifting cultivation (or bush fallowing), ley systems, runoff and irrigated agriculture and dryland farming.
— Monitoring of the demographic and socio-economic conditions with a view to achieving a better

understanding of the relationship between the natural potential and its current exploitation. This would help in improving environmental management in crop production systems in the ASAL.

— Monitoring landuse policy and planning in the ASAL with a view to improving environmental management in crop production systems and solving problems of desertification and conflicts in the land use competition. This implies:

- training and education that stresses the interdisciplinary approach to desertification and the mobilisation of popular participation of the people at the grassroots level.
- legislation concerning water resources, land utilisation methods and systems of land tenure would contribute significantly to the improvement of environmental management in crop production systems in the ASAL.
- technical personnel to carry out data processing and synthesis and provide channels for communication and feed-back between national institutions for agricultural research and the policy-making authority.
- monitoring deforestation and cultivation on steep slopes and along river banks.

The most important point is how to effectively monitor desertification caused by inappropriate crop husbandry in the ASAL. Detailed inventory of land potential and of desertification "threat" map should be compiled. Also, many of the affected geographical areas of ASAL are unmapped and understudied. Control of desertification in the ASAL requires transnational measures which commonly require support from international agencies. The efforts to combat desertification is mentioned elsewhere (Ojany, 1986), particularly the work being done in Northern Kenya with the help of UNESCO and the West German Federal Government under the Integrated Project on Arid Lands (IPAL) and now, Kenya Arid Lands Research Station (KALRES), promises to provide a most appropriate method for understanding (before rehabilitating) degradation of the ASAL. Although IPAL's (KALRES) primary task was to find direct solutions to the more urgent environmental problems associated with desert encroachment and ecological degradation of arid lands, it paid no attention to the monitoring of the impact of crop husbandry on desertification. Similarly, Kenya Rangeland Ecological Monitoring Unit (KREMU) is concerned with monitoring and assessing the ecological degradation of the rangelands. To date no institution or international agency is directly concerned with monitoring the impact of crop husbandry on desertification in the ASAL. Because of this difficulty, we have proposed what we consider as the appropriate monitoring mechanism required for improved environmental management in crop production systems. We need efficient methods of monitoring and assessing desertification in the ASAL. But the question is who can carry out the monitoring mechanisms in the ASAL efficiently?

REFERENCES

Baker, R. (1986), "Environmental degradation in Kenya: Two conflicting explanations," in *African Environment* 17-18-19, enda Dakar, pp. 17-43.

Brown, L.H. (1963a), *The Development of Semi-Arid Areas of Kenya*. Ministry of Agriculture.

Brown, L.H. (1963b). *A National Cash Crops Policy for Kenya (Parts I and II)*. Ministry of Agriculture and Animal Husbandry, Nairobi.

Burley, J. and Von Carlowitz, P. (1984), *Multipurpose Tree Germplasm, Proceedings of a planning workshop to discuss international cooperation held at the National Academy of Sciences*, Washington. International Council for Research in Agro-forestry, Nairobi, Kenya.

Chanda, R. (1980), "Desertification in Africa: A Result of reduced resilience of arid ecosystems," in *The Development Problems and Prospects of Semi-Arid Areas in Eastern Africa. Proceedings of Workshop*, OSSREA, Addis Ababa, Ethiopia. pp. 1-12.

Dresch, J. (1975), "Reflections on the Future of the Semi-Arid Regions", in *African Environment: Problems and Perspectives: Special Report 1*, edited by Paul Richards. International African Institute.

Eckholm, E. (1984), "Poverty, population growth and desertification". *Desertification Control*, 10:37-41.

Elgabaly, M.M. (1977), "Water in Arid Agriculture: Salinity and Waterlogging in the Near-East Region", in *A Journal of the Human Environment: Research and Management. AMBIO*, 6:36-39.

El Houri Ahmed, A. (1979), "Effects of land use on Soil characteristics in the Sudan", in *Soils Research in Agro-forestry*, edited by H.O. Mongi and P.A. Huxley. ICRAF, Nairobi, Kenya p. 1-13.

Felker, P. (1978), "State of the art: *Acacia albida* as a complementary permanent intercrop with annual crops". University of California, Riverside, California, USA.

Grove, A.T. (1974), "Desertification in Africa", *Paper presented at the Alice Springs Field Meeting of the IGU Working Group on Desertification in and around Arid lands.*

Heady, H.F. (1972), *Range Management in East Africa.* Government Printer, Nairobi.

Ojany, F.F. (1986), "Desertification in Africa with special reference to East Africa". *Geographical Reports of Tokyo Metropolitan University,* 21:23-34.

Ominde, S.H. (1971), "The Semi-Arid and Arid Lands of Kenya", in *Studies in East African Geography and Development,* edited by S.H. Ominde. Heinemann.

Pratt, D.J. and Gwynne, M.D. (1977), *Rangeland Management and Ecology in East Africa.* Hodder and Stoughton.

Rachie, K.O. (1983), "Intercropping tree legumes with annual crops", in *Plant Research and Agro-forestry,* edited by P.A. Huxley. International Council for Research in Agro-forestry, Nairobi, Kenya, pp. 103-116.

Rapp, A. (1974), *A review of desertification in Africa-water, vegetation, and man, SIES Report No. 1.* Secretariat for International ecology, Sweden, Stockholm.

Redhead, J.F., et al., (1983), "The intercropping of grain legumes in agro-forestry systems", in *Plant Research and Agro-forestry,* edited by P.A. Huxley. International Council for Research in Agro-forestry, Nairobi Kenya, pp. 115-124.

Republic of Kenya (1974), *Development Plan for the period 1974-1978, Part I.* Government Printer, Nairobi.

Republic of Kenya (1979), *Development Plan 1979-1983, Part II.* Government Printer, Nairobi.

Republic of Kenya (1981), *Sessional Paper No. 4 of 1981 on National Food Policy.* Government Printer, Nairobi.

Republic of Kenya (1984), *Kenya's experiences in combating desertification: Kenya's country position paper.* Ministry of Environment and Natural Resources, National Environment and Human Settlements Secretariat.

Republic of Kenya (1986), *Sessional Paper No. I of 1986 on Economic Management for Renewed Growth.* Government Printer, Nairobi, pp. 62-90.

Ruthenberg, H. (1976), *Farming Systems in the Tropics.* Clarendon Press, Oxford.

Schleiff, U. (1977), "The significance of potassium fertiliser for salt-sensitive crops under irrigation with saline water", in *Applied Sciences and Development.* Institute for Scientific Cooperation, 10:96-104.

Smith, D.L. and Obel, E. (1980), *Uses of trees and Shrubs identified by rural children in Kenya.* Mazingira Institute.

Stein, C. and Schulze, C. (1978), "Landuse and Development Potential in the Arid Regions of Kenya", in *Applied Sciences and Development.* Institute for Scientific Cooperation, 12:47-64.

Swynnerton, R.J.M. (1954), *A plan to intensify the development of African Agriculture in Kenya.* Government Printer, Nairobi.

UNCD (1977), *Desertification: Its Causes and Consequences*. Pergamon, Oxford.

UNEP (1979), *Desertification control: The Semi-Annual Bulletin on Plans and Activities, Vol. 1, No. 2*. United Nations Environment Programme, Nairobi, Kenya.

Verstraete, M.M. (1983), "Another look at the concept of Desertification", in *Origin and Evolution of Deserts*", edited by S.G. Wells and D.R. Haragan. The University of New Mexico Press, 9:213-228.

Wenner, C.G. (1979), *An outline of Soil Conservation in Kenya*. The Information Centre of the Ministry of Agriculture, Nairobi.

Woodhead, T. (1970), "A Classification of East African Rangeland II: The water balance as a guide to site potential," in *Journal of Applied Ecology* 7.

8

Problems of Desertification and Strategy for Combating Desertification with Reference to Rajasthan

H.S. SHARMA

Introduction

Desertification is the greatest single environmental threat to the future well being of the earth. It has been estimated that about one-third of the land surface of the globe is affected by desertification. This area contains about 15 per cent of the total population of the world. The problem of desertification has become most acute in drought prone state of Rajasthan during last two-three decades. The severe drought that extended from 1981 through 1987 focussed the attention of the Government and scientists on the human aspects of land degradation and led to the convening of the seminars, discussions and relief works in drought affected areas of North-western part of India.

The aim of this paper is to analyse the problem of desertification and to suggest a strategy of combating this problem. The paper is divided into three sections: Section one deals with theoretical aspect of desertification and its causes and processes. Section two focusses on the extent, types and causes of desertification with reference to Rajasthan and section three outlines strategy for combating desertification in Rajasthan.

Methodology

The present study is based on review of reports, research monographs and collection of data from Government departments. The data concerning physical and environmental hazards were computed and gathered under UGC Career Award Scheme. The study under this scheme was based on field investigation and interpretation of landsat imageries. The data regarding population statistics and landuse were obtained from Census Handbooks and other revenue departments and Government offices. A map of desertification hazards has been compiled depicting possible physical and biological indicators of desertification in Rajasthan.

SECTION I

Some Theoretical Considerations

Desertification is a term that has been in use since at least 1949 when Auberville, a perspective and well informed botanist and ecologist published a book on Climates, Forests, etc. desertification de L Afrique Tropical. He defined desertification as the change of productive land into a wasteland as the result of ruination by man—induced soil erosion. He associated it with the humid and sub-humid tropics where he worked. Desertification is a term employed rather loosely specially with reference to the deterioration of arid and semi-arid lands that are used beyond their capacity for sustained production by cultivators and Herman etc.

Despite the fact that a world conference has been held on the subject, there is no generally accepted definition of 'desertification'. In the International Conference held at Nairobi in 1977, desertification has been defined as the 'diminution or destruction of the biological potential of the land, which leads ultimately to desert like conditions and is an aspect of the widespread deterioration of eco-system under the combined pressures of adverse and fluctuating climate and excessive exploitation'.

Unfortunately, there has been much confusion over the meaning of *desertification.* The largely invalid concept of cxpanding deserts and advancing sand dunes has become a more permanent image in the public eye than the less visible and much more serious phenomenon of land degradation in drylands which is addressed here. This is the issue that affects so many people and is largely man-made. If fully recognized and tackled, it should be resoluable by man.

Desertification, as defined by UNEP, is land degradation in arid, semi-arid and dry sub-humid areas resulting mainly from human activities. This definition was modified by the 1992 UN Conference on Environment and Development (UNCED), to read as follows:

> Desertification is land degradation in arid, semi-arid and dry sub-humid areas resulting from various factors, including climatic variations and human activities.

This definition has been internationally negotiated and approved at UNCED as the operational standard for Agenda 21. Unfortunately, however, it does include the three separate elements of short-term drought, long-term climate fluctuations and land degradation induced by human beings. Each of these is different and needs to be addressed in different ways, although there are interactions.

Droughts are natural phenomena which recur periodically and can be prepared for. Severe and prolonged drought in degraded dry lands results in increased demands on the production of humid lands. In case of prolonged drought in certain key areas, global food security is already at a precarious state and this situation will become critical if desertification continues unabated.

Desertification and Droughts

A common misapprehension about desertification is that it spreads from a desert core, like a ripple on a pond. The truth is that land degradation can and does occur far from any climatic desert; the presence or absence of nearby desert has no direct relation to desertification. The process beings as a patch on the landscape where land abuse has become excessive. From that patch which might be around a watering point or in a cultivated field, land degradation spreads outward if the abuse continues. Ultimately, the patch may merge into homogeneous area, but that is unusual on a large scale.

The second misconception is that droughts are responsible for desertification. Ironically, association of the Sahelian drought with desertification had led to the common misconception that drought causes desertification. In truth droughts do not cause desertification, but it only exaggerates the harmful effects of improper land management.

The researches all over the world indicate that desertification is caused by overgrazing, excessive woodcutting, land abuse, improper soil and water management and land disturbance; its effects appear as reduced productivity of land and environmental degradation. At UNCOD the data of the conference were reviewed and three possible causes of desertification have been defined:

1. Climatic change resulting from changes in global atmospheric circulation;
2. Mismanagement resulting from an inexorable build up of pressure on resources reducing large areas to desert like conditions, i.e. the same effect on I but caused differently.
3. Biogeophysical feedback relating II to I by which means it is hypothesised that changes in landuse are able to affect changes in climate. Several theories have been proposed by scientists to seek a feedback mechanism between landuse and climate (The *dust theory*—inverse relation between dust and rainfall; The *albedo theory* direct relation between deforestation and reflectivity; the *moisture theory*—reduced vegetation and reduced soil moisture contribute diminution of rainfall; the *Carbondioxide Theory*—relates that high percentage of carbondioxide increases air temperature.

SECTION II

Desertification in Rajasthan

Rajasthan has undergone a propound ecological change during the last two-three decades. The State embraces a huge area or approximately 3,42,262 square kilometers and of population of 3,41,08,292 (1981) and 4,40,05,990 (1991). It is the second largest state of India in terms of area and occupies 9th place in terms of population (1991). The Aravallis divides the State into two halves and they have undergone massive deforestation during the last 20 years or so. The north-western eleven districts consist of the Thar Desert comprising various types of sand dunes and rocky plateau. The state is frequently drought prone and the occurrence of droughts has rendered the problem of desertification very acute.

Overgrazing, soil erosion and salt damage to irrigated land are long standing problems of Rajasthan. Though waterlogging and salinisation are of relatively recent origin, ameliorative measures to reduce land degradation have been undertaken by Government in the effected areas. The problem is an immense one in terms of the area involved and the amount of degradation that has occurred.

Desertification Characteristics: Historical Perspective

Man first placed his imprint on the landscape of Rajasthan thousands years ago. Historical records and evidences make it clear that the destructive impacts of man was already apparent five thousand years ago in the Ghaggar and lost Saraswati plains. It was then what has been called the first period of accelerated man made erosion began. Wood cutting and cultivation of slopping lands of the Aravallis and sand dunes were responsible for most of the land degradation that occurred in the Aravallis and around former fertile plains of the Ghaggar, Chambal, Luni and Banas rivers of Rajasthan. Sedimentation of canals, dams, waterlogging and salinisation combined of raise havoc in most of the river valleys of Rajasthan.

In recent years grazing pressure seems to have become a much greater problems than it was before. There has been an expansion of human and livestock population between 1971-1981. Cultivation has been extended into pasture lands. Wells have been constructed in arid and semi-arid parts and improved transportation facilities have made formerly remote areas accessible to livestock. As a consequence, cropping has become a high risk land use in the climatically marginal areas that were formerly the best pasture lands. Destruction of vegetations on the Aravallies, sandy lands and upland has led to well spread soil erosion, depletion of soil fertility.

Location and Type of Desertified Lands

The process and characteristics of desertification differ under different morpho-climatic zones of Rajasthan. Desertified lands are mainly distributed through western districts which generally fall under arid and transitional between arid and semi-arid morphogenetic regions and the vast areas to the north-east of Aravallis, extending from the Sambhar inland basin to the border of Haryana and U.P. It is also found in relatively small areas of south-eastern uplands parts of Rajasthan and the eastern plain of the Banganga and Chambal and its tributaries. The desertified lands may be classified into four major areas according to origin and morphogenetic characteristics:

Desertified Lands in Arid Morphogenetic Region

The three hundred millimeter isohyet forms the boundary of this region. The districts of Barmer, Jaisalmer, Jodhpur, Bikaner, part of

Jalor, Ganganagar fall under this region. The area is mostly dunal except parts of Jodhpur and Jalore. The very severe desertified lands are concentrated in western part of Jaisalmer and Barmer where 80 to 100 per cent area is under sand dunes of different heights and types. As we proceed towards north-east and south-east of this complex dunal country, strong to moderate sheet and gully erosion has rendered the rocky plateau under serious threat of desertification. In the Indira Gandhi Canal Command Area, the problem of Salinization and waterlogging has cropped up. Uneconomic and intensive use of canal water in Ganganagar and Bikaner have developed some alarming problems. In these areas due to sandy terrain and seepage losses from the main canal, distributaries and field channels, the ground water table is rising at an alarming rate of 0.43 to 0.83 meter per year. In the Ghaggar flood area round Suratgarh, Borpal and Manek-thar areas water table has risen by 5-6 meters within past 20 years. Due to this rise in water level 6920 in mcm water is being added to the aquifer annual leading to the development of soil salinity (kallar) in many places around 18 reservoirs of the Ghaggar Diversion Scheme. A recent study conducted by Dhiman on the basis of remote sensing technique reveals that the critical marshy lands and salinized land are increasing at a faster rate since 1972-1986. The studies of CAZRI in Nachna area reveal that rise of water table has also caused seepage problem where canal water has only been introduced recently. It has been estimated if in the canal irrigated area, the present rate of ground water is allowed to continue, then, more than 40 per cent of the canal command area is likely to develop drainage and salinity problem within 50-100 years. Canal irrigation and subsequent agricultural practices on stablised dunes and marginal area of Ganganagar, Bhadra and Bikaner are causing fresh reactivation of sand dunes and formation of new sand dunes.

In contrast to above mentioned areas of the canal command of arid morphogenetic region water table has gone down to 15 to 20 meters due to consecutive droughts from 1982-87 and human and animal consumption. The growth rate of human population and animal population during 1971-81 has been reported highest from 42 to 47 per cent which is highest so far as the deserts of the world are concerned.

Desertified Lands in Transitional Morphogenetic Region

The area between the Aravallis and 300 mm. isohyte is under severe desertification hazard. The area is mostly characterized by moderate wind, sheet and gully erosion. The strong wind and gully erosion hazard has been observed between Nagaur and Churu. Because of intense human and animal pressure the reactivation of dunes has set in, forming new barchan dunes in the vicinity of Chur, Jhunjhunu and Nagaur. Strong sheet erosion is perceptible in western slopes of the Aravallis due to largescale deforestation and cultivation on piedmont marginal areas. The growth rate of population in these areas is 27 to 32 per cent between 1971-81 which is highest in Chur and Jalore districts. In some of the areas of this morphogenetic zone disertified lands are found around water bodies in the lower reaches of the water courses. This type of desertified land is mainly linked to the changes of river courses and to exhaution of water resources. The abandonment of Khadins in this region is usually a result of either decrease of water volume or drying up of water resources due to consecutive droughts. Other causes of desertification in this region include the reactivation and shifting sand on the edge of inter-dunal flates due to destruction of vegetation on sand mounds through collection of firewood and overgrazing.

Desertified Lands in Semi-arid Morphogenetic Region

The areas adjacent to desert margin like semi-arid region of Rajasthan is vulnerable to desertification hazard. The area is characterised by widespreads and sheets and obstacle dunes on the windward sides of the isolated hogbacks and hills. A particularly larges and field occurs in the Sambhar area, west of Jaipur and extended up to Jaipur. Dunes extends a considerable distance east and north of Jaipur up to Kotputli. A large number of obstacle dunes are found at Banada, Jhir, Dausa, Lalsot, Karauli, Gangapur, Newai, Acharol and Tonk. This appears to mark the easternmost extension of the aeolian features in Rajasthan. The strong to very severe sheet and gully erosion has been noticed in this zone which has been accelerated during last 20-30 years as a raesult of large scale deforestation from the Aravallis, over cultivation and excessive biological stress. The Jaipur district records highest growth rate of population between 1971-1981 in this semi-arid zone. In Sawai Madhopur district also the growth rate was 27-32 per cent. The consecutive droughts of previous years have also accelcrated the effects of desertification. The water table has gone down by 20 meters. The surface reservoirs have dried up. The river course has been silted due to accelerated sheet erosion from the Aravallis. The introduction of canal irrigation in this semi-arid morphogenetic region especially in the Morel Command Area, Parbati Command Area, the saline soils are developing causing decline in productivity of soils. The systematic mapping of the Moral Dam Command Area on the basis of landsat imageries reveal nearly 1153 hectares of the command area is under threat of salinity which constitute nearly 2.91 per cent of the area.

Desertified Lands in Sub-humid and Humid Morphogenetic Regions

The nature and processes of desertification in this morphoclimatic zone is different and desertified lands are confined in patches where deforestation has been maximum. The Aravalli hills, Chittor hills, uplands of tribal Banswara and Dungarpur areas are under threat of desertification. The cutting of forest for fire wood needs and clearance of forest for construction of Mahi, Bajaj Sagar Dam and series of dams along the Chambal resulted into strong to severe sheet erosion. The growth rate of population of these areas has been reported from 20-37 per cent. The parts of Kota and Banswara districts have reported 32-32 per cent growth rate during 1981-91. Overgrazing is another cause of desertification. The nomads of eastern Rajasthan generally cross these areas for search of fodder and good pasture lands in adjoining districts of M.P. Therefore, these areas are under the grip of overgrazing causing serious ecological and socio-economic problems in these areas.

Causes of Desertification in Rajasthan

There are four views prevalent on the problem of desert expansion and desertification in Rajasthan. These are as follows:

Stationary Theory

The scientists of CAZRI held the view that Great Indian Desert is stationary since its formation some time in Mid-Miocene period. Some temporary modifications in the climate may, however, been experienced during the glacial period but this did not appreciably affect the desert condition. According to them, the climatic boundary of the desert is neither expanding nor deteriorating but has been occupying essentially the same position.

Sand Spread Theory

Another alarming view is that desert has been spreading in the form of sand movement towards the north-eastern part of Rajasthan through the Aravalli gaps.

Man-induced Sand Reactivation Theory

The third vie holds that the existence of fossilized dunes and sand sheet are not result of current aeolian activity. They were deposited in the former arid phase and because of the increasing human and animal pressure in the form of cultivation and urbanisation in Jaipur have been reactivised causing initiation and intensification of desertification hazard in this part (Sharma, 1988).

The fourth view attributes desertification to the processes of modernisation, as the expanding horizon of new science and technology aimed of maximum output from land or returns from these investments (Hooja, 1985).

In Rajasthan, as in other parts of the world, the process of desertification is the result of a combination of several factors, some of which are natural and completely beyond human control (such as droughts), while others are the result of human and animal interaction with the environment.

Rainfall Trends/Drought

Rainfall Trends in Rajasthan

The study of rainfall trends of the last 100 years in Rajasthan shows that certain stations have recently experienced apparent downward trend in rainfall, whereas others have shown a high variability from year to year, but no constant trend has been noticed. Now, the question arises, is there any widespread evidence that rainfall in the Indian desert and adjoining areas is decreasing? If so, is a fluctuation in progress, or is the decrease a lasting change? On a very long time scale there have certainly been great changes in climate specially within the Quarternary period. During the past 10,000 years Rajasthan has seen alternating between arid and humid phases resulting in accumulation of sand dunes especially during the arid phase.

There is thus plenty of evidence of past desiccation of climate that led to the spread of the desert surface in the north eastern part of the present desert boundary. Is there evidence of a similar desiccation in progress today? Is the widespread desertification now in progress due to such a downward trend of rainfall? There is strong evidence to suspect that more and more humid and sub-humid zones of the country, which had earlier experienced no significant moisture stress, are now under the intensification of man-induced desertification process.

The unprecedented drought situation that has occurred within the past three four years in Rajasthan is not a new thing. There are plenty of references of such a situation in the Vedic literature and in the recorded history of India. The droughts of 1877, 1900, 1918, 1940, 1965, 1972, 1979, 1984-85 and 1986-87 are well known because of their disastrous consequences. The most notable characteristics of such droughts are a random distribution in time, spatial coherence over large and in some regions, persistence for years at an end.

Climate Desertification Linkages

The prolonged drought for four consecutive years has prompted the question whether the climate is being affected by human action? There are two mechanisms whereby this might happen. First overgrazing, unwise use of land and the stripping of forest cover, all tend to make the surface more reflective to solar radiation in this part of the country. Dynamic modelling indicates that such raised albedos tend to diminish rainfall and further potential disastrous positive feedback. Second, in regions like Rajasthan, far from the ocean, much of the rainfall is caused by re-evaporated soil moisture. Faultry landuse practices have reduced the storage capacity of enhanced evaporation.

Population Pressure

Successive census reports have revealed that population of Rajasthan has been growing at an alarming rate. It has registered more than double over the past thirty years, having risen from 1.59 crores in 1951 to 3.41 crores in 1981. The highest growth rate, however, has been recorded during 1971-81. In absolute terms, population has increased by 238 lakhs during last 80 years. In 1981, Rajasthan has recorded a growth rate of 32.97 per cent as compared to the national growth rate of 24.66 per cent which has declined to 28.44, still, higher than the national average (23.85). In particular, the western arid districts have registered an abnormal growth. The alarming growth rate led to an adverse environmental situation enumerated as follows:

1. Reduction in man-land ratio which is normally devoid of adequate natural sustenance. There is a corresponding shikage of cultivated land.
2. Rise in net sown area at the cost of grazing land.
3. Over dependence on agriculture has led to land degradation due to non-availability of adequate cultivated lands.
4. Urbanization and industrialization in Kota and Jaipur, Banswara and Dungarpur districts have further aggravated the man land situation and reduction of forests for fuel and housing material.

Animal Pressure

Similar to increase in human population, there has been a rise in number of livestock population (Table 8.1). This has led to overgrazing, deforestation and exploitation of the woody bio-mass of the region.

Table 8.1: Growth of Livestock in Rajasthan.

S.No.	*Type of Animal*	*1956*	*1961*	*1966*	*1972*	*1977*	*1983*
1.	Cows	+12	+09	-09	-05	+03	+05
2.	Buffaloes	+13	+17	+05	+07	+11	+19
3.	Sheep	+37	-16	+20	-03	+16	+35
4.	Goat	+58	-08	+28	+18	+02	+25
5.	Camel	+33	+25	+20	+16	+01	-53
6.	Pig	+32	-02	+16	+40	+11	+37
7.	Poultry	+39	+75	+14	+50	+25	+46
8.	Total	+26	+02	+11	+04	+06	+23

Landuse Change

Due to increased human and livestock population, vast area of Rajasthan have been brought under cultivation often extending to marginal lands, piedmont slopes of the Aravallis and other fringe areas, where the physical conditions do not warrant farming. The percentage increase in various land use categories 1971-81 in shown in Table 8.2.

Table 8.2: Landuse Change in Rajasthan.

Landuse	*Percentage increase over 1971-81*
1. Forest	33.2
2. Land put to non-agricultural use	13.6
3. Barrani and unculturable land	-64.4
4. Permanent pasture	7.7
5. Culturable waste	61.1
6. Current fallow	-7.3
7. Other fallow land	-1.4
8. Net area sown	-3.3
9. Double cropped area	49.5
10. Total cropped area	13.5

Source: Census Handbook, 1981.

The official data shows an increase of forest area but the NRSA sponsored study of comparative data for 1972 and 1982 generated through satellite imagery has revealed that degradation of forests of the Aravalli and other parts of Rajasthan is taking place at an alarming rate. The total forest area in the state is 31753.57 square kilometers or 9.27 per cent of the total geographical area. If the rate of depletion will continue, the Aravallis will be barren by the 21st century.

Massive investments on various irrigation projects in Rajasthan have been made but the returns are not commensurate. Introduction of irrigation farming without adequate drainage measures has given rise to accumulation of minerals and salts in the soil and near surface formation in a tract having naturally impeded drainage and heavily charged with minerals and salts. All these factors have led to diminishing returns from land and intensification of man-induced desertification processes.

Development of new settlements on stabilised dunes, urbanization and construction of roads without regard to drainage channels (as in case of Jaipur city), canal without regard to physical constraints have led to rapid mobilization of stable sand belts and consequent desertification by removal of vegetation cover and wind action.

SECTION III

Strategy for Combating Desertification

The problem of desertification in Rajasthan is so widespread that massive efforts are needed to improve degraded lands. There are number of institutions and government departments engaged in research, surveys and monitoring activities of arid-zone and drought prone areas of Rajasthan. In the past some sporadic attempts have been made to develop the arid zone by former princely states. There is no evidence of any organised and systematic attempt having been made before independence to tackle the problem of desert and semi-desert areas in a comprehensive manner. In 1951-52, the need to conserve and improve the resources of the desert region was recognised and a symposium was convened by Indian Science Academy. Besides this, an ad-hoc committee of experts was appointed by the Union Government. Accordingly, Desert Afforestation Centre was set up at Jodhpur in pursuance of the recommendations. This programme was renamed in 1957 as the Desert Afforestation and Soil Conservation Station. At the same time, arid zone project was initiated by UNESCO to initiate and intensify research on problem of arid zone in different parts of the world. In the pursuance of UNESCO project recommendation, the Desert Afforestation and Soil Conservation was further reorganised in 1959 as the Central Arid Zone Research Institute. In 1966, a Desert Development Board was set up. In September 1972, the activities of the Desert Development Board were reviewed and, finally, in Fifth Five Year Plan, it was merged with Drought Prone Area Programme (DPAP). The Desert Development Programme (DDP) was launched in 1977-78. The main objectives of the programme are to check the process of desertification and at the same time to develop and utilise the available resources of the region. The programme covers arid region comprising eleven districts and 84 blocks. The following major activities have been taken up under this programme:

1. Afforestation with special reference to shelter belt plantation grassland development and sand-dune stabilization.

2. Ground water development and utilization.
3. Contribution of water harvesting structures.
4. Rural electrification for energing tubewells and pumpsets, and
5. Development of agriculture, horticulture and animal husbandry. The Task Force recommended the following programmes in Seventh Plan:
   1. Promoting more productive dryland agriculture on the basis of soil-water-climate resources of the area.
   2. Development and productive use of water resources of the area.
   3. Soil and moisture conservation including production of proper land use practices.
   4. Afforestation including farm forestry.
   5. Livestock development including development of pasture and fodder resources.
   6. Watershed and mini-watershed should be the basis of planning. In Sixth Five Year Plan about Rs. 6700.00 crores were allocated, out of which nearly Rs. 77.6 was utilized.

Drought Prone Area Programme (DPAP)

A special programme was launched in India in 1970-71 for the development of areas chronically affected by drought. The objectives of this programme are of developing the land, water, livestock and human resources of drought prone areas of the state. In Rajasthan, DPAP was initially started in 1974-75 in 72 blocks of 10 districts, but, gradually was extended so as to cover 79 blocks spread over 13 districts. In 9 districts covering 61 blocks, DPAP and Desert Development Programme (DDP) were running simultaneously. However, to rationalise DPAP/DDP programmes since 1982-83 the Government of India separated 61 blocks in which only one programme, namely, DDP is in operation, while DPAP continued in southern hills and semi-arid districts of Rajasthan. In Seventh Plan 54 new blocks of some districts of south-eastern parts of Rajasthan were included. In Seventh Plan, it was emphasised that DPAP will continue which would *inter alia* include increased stress on ecological balance and increasing the per capita through effective utilization of scarce water, conservation of scanty rainfall and arresting its runoff in drought prone areas.

Despite considerable contribution and prodigious efforts of the Government, significant breakthrough has not yet been achieved. A major reason for this may be the lack of integrated approach, and most of the efforts for controlling desertification has been sectoral in nature without adequate cognizance of collateral issue.

Conclusion

In Rajasthan, desertification is recognised as an important national problem. The strategy of combating desertification must be based on ecological characterization of different morphogenetic regions and an application of control measures suitable to each natural zone by taking into account the degree of desertification involved. Prevention of further deterioration should always be the first consideration whether the land is in good or poor condition, because prevention is always cheaper than reclamation. If priorities are to be established on an economic basis it is better to concentrate first on protecting and improving the good land rather than the poor land. For example, most of the Indira Gandhi Canal Command Area, Chambal Command Area, Morel Command Area, Jawai Canal Area, etc. are not now salinized or waterlogged; much of it can be kept that way with only the application of good field water management practices. The strongly developing desertified land and seriously desertified lands should be the focal point of desertification control as they constitute a great threat to agricultural production: effective measures include building system of shelter belts, planting forest in the depression among dunes and growing sand, fixing plants and combining forestry with agriculture. In the developing desertified lands which have not yet reached the serious stage and the land can still yield certain amount of biological production, this type of land is often not given adequate consideration. Therefore, the measures should be to:

1. Prompt action to reorient cropping methods and control dry farming techniques.
2. Improvement of appropriate control measures, such as establishment of wind break networks, reducing the pressure of grazing, closing off natural pasture lands and protecting natural vegetation. The slightly desertified lands could easily become developing desertified land under incorrect utilization.

It is important that adequate consideration should be given to this type of land and necessary preventive measures be taken. These should be complemented by planned landuse based on the land capabilities in the region. Grazing should be regulated in accordance with the carrying capacity of pastures.

3. There has to be a comprehensive water harvesting management distribution policy for Rajasthan. Otherwise, the state will stagger from crisis to crisis. Such an approach to utilization would stop the advance of desertification, reverse the process and, therefore, ensure sustained production.

REFERENCES

Dhabaria, S.S. (1984), "Intensification of Desertification Hazard Through Sambhar Gap: A Problem of Environmental Degradation in Indian Arid Zone", *18th International Symposium on Remote Sensing of Environment*, Paris.

Dregne, H.E. (1983), *Desertification of Arid Lands*. Harwood Academic Publishers, London.

Hooja, B. (1985), "Reforming Desert Development Administration and Strategies in Rajasthan – An Extensive Approach for Desert Development and Environmental Preservation". *Indian Journal of Public Administration*, Vol. XXXI (3): 910-32.

Jain, J.K. (Ed.) (1988), "Combating Desertification in Developing Countries", *Scientific Review on Arid Zone Research* (4), Scientific Publishers, Jodhpur.

Roy, B.B. and Pandey, S. (1971), "Expansion or Contraction of the Great Indian Desert", *Proceedings of the Indian National Science Academy*, Vol. 36(6): 331-34.

Sharma, H.S. (1988): The Problem of Desertification in Rajasthan, in V. Vidhyanath, and R.R. Rao, (eds.), *Development of India's Resource Base: Patterns: Problems and Prospects*, pp. 358-66. Gyan Publishers, New Delhi.

UNCOD (1977), *Plan of Action*, Nairobi.

Zhanda, Zhu and Shu, Liv (1981), Desertification and Desertification Control in Northern China, *Desertification Control*, Number 5, UNEP.

UNEP (1992), Status of Desertification and Implementation of the United Nations Plan of Action to Combat Desertification, Report of the Executive Director, February.

UNDP/UNEP (1992), Draft Report on the Expert Meeting on Desertification Land Degradation and the global Environment Facility, Nairobi, November.

9

Impact of Desertification on Integrated Management of Resources in Kachchh District, Gujarat

AMAL KUMAR SEN

The Indian costal desert is concentrated along the Gulf of Kachchh and extends into the interior of Gujarat and into a small portion of Rajasthan. The oceanic influence is extended in the north up to the head of the Luni delta. Geographically, it is concentrated mostly around the Rann of Kachchh. The annual precipitation of the region is below 500 mm. and the coefficient of variability of the annual rainfall (based on the data of 1901-50 rainfall) is above 60 per cent. It means the degree of rainfall reliability here is exceptionally low. The coefficient of rainfall variability (based on the 1901-50 data) and rainfall, temperature and humidity characteristics (standard normals based on the data from 1931 to 1960) of this area are presented in Table 9.1.

The Indian coastal desert conforms to Meigs' (1973) hot subtropical type so far as the climatic aspects are concerned. But from the locational point of view, it belongs to the Tropical Belt. Meigs' map (1973) includes this area within his "Kathiawar Makran Coastal Deserts" of Iran, India and Pakistan. But his delimitation of coastal deserts in India and Pakistan cannot be universally accepted, particularly the northern limit of the boundary up to Sukkur in Pakistan and Barmer, Tilwara and Jalore in India. Considering the aridity indices, drought factors, littoral locations and associated oceanic influences, the coastal desert of India extends approximately from 22° N to 25° N to 68° E to 73° E.

From the administrative point of view, the Indian coastal desert comprises a part of the Sanchore tehsil of Jalore district (the Luni delta) which includes 13 villages and the southern tip of the Chohtan block (14 villages) of the Barmer district in Rajasthan, Bhuj (Kachchh), Jamnagar, parts of Rajkot (Maliya tehsil) and the Banaskantha or Palanpur (Santalpur or Varh, Tharad and Vav tehsil) districts in Gujarat. The total area of this coastal desert is about 65,520.8 km². This coastal desert includes the Rann of Kachchh which covers about 9,000 km² (Sen and Shankarnarayan, 1977).

Table 9.1: Rainfall, Temperature and Humidity Conditions in the Indian Coastal Desert.

Station	*Normal Annual Rainfall in mm.*	*Coeffi-cient of Rainfall Variabi-lity (in %)*	*Mean Monthly temperature (°C)*		*Average relative humidity(%)*			
			Warmest Month	*Coldest Month*	*0830 hrs.*	*1730 hrs.*	*0830 hrs.*	*1730 hrs.*
Bhuj (Arid)	322	62.29	32.1	18.1	66	33	45	24
Jamnagar (Arid)	471	88.38	31.5	18.5	70	54	60	37
Rajkot (Semi-Arid)	589	38.08	32.6	19.4	72	27	51	22

Kachchh Area with Special Reference to Banni—Resource and Desertification

The Kachchh district is a crescent-shaped area and extends from 22°44′-24°11′ N and 68°09′ E to 71°54′ E. This arid tract is a unique geographical location surrounded by water and wasteland. The total area of the district is about 2,334 sq. km. with a total annual rainfall of about 320 mm. having 14 rainy days on average in a year. Two and three consecutive years of rainfall with less than 80 per cent of the normal is common in the Bhuj area. Mean annual wind speed in Bhuj and Mandvi is 11.3 and 23.3 km/h. respectively, mean number of days of duststorm is 0.6. Here, the normal climatic features of desert are modified to a considerable extent along the littoral—high humidity accompanies the high temperature due to direct evaporation from sea water. Such an area represents a hot coastal desert (Meigs, 1973). The study of land degradation and desertification hazards are mainly conducted by analysis of landsat imagery (MSS, FCC and TM data for 1982, 1984, 1986 and 1988).

Banni is a low-lying tract of about 2,000 sq. km. that stretches to the SW of Pacoham. This extends from Lakhpat to Kheowada running almost parallel to the coast of Kachchh for about 110 km. (Gulf of Kachchh). The terrain includes about 40 villages (Vandh or Jhil) and 70 houses (Phunga) of cattle breeders. About 780 sq. km. includes area of superior grassland—a narrow belt of low-lying grassland stretching towards the west in a row of capes, cliffs and promontories. The margin is low throughout and can hardly be distinguished from the Rann. Banni (wasteland) is apt to be covered (Patel, 1971) in times of high flood and a popular belief is that it owes its origin to a bar of sand from the rivers which flow north. But it is doubtful whether the short north-flowing consequent streams like the Gagansar, Bhaward, Kali, Godhatad, Dhidud, etc., are capable of bringing so much detritus to develop such a long stretch of land. Study of Landsats and analysis of CCT indicates that the junction of the mainland and the Banni is a faulted one with a downthrow to the north. Intertrapped beds between two successive lava flows is noticed towards the east by an extensive system of trap dykes and sills and other irregular masses of large dimensions (Wadia, 1973) which is the vicinity of Banni become very complex. The intrusion forms part of the Deccan Trap series. The feature distinguished through the study of bands 4 (Green) +5 (Green) +7 (Blue) and 4, 5, 6, 7 through green filters of Landsat data along with conjunctive use of available information and 5 per cent ground truth suggest that the Banni is of volcanic origin—mud flows (Sen and Shankarnarayan, 1986) and resulting mud flat.

But other possibilities are there. Banni is so low that it can hardly be distinguished from the Rann. It is evident in a general uplift of the region to the north and subsidence on the south. The old sediments are highly folded and faulted. There are three major faults close to the Banni and along the central hilly tracts with E-W trend and a down throw to the north. Hence, this is structurally a weak area under tension and compression. More than 75 tremors of minor and major intensities have been felt since 1880 to date of which the 1898, 1903, 1940 and 1956 ones are comparatively severe. This with the present structural lineaments suggests that the Banni area is slowly rising from the sea level and, hence, it owes its origin to neotectonic movement-raised mud flats (Kar, 1988).

Banni is barren and on the whole treeless. Besides the sandy flat and mud slack of this area three elevated areas are mapped between Pachham/Khavada (460 m.), Khadir (87 m.) and Bela (69 m.). The raised areas are commonly known as bets and are indicated in the map (Fig. 9.1) as Banni terraces. The elevated areas are formed due to heaped up sand during the summer by wind erosion/deposition. This configuration has effects on the desertification hazard in the area.

Desertification and Land Degradation

Figure 9.2 indicates the status of desertification, desertification processes involved and its relation with geomorphic units.

The high level plateau (2) and pediplains and pediments (4) are virtually colluvial plains, with moderately deep fine loamy sandy loam soils with *Euphorbia, Z. nummularia, Acacia nilotica* and *A. lencophloea* are predictably free of any desertification hazards. Very slight land degradation occurs due to wind erosion/deposition along the pediplains and pediments. Isolated hills and ranges (1) and dissected plateaus (3) are rocky uplands with variable thin soil on hills and moderately deep soil on upland supporting *Acacia senegal, Salvadora oldoides* cover are subjected to severe desertification hazards due to wind and water erosion. Rock expansion and overgrazing and cultivation on marginal lands are the other desertification processes involved.

Coastal erosional plains locally lateritized (5) and older alluvial plain of unconsolidated sand, silt and clay (6) are good agricultural lands having loamy soil *Azadirachta Indica, Acacia nilotica* and *Salvadora oldoides* are also practically free from any desertification hazards. Coastal erosional/alluvium plains and alluvial plains have maximum potential for cropping.

Sand and silt with coastal dune (9) occurs in isolated patches rarely attaining a height of 10 along the southern coast extending from Lakhpat to Mandavi which are subjected to severe to very severe desertification hazards caused by tidal effected, sand deposition (marine) and wind erosion. Commercial plantation and cultivation on these land aggravate desertification hazards. wind-blown sand are dull white to buff, medium to coares grained and vary in slica content from 85 to 92 per cent. The sand dunes slowly migrate inward and are a menace to desertification hazards.

Lagoons infilled with mud (10) also occur along the southern coast mostly in isolated patches except two little extensive areas flanking the waterlogged saline flat towards the east. These are tidal

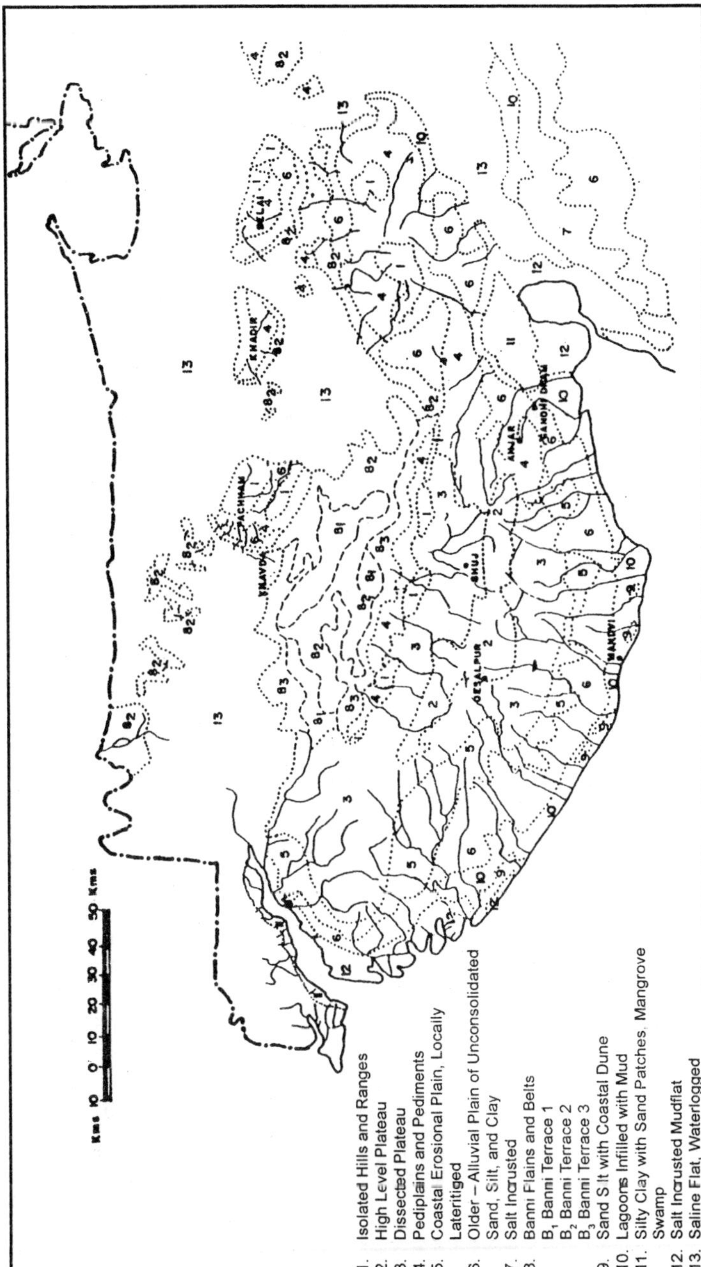

Fig. 9.1: Geomorphological units of Kachchh.

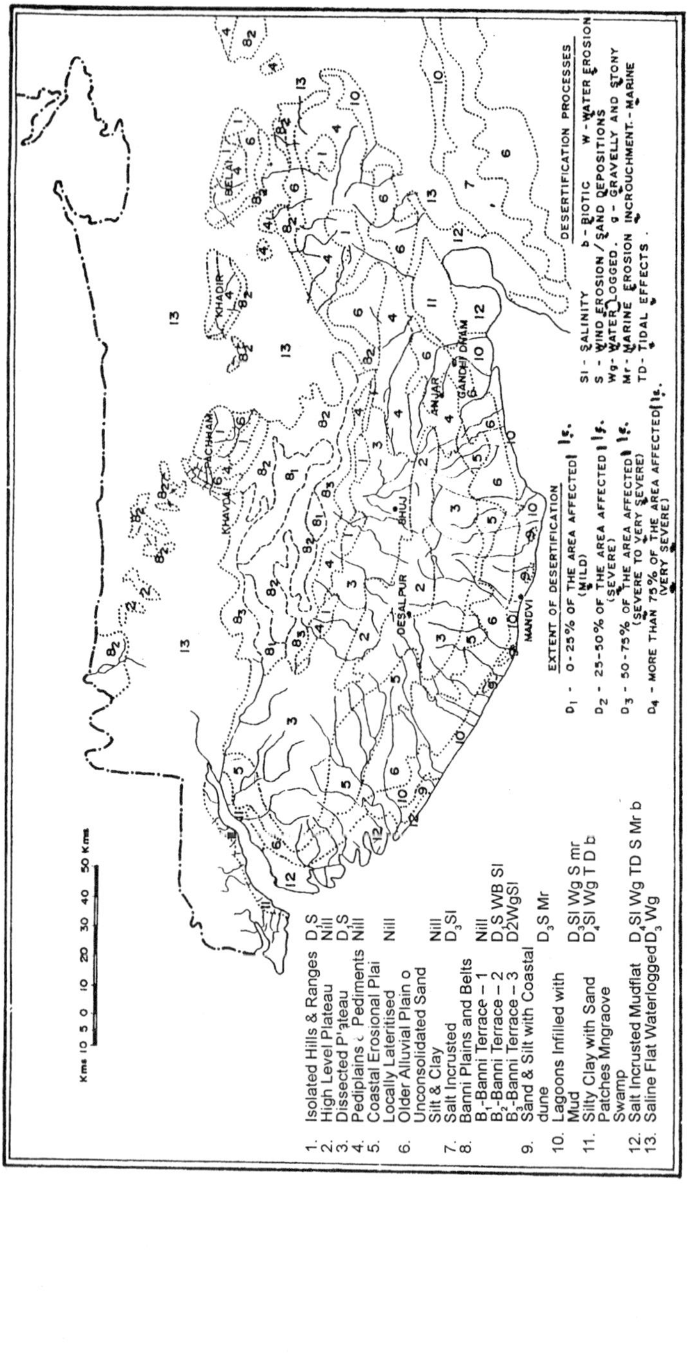

Fig. 9.2: Desertification hazards and Geomorpholigical units (Kachchh District).

mud flats having deep to very deep fine loamy saline soils at base (silty area) and deep coarse sandy soils on the sand bars (wind blown) with predominance of halophytes like Sureda sporobolus. This unit is affected by salinity, Waterlogging, marine and wind erosions and subjected to severe to very severe desertification hazards. These are mainly marine alluvial fans.

Salt incrusted mud flats (12) occur in between units 9 and 11 along the southern littoral (coastal) tract. These are highly saline isolated alluvial fans dominated by munds and like the previous unit are subjected to marine erosion under tidal influence. High salinity marine and bare soils without surface vegetation cause severe desertification hazards in these pockets. This is clearly evident between the Gandhidham-Anjar and Bachau tract.

Silty clay with sand patches (11) mangrove swamp are salty and swampy stretches with small packets in raised grounds near Lakhapat (Kateshwan) and Gandhi Dham. The area is subject to be affected by marine erosion, tidal effects and high salinity hazards. Accordingly, desertification hazards are likely to be very severe. The zone consists of swamps and is vegetated with mangroves like *Avecenia officinalis, Casuarina equisetifolia, Cassia auriculata, Sueda maritiama, Euphorbia tirucalii*, etc.

The saline flat waterlogged (13) is the level bed of the Rann. Almost flat the tract is flooded during the period of S.W. winds and rains. Most of the Rann is dry by the middle of November. These are dominated by fine silts and marine sands which are susceptible to be carried by wind during the summer and wind erosion/deposition causes raised levels by heaped up of sands. The waterlogged areas are deeply crusted with salt. Wind erosion and salinity are the active forces (natural) of severe to very severe desertification hazards. In general no sign of life breaks the weary lonliness.

Desertification in Banni Area (No. 8 in Map 1)

Banni means wasteland. The subsoil water is brackish and as a result potable water is scarce in this area. Banni, however, is a great centre of cattle breeding, where good quality grass grows. The Banni with its raised mud flats (8_1, 8_2, 8_3) is vast sandy tract with sparsely vegetated plants like *Prosopis juliflora, Acacia araluica, Salvadora persica, Prosopis cineraria*, etc., and grasses like *Andropogon annulatas, Aristida imberdis, Cynodon dactylon, Eleusine indica,*

Eleusine aristida, imberdis, Cynodon dactylon, Eleusine indica, Eleusine aristida, Aeluropus villosus, etc. (Patel, 1971). Raised mud flats of Banni (8_2) with fine loamy and deep coarse loamy soils are subjected to land degradation due to the effect of wind and water erosion (severe). The other rained mud flats (8_2) in the north are being gradually uplifted and tilted leading to leaching of salt, resulting in less salinity in the uplifted part and more in the lower dip slope zone. Salinity may affect or cause severe desertification hazards if biotic inference occurs. Alluvial fan (8_3), parallel along the saline flat (1_3) is subjected to waterlogging during the rainy season and these are practically wastelands. Eighty two lands, around the southern part of Ropar and Bachau, do not have wells due to high salinity cause by intrusion of sea water along the coastal area. Salinity is thus a process of desertification in the Banni area. The raised part of Banni (8_1) is almost free from desertification hazards being covered by good grassland. Due to high salinity waterlogging and wind erosion two types of watelands—saline waste and saline waste with scrub—can be mapped in the Banni area. The digital analysed Landsats and Map 2 indicate the status and changing pattern of desertification hazards in this terrain.

Ecosystem, Vegetation and Grasslands

Based on the ecosystem consideration of the tract the vegetation has been grouped by Rao and Agrawal (1964) under two heads, viz. (a) strand and (b) salt-marsh.

The strand communities are again studied under two sub-heads: (a) sandy strands, and (b) rocky sandy strand. The sandy strand communities are composed of either halophytes or psammophytes. Since the habitat is more or less uniform over vast distances, the same species with a wide range of distribution can be seen almost everywhere on tropical beaches. Ipomoea pescaprae forms the dominant community on the sandy shores. Rao and Aggarwal (1964) have identified nine community types under the sandy-strand ecosystem. These are: (1) Hydrophylax maritima, (2) Ipomoea pescaprae, (3) Asparagas plumosus, (4) Halopyrum mucronatum, (5) Cyperus conglomeratus, (6) Cyperus aristatus, (7) Sporobolus termulus, (8) Leptadenia pyrotechnica, and (9) Cynodon dactylon. The first two are found all along the coast and the second one is an excellent strand binder. The third type if found on the sandy bars, is often grazed. In the coastal area, the fourth type is a consipicuous component of the strand flora. The fifth, sixth and seventh communities exhibit dense patches on dry sandy spots and are found associated more or less with all types of strand vegetation. The eighth community type is found growing in abundance on the sand dunes. The last community, a ubiquitous grass, is found throughout the coast, adjoining the strand vegetation.

The sandy-strand habitat is composed of water-laid loose sands which have given rise to sand hills or sand dunes.

The rocky-sandy strand habitat in the coastal desert is characterized either by the presence of sand in the rock crevices or by the presence of a thin mantle of soil over the underlying rocks. This ecosystem is marked by the following predominent plant communities: (1) *Capparis cartilaginea*, (2) *Fagonia cretica*, (3) *Indigofera cordifolia*, (4) *Kick-xia ramosissima*, (5) *Lepidagathis trinervis*, (6) *Sericostoma pauciflorum*, and (7) *Enicostema verticullatum* mixed species.

The salt-marsh vegetation occurs in the littoral sub-littoral belt and in the Luni Delta. The flora comprises mainly the following species: (1) *Avicennia alba*, (2) *Suaeda nudiflora*, (3) *Aeluropus lagopoides*, (4) *Fimbristylis cymosa*, (5) *Scirpus maritimus*, (6) *Urochondra, setuluso* and (7) *Atriplex stocksii*. The first is met all along the fringes of the inland creeks and shallow muddy areas. The second is the characteristic feature of saline areas. Others are found in the interion areas.

Behind the strand habitat, salt-marshes, slacks and mud formations there lies the semi-arid coastal plain.

In short, the dominant floristic species of the Indian coastal desert are: *Capparis decidua*, *Zizyphus jujuba*, *Prosopis cineraria*, *Avicennia officinalis* and *Rhizophora mucronata*.

Geology

The Indian costal desert has the following geological formations:

1. *Recent subrecent*: Alluvium, blown sand, milolites, tidal flats, raised beaches.
2. *Tertiary*: (1) Pliocene—Dewarka beds (2) Miocene—Gaj beds.

The Deccan trap encompasses the Kachchh area to a little extent. The saline marshes predominate in the Rann of Kutch and its adjoining areas. The major portion of the coastal sands is of marine origin composed of shells of marine microfauna and is very rich in $CaCO_3$.

Soils

Soils of the Indian coastal desert have been described by Rao and Aggrawal (1964) as follows:

(i) *Soils of sandy strand*: These soils are sandy to loamy-sand desert soil. Their pH exhibits moderate alkalinity. Organic matter ranges between 0.02 and 0.64 per cent. The values of total dissolved solids and sodium chloride point to the effect of salt-spray, but not to the direct inundation with sea-water.

(ii) *Soils of the rocky-sandy strand*: The Soil cover over the underlying rocks has been formed by the weathering of rocks in situ. The soil is mixed with conglomerates. In some places, water-laid or wind-blown sands are present in thin layers over the underlying rocks, and hence become mixed with the soil from the weathering rock. The soil is loamy sand. The pH shows mild to moderate alkalinity. Organic matter ranges from 0.22 to 0.81 per cent. The contents of the total dissolved solids and sodium chloride indicate the effects of salt spray only. Soils are highly calcareous.

(iii) *Soils of the salt-marsh*: Soils are saline and those from salt pans are sandy loam, loam or silt loam with a mild degree of alkalinity. Organic matter is much higher. Dissolved solids and sodium chloride are very high as a result of the direct influence of sea-water.

(iv) *Soils of slack and mud formations*: These soils are loamy sand to sandy loam to sandy clay loam with mild to moderate alkalinity. Organic matter of the surface soils varies within wide limits (0.26 to 4.69%) depending upon the age of these slacks and mud formations. The content of dissolved solids and sodium suggests that the soils are not under the direct influence of sea-water. All the soils are highly calcareous (40.97 to 56.58% $CaCO_3$).

Groundwater Resources

The geology and water-bearing formations of the area have been earlier reviewed by Adyalkar (1964). In Kachchh, rocks of Patcham and Chari series comprise calcareous shales and limestone and the quality of water is, therefore, saline. The Umia series were deposited only under fluviation and estuarine conditions and partly under marine conditions. The marine formation of the series yields only saline water and hence, there is no development of ground water in the fossiliferous sediments of the series. In the Bhuj sandstone groundwater occurs under both confined and unconfined conditions. The static water level in the dug well ranges from 2 to 30.5 m bgl and that in the dug-cum-bore wells and tube-wells tapping confined aquifers from 3.4 to 31.4 m bgl. Over 9,000 to 26,000 million litres of groundwater have been utilized for the purpose of irrigation in the lower and upper Bhuj sandstones yield about 13.55 lps. The quality of water is generally good. In the Deccan trap in Kachchh, groundwater occurs along the joints, fractures and secondary partings, which an generally limited in depth up to 50 m. The static water level in the dug well ranges from 4 to 12 m., with yield capacities of 0.05 to 0.20 lps. The yield capacities of the tertiaries in Kachchh ranges from 0.06 to 0.24 lps, with slight to moderate saline water. Irrigation with groundwater is a universal local practice.

Prospects for Development

Considering the limitation of resources and the constraints imposed by the arid environment, the Indian coastal deserts have restricted choices for development. The resources of the region have not yet been fully assessed. For this assessment, integrated basic resources survey in the region should prove useful. The development planning of the Indian coastal deserts may be recommended in the following order of priority.

(i) *Rangeland Development and Diary Farmint*: The existing permanent pasture lands and other available grazing grounds should be fenced, and controlled grazing should be introduced.

(ii) *Development of Mineral Resources*: The available minerals, such as salt, bauxite, calcite and limestone, should be properly extracted. The modernization of the mines will increase the output and income per head. In vestigations on soil and natural gas should be continued.

(iii) *Fishery*: The coastal area of the Gulf of Kachchh contains good fishing grounds. This particular industry has a good prospect in the Kachchh.

(iv) *Urban and Industrial Development*: The planning of urban development in the coastal arid zone should be considered from the following viewpoints: (a) development of ports, e.g. Mandvi, Kandla, (b) industrial development of settlements, e.g. Naktharana, Gandhidham, Anjar and Bhuj.

(v) *Tourism*: There are great prospects to develop tourism and its associated industries in this arid zone. Seaside resorts can be developed along the Gulf of Cambay and the Gulf of Kachchh. The usually sunny weather and the scenic beauty of the regions may also help to attract tourists. The historical and architectural significance of places may also help to develop tourism. The prospects of having a view of the desert and of the Rann of Kachchh should provide much scope for the development of a viable tourism industry.

With the development of tourism, it is almost certain that the local arts and crafts will find avenues for a fuller expression.

Game sanctuaries can also be developed in the Rann of Kachchh area and in Kachchh district. The Ran of Kachchh receives many migratory birds, e.g. the flamingo from coastal Africa, Asia and other countries. Other interesting birds, e.g. florican, imperial sand-grouse and Houbara are also found in large numbers.

Wildlife sanctuaries can also be developed in the area, particularly for wild ass and Blue Bull. Jamnagar and Okha can be developed as transport centres for the lion sanctuary in the Gir forest. Tourism and the resort industry have developed the economy of many coastal deserts in countries such as Peru, Chili and Israel. There is enough scope to develop the Indian coastal desert, too, in this respect.

(vi) *Agriculture*: Agriculture is still the main occupation of the people. But owing to extremely arid conditions, high salinity, limited irrigational facilities and marshy land, it is doubtful if it can be developed to the desired level. Dry-farming techniques may be tried to develop agriculture. For the development of this coastal desert, as in other coastal deserts of the world, agriculture should necessarily receive the lowest priority.

Acknowledgements

Sincere thanks are due to Sri D.K. Das Sharma, Director, G.S.I. (W. Circle, Jaipur, now Director Palaeontology, S. Circle, Hyderabad) and Sri Amal Kar, Senior Scientist, CAZRI, for their help. The Geomorphological map is partly based on Sri Das Sharma's work and is duly acknowledged.

REFERENCES

Adyalkar, P.G., (1964), "Palaeogeography, sedimentological framework and groundwater potentiality of the arid zone of western India", *Proc. Symp. Prob. Indian Arid Zone,* Jodhpur (23rd Nov. to 22nd Dec. 1974), pp. 4-17, Ministry of Education, Govt. of India, New Delhi.

Ananda Rao, T. and Aggrawal, K.R., (1964), "Ecological studies of Saurashtra coast and neighbouring island", *Proc. Symp. Prob.* Indian Arid Zone, Jodhpur, pp. 31-42.

Kar, Amal, (1988), "Possible Neotectonic Activities in the Luni—Jawai Plains, Rajasthan", *J. Geol, Soc. India*, 32(6), pp. 522-526.

Meigs, P., (1973), "World distribution of coastal desert" in *Coastal Deserts—Their Natural and Human Environments* (eds. Amiran, D.H.M. and Wilson A.W.), Pt. 1, Chapter 1, The University of Arizona Press, Tucson, Arizone, pp. 3-12.

Patel, G.D., (1971), *Gazetter of India,* Gujarat State, Kachchh District, (ed.), Govt., of Gujarat, Ahmedabad, p. 623.

Sen, A.K. and Shankarnarayan, (1986), "Analysis of space photographs for mapping land resource unit of Indian coastal desert", *Proc. Int. Seminar on Aer. Phot. Int. and Remote Sensing for the Developing Countries,* No. 1, Vol. 2, T. 5-P/18, pp. 1-9.

Sen, A.K. and Shankarnarayan, K.A., (1977), "The Extent, Resources and Prospects of the coastal desert of N.W. India", (Ch. 10) in *Desertification and Control,* I.C.A.R., pp. 81-88.

Wadia, D.N. (1973), *Geology of India,* (Fourth Edition), Tata McGraw-Hill, New Delhi, p. 508.

10

Land Degradation Due to Channel Shift

A Case Study of the Kosi Embankment Area

RAMASHRAY PRASAD

Introduction

The Kosi river is notorious for its flood during monsoon period when there is high discharge and associated sediment in the flowing water. This sort of behaviour of rivers in the plain is common as it is based on natural physical laws of rising discharge and accompanied silt load; but the nature, amount and magnitude of the Kosi is far more than other rivers of the world—only comparable to the Hwang Ho (Yellow) river of China. The spread of water of the river (width) is changing with time and space. During high flow period of flood, discharge is enormous and, thus, the width of flowing water widens but with decreasing amount of water in the channel, width becomes narrower.

Most of the dynamic river regime of the world which create flood in the plain, the greatest complex one is found in shifting of the courses and this process is clearly observed in the lower Kosi basin. Shillingford (1895) considers the shift of the Kosi channels to be back and forth just like a pendulum from the points of its debauch in the plain at Chatra. A line joining Bhelhi village in the west and Purnia in the east gives a marked shift in the courses of the Kosi. It is almost 115 km. in a period of 214 years (GFCC, 1983) from 1736 to 1950 with an average rate of shift up to 0.54 km./year. Before 1883, the rate was only 0.25 km./year which got accelerated to 1.14 km./year, on an average in the post 1883 period. During 1923-33, it went up to 2.64 km./year (Prasad, 1999). In this process of chanel shifting, a vast tract of north Bihar plain was used to be devastated by the Kosi before 1963, the year of completion of Bhimnagar barrage. The nature and behaviour of the Kosi remains the same what it was during pre-barrage period (Prasad, 1987). Still, it has not got the stability and, thus, is shifting within the embankments, except some breaches. In this process, agriculture land is destroyed and, hence, a proper planning and river monitoring should be initiated to check the land getting depleted.

Causes of Channel Shift

The Kosi, a large braided river is characterized by wide channels, rapid shifting of bed material and displacement of the position of river courses (Leopold, Wolman and Miller, 1964). Rao (1956) suggested that the causes of channel shift are mainly five: (a) valley slope and gradient, (b) bed load and coarse silt, (c) discharge, (d) bed resistance, and (e) transverse slope. He has taken all these factors into account and concluded that these variables occur in conjunction and are not readily studied in isolation. All these factors supporting for vigorous shifting of the course are found to be associated with the Kosi. Rao (1956), therefore, summed up that, to reduce the lateral shifting of the river below Chatra, it is necessary to flatten the gradient or reduce or trap the coarse silt and if possible, to reduce the intensity of peak flood flow. The most important contributory factor for side sway of the river appears to be due to excessive coarse sediment in the river.

However, Wells and Dorr (1987) are of the view that the channel shift is caused by the high discharge associated with enormous amount of coarse silt, although, tectonically fragile and seismically weak conditions provide loose and shattered material to be carried away with high flow during monsoon period. Slope being very high in upper catchment, they are easily transported down. The slope and bed of the channel in plain is not capable of performing the work as it is done in the upper part. Hence, deposition takes place and chokes the mouth/valley of the stream. Again, when relatively bigger discharge occurs, the water gets spread over a large area and spills over towards lower side, a small carving takes place. With the passage of time, in a few years, this newer one gets deepened and widened progressively. The newer one becomes the main channel with high velocity and discharge, and the earlier one gets abandoned. This man channel is dominant for some years and after that in suitable conditions

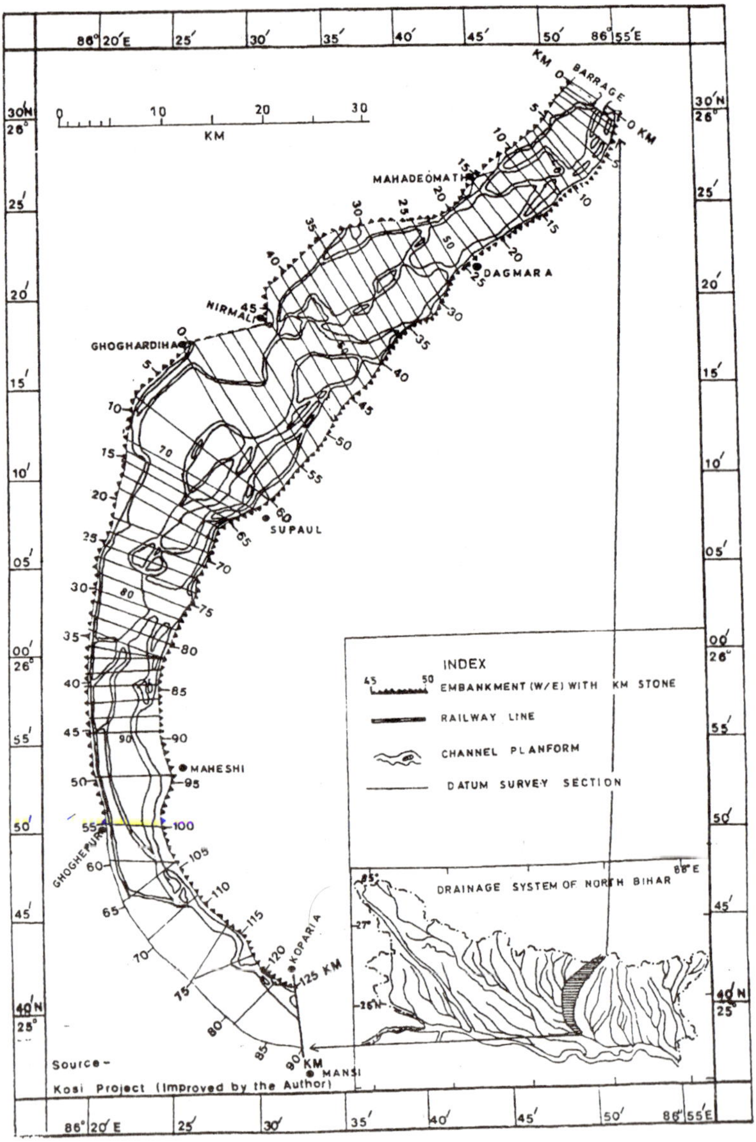

Fig. 10.1: Kosi Embankment Area.

as stated above, another channel is carved out leading for further shift of the channel and abandonment of the old flowing (Prasad, 1999). The Kosi fan is flat. Strictly, then, climatic floods by monsoonal depressions turn the river Kosi into a 15 to 30 km. wide flowing sea (Inglis, 1967).

Objective

The main objective of this study is to throw light on the problem of degradation of agricultural land due to channel shift of the Kosi. Steps have also been taken to suggest some remedial measures for the proper management of land, so that the area gives a very good agricultural yield.

Study Area

The area under this study is within the two embankments—eastern and western—of the Kosi downstream the Bhimnagar barrage in the north and Mansi-Koparia railway bridge in the south. The entire area encompassed by these limits is 1251 sq. km. with a varying width of 6 km. to 17 km. and that of the length is 125 km. The geographical coordinates of the area is 25°43′ N to 26°36′ N latitude and 86°20′ E to 86°55′ E longitude (Fig. 10.1). In terms of the fluvial processes of the Kosi, the entire area is not behaving uniformly. In other words, scouring and filling tendency of the river in different parts varies considerably based on which it is divided into four 'Channel Reaches'. It is given in the Table 10.1 along with its reach-wise characteristics during pre- and post-barrage (pre- and post-1963) period.

From Table 10.1, it is quite evident that two upper reaches were getting degraded in both pre- and post-barrage period. Reach I experienced a bit higher scouring (3.56 cm/year) in pre-barrage but in post-barrage period it has decreased (0.83 cm/year). Prior to 1963, Reach II was getting degraded at a rate of 0.37 cm/year but after that it has gone up to 1.86 cm/year. Reach III was witnessing an aggradational tendency with a tune to 9.56 cm/year before 1963 and in post-1963 it is rising at a rate of 6.36 cm/year. For Reach IV information about aggradation/degradation is not known for pre-barrage days but during post-1963, it is aggrading with an alarming rate of 12.03 cm/year (Desai, 1982). Author has calculated for entire area and found that the average aggradational tendency is

3.04 cm/year after the barrage were operational but prior to that, for Reach I, II and III, it works out to be 2.06 cm/year. If Reach IV is excluded for the comparability, it is only 0.73 cm/year deposition of sediment in post-barrage period. From these observations it is evident that the barrage and embankments have played a leading role in reducing the rate of deposition. It is because, discharge is made constricted within the embankments giving a higher velocity and, thus, transporting the sediment load downstream which yield lower deposition and, hence, lower rate of rising the embankment area.

Table 10.1: Four channel reaches of the study area.

Reach	*Place along Eastern Embankment*		*Distance from Bhimnagar Barrage (in km.)*		*Total Area in sq. km.*	*Avg. Scouring/ Filling/Year (Depth in cm.)*	
	From	*To*	*From*	*To*		*Pre-1963*	*Post-1963*
Reach I	Bhimnagar	Dagmara	0.00	25.25	156.0	–3.56	–0.83
Reach II	Dagmara	Supaul	25.25	62.82	545.2	–0.37	–1.86
Reach III	Supaul	Maheshi	62.82	94.25	293.2	+9.56	+6.36
Reach IV	Maheshi	Koparia	94.25	125.00	256.6	NA	+12.03
All Reaches	Bhimnagar	Koparia	0.00	125.00	1251.0	+2.06	+3.04

Note: The average scouring/filling/year for entire study area during post-1963 is 3.04 cm., while for 3 upper reaches it only is 0.73 cm.

Data Source

This study is based on the configuration plan, i.e., the lying of channel in the area on a map. The datum level survey is conducted every year after monsoon months in December/January when there is low flow and it is easier to transverse, even, with smaller boats. Altogether there are 98 cross-sections with well defined azimuths along which survey is conducted. A total of 32 cross-sections lies upstream to the barrage which is not the area under study here. Within this study area cross-sections are numbered from 33 to 98, i.e., 66 cross-sections. During this course, a map is prepared in terms of channel and non-channel area within the embankments by the Kosi Project Authority. A shift of channel course is identified from one flood season to another. Because of this a large area lying out of the bed is occupied by the bed of the Kosi in another year's flood due to oscillation of the river 'thalweg' and width. Therefore, channel configuration plans have been collected from the Research and Investigation Division of the Kosi Project Authority at Birpur, Bihar. These plans are from 1975 to 1986 barring 1985. Plans for other than these years are not available.

Methodology

Here, method adopted is very simple. It is based on configuration plans which are prepared at a scale of 1:50,000, thus, the size runs into a few metres in length. These plans have been reduced through electronic copier and area under the channel and above the channel has been computed very accurately by planimeter for all years mentioned above. To bring the map into reproducible book size, it is further reduced. The plans of four years namely; 1975, 1978, 1983 and 1986 have been superimposed on a single base to give a visual glance in terms of the shifting of channel bed which are shown in Fig. 10.2 to Fig. 10.5. However, it is found that the area under channel varies considerably on year to year basis without giving any trend for projection, hence, considering the total area, year-wise percentages have been computed for both sets—area occupied by channel and area above the channel which could be used for agriculture. Year-wise addition of both percentages is 100 and it gives the idea about the changing amount of land on yearly basis because of channel shift. The percentages given in a column could, even, be compared in terms of variations in the area. The ratios have also been found out between channel area and area above the channel within the embankments.

Results and Discussion

It is quite apparent that the river bed or the area occupied by the channel is not available for agricultural purposes except fishing. If the channels are stable, there is not much problems to the agricultural land. Area which is outside the channel can be utilized for crop growing according to the suitability of the soil and its fertility. But, when there is no stability of channel, area which is available for crops in one year turns into stream bed in the other year and, thus, it is eaten up by the shifting of the channel bed. A part of the area under channel comes out from the bed of stream but that is not at all suitable for crop growing as it is devoid of humus and, thus, fertility. It is the deposition of sediment from coarse sand to fine silt brought down by the Kosi river from its large upper (mountainous) catchment (more than 61,000 sq. km.). Apart from the shifting of the Kosi channels within its embankment, the variation in area occupied by channel in different years is quite significant. Hence, it is better to study the nature and amount of variations in channel and non-channel area on year to year basis reach-wise.

Reach I

The channel planforms of 1975, 1978, 1983 and 1986 have been superimposed which is given in Fig. 10.2. From this plan, it is quite evident that the entire embankment area is completely governed by its channels. Channel shift is an usual phenomenon on year to year basis. Every year the channel occupy its different position, sometimes, it is partly and sometimes in the entire reach. It is also interesting to note that the major and dominating channel lies towards the eastern half of the area. The Kosi is a very interesting example of braiding. There are many branching and sub-branching of channels within the embankment. Protection measures are taken up for both the embankments but more emphasis is given to eastern as it is under severe attack of flowing water at the high stage of the river.

The area under channel and above water level (agricultural land) shown on the plan has been computed with the help of planimeter. It is given in the Table 10.2, it is quite clear that, on an average (24.36%), one-fourth of the area is under the channel cover and that of remaining three fourth (75.64%) above the channel at the time of survey in December/January. Channel area in this reach vary from 16.60 per cent in 1975 to 49.62 per cent in 1978. In 1981, the percentage of channel area is 18.01. There are only two instances when the percentage is below 20 and that of only once above 30 i.e., that is 1978. Other years are represented between 20 to 30 per cent, to be very precise, it is 21.35 to 29.74 per cent. It works out to be an interval of only 8.4 per cent.

For above eleven years, the area not covered under channel

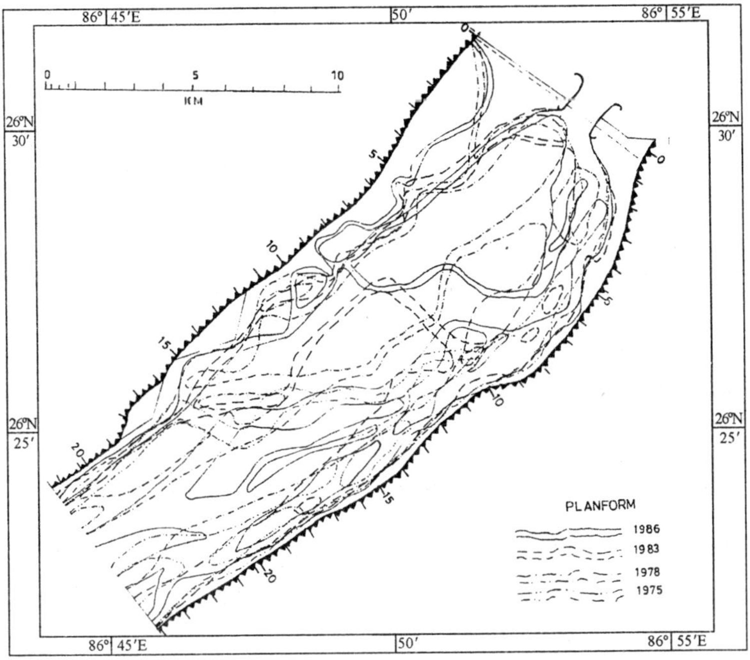

Fig. 10.2: Composite Channel Planform: Reach I.

(non-channel area or agricultural area) is also given in the table and its percentage. Although the percentage itself is capable of standarding the comparison, but to see the ratio between the channel and non-channel area at the time of survey is also given. It represents the area above the channel in multiple of the area under channel. The smallest ratio is 1.02 times in 1978 and 5.02 times in 1975—an interval of four times between them. The average ratio of eleven years works out to be 3.11 times and its standard deviation of channel and non-channel area is 13.1 and that of coefficient of variation is 33.68 per cent. Column VI very clearly shows the year to year changes in the agricultural land which may be used to grow some crops. From this table it is clear that the variation in land which may be utilized for agriculture is great. From 1977 to 1978, 43.8 sq. km. area is consumed by channel and in the next year 39.8 km. is again exposed from the channel bed. Thus, the variation from one year to another gives a fluctuating behaviour of area within the embankment.

Table 10.2: Variations in channel and agricultural area: Reach I

Year	*Channel Area (sq. km.)*	*% to Total (Channel)*	*Area Above Channel (sq. km.)*	*% to Total (Non-Channel)*	*Yearly changed in Agr. Area (sq. km.)*	*Ratio (Channel Area to Non-Channel Area)*
1975	25.9	16.60	130.1	83.40	-	5.02
1976	35.7	22.88	120.3	77.12	-9.8	3.37
1977	33.3	21.35	122.7	78.65	+2.4	3.66
1978	77.1	49.42	78.9	50.58	-43.8	1.02
1979	37.3	23.91	118.7	76.09	+39.8	3.18
1980	36.2	23.21	119.8	76.79	+1.1	3.31
1981	28.1	18.01	127.9	81.99	+8.1	4.55
1982	36.5	23.40	119.5	76.60	-8.4	3.27
1983	35.8	22.95	120.2	77.05	+0.7	3.36
1984	35.6	22.82	120.4	77.18	+0.2	3.38
1986	46.4	29.74	109.6	70.26	-10.8	2.36
Average	38.9	24.36	117.1	75.64	-	3.11

Reach II

The channel planform of Reach II is superimposed on Fig. 10.3 for the year 1975, 1978, 1983 and 1986. This reach is the largest among four reaches in terms of its area. The entire reach is vulnerable during rainy season when heavy discharge of water and sediment come down. Sediment deposition within the old channels and carving a new nearby the older one is not a new event for this river. Thus, the channels are not at all stable from year to year observation. No clear-cut pattern in the shift can be stated with high level of confidence. If at all it is to be stated, it is the unpredictability of the channel change from year to year basis within the embankment. The entire area is crissed-crossed by channels in different years. It is only for four years of representation. If the number of years are increased, probably entire area will be occupied by river channels barring a few pockets and spots. But one thing is quite evident that in each year the main dominating channel is lying towards eastern half.

Table 10.3 given below represents the information for Reach II which is between cross-section nos. 50 to 70. The total area of this Reach is 545.2 sq km. The area is greater on account of its larger width. Here, width being more, channels become narrower in comparison to the spacing between the embankments. Because of

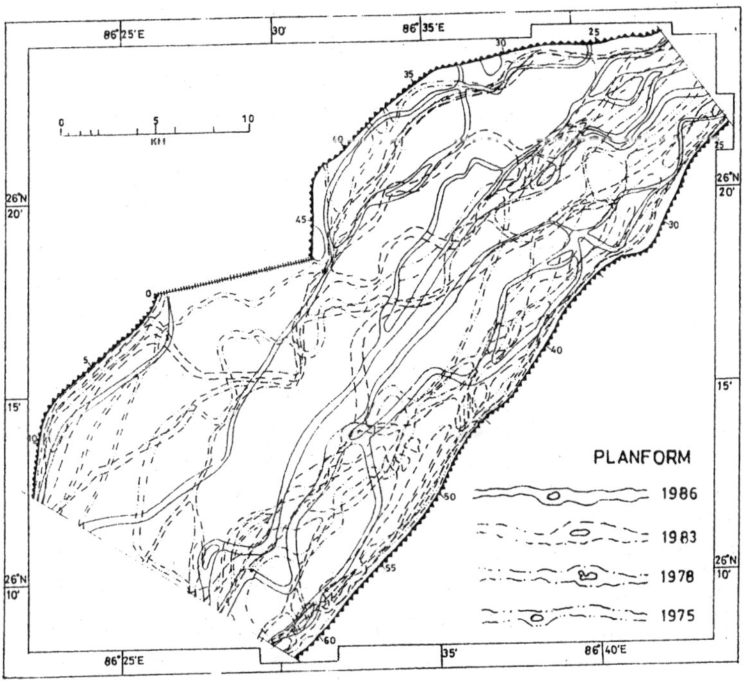

Fig. 10.3: Composite Channel Planform; Reach II.

which the area under channel cover comes down and the percentages too.

From this table, it is quite apparent that the area under channel is marginalised drastically in comparison to that of Reach I. Here, the average percentage of channel area to that of total area of the reach is only 15.13. It has never exceeded 18.34 per cent (1983) and that of lower one is 12.28 per cent (1979) with an interval of about 6 per cent. The standard deviation works out to be 8.8 and that of coefficient of variation is 10.67 per cent for channel and non-channel area. The radio between these two is 4.45 times in 1983 and maximum of 6.76 times in 1977 with an average of 5.61 times for given eleven years. In comparison to Reach I, standard deviation in this reach is smaller conveying the idea of a little more stability of channel area. The horizontal movement is still an usual phenomenon (channel planform described above) but the amount of area lying under channel remains almost identical over different years otherwise it does not show any channel stability in terms of the shifting character. Maximum area engulfed by the channel is 25.8 sq. km. during 1983 in comparison to 1982 but in the next year (1983) 21.8 sq. km. comes out from the belly of the channels. The minimum change is only 0.8 sq. km. from 1979 to 1980.

Table 10.3: Variations in channel and agricultural area: Reach II.

Year	*Channel Area (sq. km.)*	*% to Total (Channel)*	*Area Above Channel (sq. km.)*	*% to Total (Non-Channel)*	*Yearly changed in Agr. Area (sq. km.)*	*Ratio (Channel Area to Non-Channel Area)*
1975	94.5	17.33	450.7	82.67	—	4.77
1976	86.0	15.77	459.2	84.23	+8.5	5.34
1977	70.3	12.89	474.9	87.11	+15.7	6.76
1978	75.0	13.76	470.2	86.24	-4.7	6.27
1979	83.3	12.28	461.9	84.72	-8.3	5.55
1980	84.1	15.43	461.1	84.57	-0.8	5.84
1981	87.2	15.99	458.0	84.01	-3.1	5.25
1982	73.9	13.55	471.0	86.45	+13.0	6.38
1983	100.0	18.34	445.2	81.66	-25.8	4.45
1984	78.2	14.34	467.0	85.66	+21.8	5.97
1986	74.8	13.72	470.4	86.28	+3.4	6.29
Average	82.5	15.13	462.7	84.87	—	5.61

Reach III

The channel outlay of Reach III for the year 1975, 1978, 1983 and 1986 have been superimposed on Fig. 10.4. It is downstream from the Reach II and it is narrowing down towards southern part. More or less the same type of pattern in channel shift is observed here, too, what it is there with Reach II, but with relative stability in the channel from one year to another under study. By giving a glance on Fig. 10.4 it is very evident that the main and wider channel, to some extent, has stablized itself in the centre. Although, equally important channel in 1986 has also appeared near the eastern embankment. Central stability is only the relative one if it is compared with Reach I and II. Because, the lateral migration in the channel is still going on but at a smaller scale. It has not stabilized entirely. In other words it can be said that the channel in this reach has got temporary and relative stability unusual to the earlier two reaches.

Table 10.4 stands for Reach III which extends from cross-section nos. 71 to 90 with a total area of 293.2 sq. km. Out of the total area

Table 10.4: Variations in channel and agricultural area: Reach III.

Year	*Channel Area (sq. km.)*	*% to Total (Channel)*	*Area Above Channel (sq. km.)*	*% to Total (Non-Channel)*	*Yearly change in Agr. Area (sq. km.)*	*Ratio (Channel Area to Non-Channel Area)*
1975	70.3	23.98	222.9	76.02	—	3.17
1976	51.7	17.63	241.5	82.37	+18.6	4.67
1977	54.6	18.62	238.6	81.38	-2.9	4.37
1978	58.0	19.78	235.2	80.22	-3.4	4.06
1979	59.4	20.26	233.8	79.74	-1.4	3.94
1980	93.0	31.72	200.2	68.28	-33.6	2.15
1981	64.5	22.00	228.7	78.00	+28.5	3.55
1982	62.2	21.21	231.0	78.79	+2.3	3.71
1983	105.0	35.81	188.2	64.19	-42.8	1.79
1984	52.4	17.87	240.4	82.13	+52.2	4.60
1986	52.5	17.91	240.7	82.09	+0.3	4.59
Average	65.8	22.44	227.4	77.56	—	3.46

of 293.2 sq. km., 105 sq. km., is under the channel in 1983 at the higher end and 51.7 sq. km. in 1976 at the lower end which works out to be 35.81 per cent and 17.63 per cent respectively. Thus, the difference between maximum and minimum is of 18.18 per cent. Out of the eleven years under study it has exceeded 30 per cent only in two occasions (1980 and 1983) and only three times show less than 20 per cent during 1976, 1984 and 1986. The percentage during other remaining years are between 20 and 30. The average for the entire period is 65.8 sq. km. under channel and that of non-channel area is 227.4 sq. km. It is 22.44 per cent and 77.56 per cent respectively to the total area of the reach. The ratio varies between 1.79 times during 1983 to 4.67 times during 1976. The average ratio for entire period is 3.46 times. The standard deviation for channel and non-channel area works out to be 16.8 and that of coefficient of variation is 25.53 per cent. From 1982 to 1983 the channel area has increased by 42.8 sq. km. and in the subsequent year 52.2 sq. km. area has come out of the river bed.

Reach IV

The superimposition of channels for the Reach IV has been done on Fig. 10.5 for all three years only. This is the lowest (last) reach which

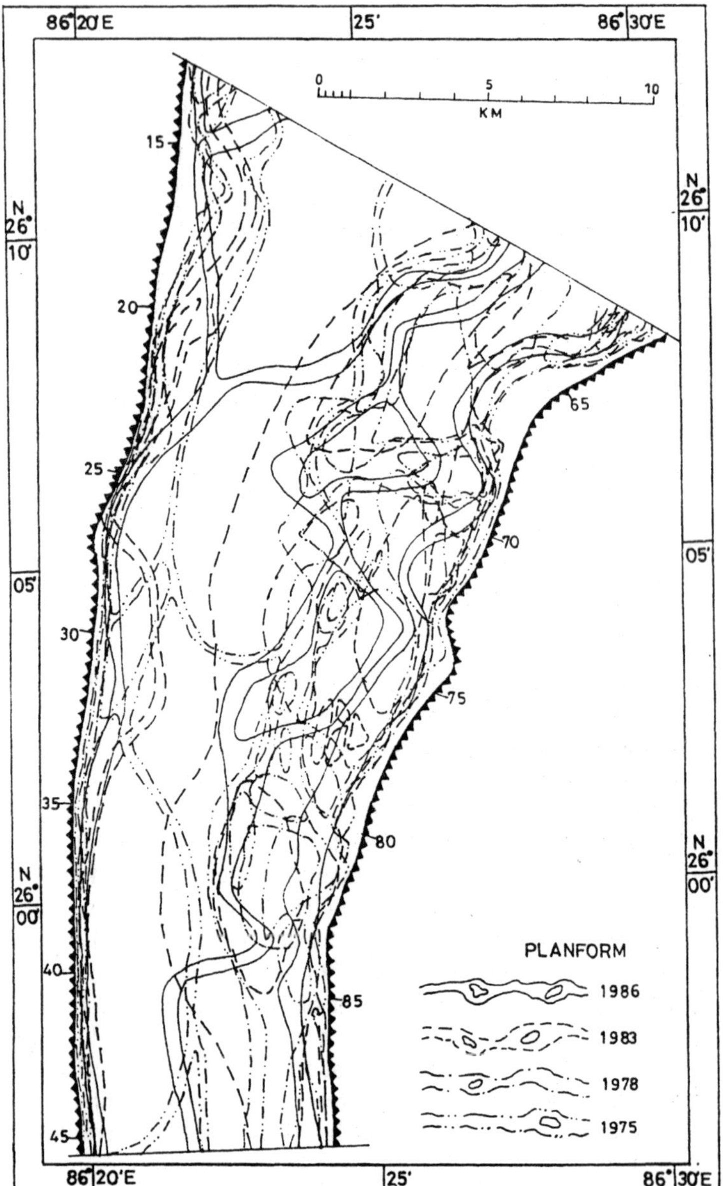

Fig. 10.4: Composite Channel Planform: Reach III.

is getting terminated by the Mansi-Koparia railway bridge in the south. Western limit along a major part has been left unembanked to facilitate the water to come into the Kosi brought down by many rivers and channels from north-western side. By giving a look on thc superimposed plan of channels, it is very clear that the channel has got its stability and it is lying in the eastern part near the embankment. Among these four reaches the last one is the most stable. With slight horizontal change in the position of channel, it is existing at the same location for over many years. Criss-crossing of channels like Reach I and II is not observed. In other words, it is apparent that the channel during 1978 and 1983 is identical with marginal change of position while the channel planform for 1986 is not available.

The area under channel and above channel in Reach IV has been shown in Table 10.5 for different years separately. For Reach IV the average channel area for those eleven years is 43.9 sq. km. and that of non-channel is 212.7 sq. km. whose percentages are 17.11 and 82.89 respectively. The maximum channel area is 62.2 sq. km. (24.24%) in 1978 and the minimum is 29.1 sq. km. (11.34%) in 1984. The different between these two is to an extent of 33.1 sq. km. which is equivalent to 12.9 per cent between the maximum and minimum. These are only

Table 10.5: Variations in channe and agricultural area: Reach IV.

Year	*Channel Area (sq km)*	*% to Total (Channel)*	*Area Above Channel (sq km)*	*% to Total (Non-Channel)*	*Yearly change in Agr. Area (sq km)*	*Ratio (Channel Area to Non-Channal Area)*
1975	44.4	17.30	212.2	82.70	—	4.78
1976	49.1	19.13	207.5	80.87	-4.7	4.23
1977	47.0	18.32	209.6	81.68	+2.1	4.46
1978	62.2	24.24	194.4	75.76	-15.2	3.13
1979	57.0	22.21	199.6	77.79	+5.2	3.50
1980	51.5	20.07	205.1	79.93	+5.5	3.98
1981	33.6	13.09	223.0	86.91	+17.9	6.64
1982	33.5	13.06	223.1	86.94	+0.1	6.66
1983	30.7	11.96	225.9	88.07	+2.8	7.36
1984	29.1	11.34	227.5	88.66	+1.6	7.82
1986	44.8	17.46	211.8	82.54	-15.7	4.73
Average	43.9	17.11	212.7	82.89	—	4.85

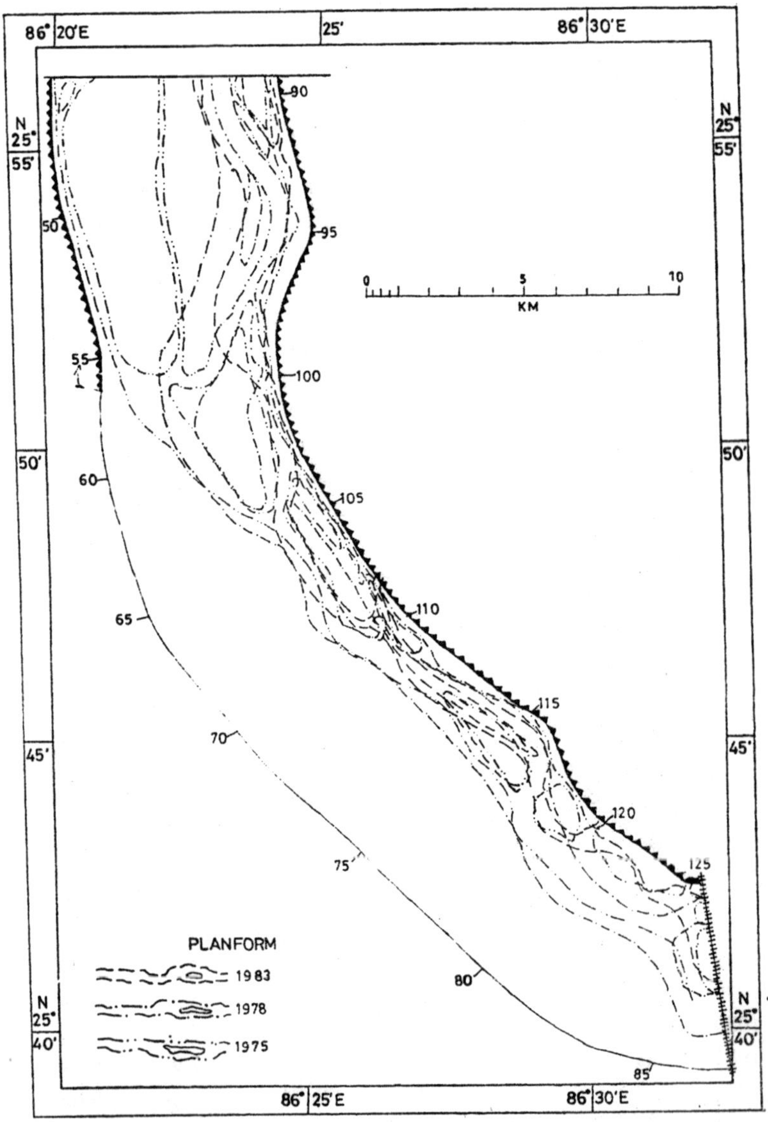

Fig. 10.5: Composite Channel Planform: Reach IV.

three occasions when the percentages of channel area has crossed 20 per cent (during 1978, 1979 and 1980). For remaining years it is always less than 20 per cent but it has never been below 11.34 per cent. The ratio between the area above and below channel varies between 3.13 times in 1978 to 7.82 times in 1984. The average ratio for entire period is 4.85 times. The standard deviation calculated for the channel and non-channel area is 10.5 and that of coefficient of variation is 23.92 per cent. The maximum area which has come out from channel is 17.9 sq. km. during 1981 in comparison to 1980 followed by the lowest one is the subsequent year (1982) which is only 0.1 sq. km. From this table it is obvious that the variation in Reach IV is comparatively smaller in comparison to other three reaches.

Complete Study Area

Table 10.6 is the addition of all four reaches together, i.e. the summation of figures discussed above into four reaches for their respective years. Out of the total area of 1251 sq. km., 272.3 sq. km. in 1978 was under the channel at the time of survey towards the maximum end and that of 195.3 sq. km. in 1984 to the minimum limit which works out to be 21.77 per cent and 15.61 per cent respectively.

Table 10.6: Variations in channel and agricultural Area: complete area under study.

Year	*Channel Area (sq. km.)*	*% to Total (Channel)*	*Area Above Channel (sq. km.)*	*% to Total (Non-Channel)*	*Yearly change in Agr. Area (sq. km.)*	*Ratio (Channel Area to Non-Channel Area)*
1975	235.1	18.79	1015.9	81.21	—	4.32
1976	222.5	17.79	1028.5	82.21	-12.6	4.62
1977	205.2	16.40	1045.8	83.60	+17.3	5.10
1978	272.3	21.77	978.7	78.23	-67.1	3.59
1979	237.0	18.94	1024.0	81.06	+45.3	4.28
1980	264.8	21.17	986.2	78.83	-37.8	3.72
1981	213.4	17.06	1037.6	82.94	+51.4	4.86
1982	206.1	16.47	1044.9	83.53	+7.3	5.07
1983	271.5	21.70	979.5	78.30	-65.4	3.61
1984	195.3	15.61	1055.7	84.39	+76.2	5.41
1986	218.5	17.47	1032.5	82.53	-23.2	4.73
Average	231.1	18.47	1019.9	81.53	—	4.41

The different between these two are 77 sq. km. (6.16%) which is not considered to be very high keeping the entire area into account. The average channel area over the entire period is 231.1 sq. km. which is 18.47 per cent of the total geographical area and remaining 1019.9 sq. km. (81.53%) is above the channel.

During the entire period under study, only three occasions have come when the percentage of channel area has exceeded 20 and these are 1978 (21.77%), 1980 (21.17%) and 1983 (21.70%). Other years represent less than 20 per cent but it has never gone below 15 per cent (to be very precise, it is 15.61%) in 1984. The ratio between channel and non-channel area varies in a limit of 3.59 times in 1978 and 5.41 times in 1984. It is taken in multiple of the channel area in respective years. The average ratio is 4.41 times. The standard deviation of the area is 26.4 and that of coefficient of variation is 11.42 per cent.

In Table 10.7, the standard deviation has been calculated with respect to the area under the channel and the area above the channel for respective reaches as well as for the total area under study. The coefficient of deviation which is the ratio between the standard deviation and its mean and expressed in percentage, has been calculated for the channel area of reaches. It can be seen that this measure of deviation is having a relationship with the area of individual channel reach in an inverse manner. Larger is the channel reach by area, smaller is the coefficient of variation. For example, reach which is the largest in area is having the minimum coefficient of variation 10.67 per cent closely followed by total area, 11.42 per cent; whereas Reach I and Reach IV which is relatively small in area

Table 10.7: Average statistics of the study area: Reach-wise.

Reach	*Mean Channel Area (sq. km.)*	*% to Total (Channel Area)*	*Area Mean Above Channel (sq. km.)*	*% to Total (Non-Channel Area)*	*Ratio (Channel Area to Non-Channel Area)*	*Standard Deviation Channel Area*	*Coefficient of variation Channel Area (in %)*
Reach I	38.9	24.36	117.1	75.64	3.11	13.1	33.68
Reach II	82.5	15.13	462.7	84.87	5.61	8.8	10.67
Reach III	65.8	22.44	227.4	77.56	3.46	16.8	25.53
Reach IV	43.9	17.11	212.7	82.89	4.85	10.5	23.92
Total Area	231.1	18.47	1019.9	81.53	4.41	26.4	11.42

are having more coefficient of variation 33.68 per cent and 23.92 per cent respectively.

Some Suggestions for Land Management in the Kosi Embankment Area

From the above discussion, it is vary clear that the land within the embankments is on the mercy of the Kosi channel shifting. Therefore, to control the Kosi is the most important aspect and it could be done in phase wise manner. The steps to be adopted in this direction are:

(a) Proper and judicious land management in the upper (Himalayan) catchment and stopping ecological degradation, and human encroachment towards fragile virgin mountain environment.
(b) Small dams along tributaries of the Kosi and big gravity dam at Chetra gorge to trap or reduce the sediment load in the plain.
(c) Narrowing down the embankment spacing through the construction of inner bund and its protection.
(d) Dredging and straightening of the braided channels to facilitate quick flow of water downstream.
(e) Through the above-mentioned steps, channel in the plain will be a bit stable and less migration of bed of the Kosi, thus, smaller amount of land degradation due to channel shifting.

Conclusion

The Kosi, a large braided river, is shifting its courses regularly. The main problem of this river is its excessive sediment load, when deposited on valley floor chokes the mouth. It does not allow to pass out the rising discharge in subsequent years, thus, on that account, the shift in the course becomes inevitable. Even after the erection of embankments and barrage, channel changes its course and a large agricultural land is annihilated. Earlier swing of the Kosi from east to west altered, and now, it is shifting towards east but within the embankments, apart from a few breaches into them. Therefore, appropriate and prudent approaches towards management of the large Himalayan catchment are unavoidable without which shift in the channel courses cannot be contained.

REFERENCES

Desai, C.G. (1982), *The Kosi River: Its Morphology and Mechanics in Retrospect and Prospect*, Vol. I. Central Water Commission, Ministry of Irrigation, Government of India, New Delhi.

Ganga Flood Control Commission (1983), A comprehensive plan of Flood Control for Kosi Basin. Ministry of Irrigation, Government of Bihar, Patna.

Inglis, C.C. (1967), "Inland Delta Building Activity of Kosi River". *Journal of Hydraulic Division, American Society of Civil Engineers Proceedings*, pp. 93-100.

Leopold, L.B., Wolman, M.G. and Miller, J.P. (1964), *Fluvial Processes in Geomorphology*. Freeman, San Francisco.

Prasad, R. (1987), Changes in Channel Configuration within the two Embankments of Kosi since 1958. Unpublished M.phil. Dissertation, Department of Geography, University of Delhi, Delhi.

Prasad, R. (1999), Dynamics of River Environment: A study of Kosi Embankment Area. Unpublished Ph.D. Thesis, Department of Geography, University of Delhi, Delhi.

Rao, K.L. (1956), Stabilizing the Regime of the Wandering Kosi. United Nations Proceedings of Regional Technical Conference on Water Resources Development in Asia and the Far East. Economic Commission of Asia and the Far East, Bangkok, pp. 308-14.

Shillingford, F.A. (1895), "On changes in courses of the Kosi River and the Probable Dangers Attacking from them." *Journal of Asiatic Society of Bengal*.

Wells, N.A. and Dorr, J.A. (1987), Shifting of the Kosi River, Northern India, *Geology*, Vol. 15, pp. 204-07.

11

Geographical Analysis of Waterlogging Induced Land Degradation in Karnal District of Haryana

R.B. SINGH and MEHTAB SINGH

Introduction

The fundamental problem which the world faces today, is the rapidly increasing pressure of population on land resources. Although a revolutionary break-through has been achieved in the foodgrains production, the farmers to a great extent, have not been able to utilise the full potential of their land due to various soil problems. In arid and semi-arid regions, low and uncertainty of rainfall leads to the introduction of irrigation in these areas which has been considered as the most effective way for improving the production but the drainage is still in its infancy stage. These failures have been basically related to the problems of rise in water-table level resulting in waterlogging and over the years salinization of soils leading to partial or complete crop failure. Waterlogging and its after effects are a global environmental problem facing many countries today. Globally, 954 million ha. of land have waterlogging-salinity problem. It has been estimated that by the year 2000 A.D., more than 2.75 million ha. of irrigated land will be lost from agricultural operation. About 80 per cent of the total irrigated area in Iraq followed by 50 per cent in Syria, 30 per cent in Egypt, 23 per cent in Pakistan, 15 per cent in Iran is affected by waterlogging problems. In Argentina, about 2 million ha. agricultural land productivity has come down due to this menace. The problem is also increasing in China, Peru, Kampuchia and Vietnam. In India alone, about 10 million ha. of cultivated land are affected by waterlogging followed by 25 million ha. affected by salinity problems. Thus, the problem related to degraded soils is a complex environmental problem for management, linked to land development policy, cropping pattern, water and soil management practices, farmers attitude and perception towards water use, irrigation, topography and drainage of the problem area.

Nature of Problem

A large part of our country's land area is not yielding the desired rate of production because of impedimental factors like soil erosion, salinity, alkalinity, waterlogging etc. and there is the degradation of productive land being converted into wasteland. It is estimated that these ceaseless onslaught to which the land is our country has been subjected, have led to the degradation of half of the geographical and about two-third of the productive areas. Haryana State is not the exception. Out of the total geographical area of 44.05 lakh ha. roughly 4.50 lakh ha. area are affected by the problem of waterlogging, salinity and alkalinity resulting in low production. The study area is part of the Ghaggar-Yamuna plain which forms a closed basin. A fairly good correlation exists among the climatic, geohydrological and soil parameters which affect the drainage problem as this region is dominated by alkali soils. The infiltration rates and surface water retention storage capacity of this soil are very low and the run-off creates the problem of drainage congestion especially during rainy season. The widely recognised increase in the area of degraded land due to these reasons has in recent years brought into sharp focus the urgent need for evolving suitable technology to reverse the process of ecological degradation. Though the problem of waterlogging has been sorted out to a greater extent in fresh groundwater areas, the subsequent problems e.g. alkalinity and salinity, are still of great concern. Suitable technologies are available. But this gigantic task cannot be done by Government alone, a people's movement will have to be sustained.

Review of the Literature: The State of Crisis

Aggarwal (1957) conducted a re-survey of usar patches in U.P., 40 years after Leather (1914) had surveyed them and noticed that the area and the intensity of usar patches had increased in ten out of twelve fields so examined. He revealed that usar patches were primarily due to soil factors mainly nearness of water table though the quality of groundwater had been good. Impedance in drainage due to the presence of an indurated layer of clay on kankar-pan in the sub-soil excelerated the process. Kulkarni (1960) calculated the rise of salt affected area of western Maharashtra which showed a sharp rise up to 1940. During this period, the water table had come up to 10 feet from ground level. Abrol and Bhumbla (1971) reported that the extent of salt-affected and waterlogged soils in India increased to about 10 million hectares. Uppal (1972) analysed the hydrographs of all the canal command areas for the period 1926 to 1958 and found the water table rising at the rate of 0.4 metre to 0.15 metre per year in all the tracts except Upper Bari Doab where a marginal fall of 0.02 metre per year was observed.

In another study, Uppal (1972) estimated the extent of waterlogged areas in undivided Punjab from 1950 to 1969. He found that waterlogged areas had increased from 8,35,959 ha. in 1950 to 1,644,705 ha. in 1962. Singh (1976) was of the opinion that the high density sub-alluvium ridge that lies burried along Hisar-Rohtak-Delhi axis obstructed the flow from Siwaliks to the south-west and might have been responsible for creating salinity. Narayana, Pandey and Gupta (1977) studied the effect of vertical drainage which showed that due to poor hydraulic characteristics of alkali soils, it turned out to be ineffective. However, in a closed basin, near Mundlana, (Sonepat district, Haryana), this technique proved to be very successful. Abrol, Chhabra and Gupta (1980) defined the alkali soils as those soils which have salts capable of alkaline hydrolysis, Exchangeable Sodium Percentage (ESP) of more than 15 and pH of saturation paste in water above 8.2. Dargan (1982) pointed out that the provision of sub-surface drainage system and conservation of rainwater within bunded saline soil is considered a convenient method for modifying the components of hydrological cycles in a desired direction for reclamation of waterlogged and saline soils. Aggarwal and Khanna (1983) in their study stated that water table is falling in fresh groundwater quality zone but in saline quality zone, water table has been rising and this factor is responsible for the degradation of soil by the process of salinization. Tyagi and Narayana (1983) estimated the seepage losses in a part of Western Yamuna Canal System and it was as high as 28 per cent. Rao, and Kumbhare (1984) presented the results of experimental studies on horizontal sub-surface drainage conducted at three typical waterlogged sites in Haryana. Kumar and Abrol (1986) indicated that besides the tolerant crops, grasses such as Diplochna fusca (Karnal Grass), *Chloris gavana* (Rhodes grass) and *Brachiaria mutica* (Paragrass) have been found highly tolerant to alkalinity.

Purpose of Study

The primary purpose of this study is to: (i) analyse the genesis of waterlogging in the study area; (ii) map out the spatial-temporal extent of waterlogged areas over three decades; (iii) measure the extent of alkalinity-salinity; and (iv) suggest methods for reclamation of degraded soils.

The Study Area

The study area selected lies on the eastern edge of Haryana between 29°09′ to 29°59′ north latitudes and 76°33′ to 77°08′ east longitudes. This area is chosen because it is a part of the Ghaggar-Yamuna plain where groundwater table is rising as well as falling especially in the study area. The study area is composed of 10 blocks, spreading into Panipat and Karnal districts of Haryana as shown in map (Fig. 11.1). The entire area appears monotonously flat. The Yamuna, Chautang and Saraswati (no defined bed) area are parts of natural drainage which tilt towards south-west. This tract is made up of new and old alluvium. The old alluvium consists of mainly interbeded lentisular and interfingering beds of clay, sand, gravels, silt, sand silty clay with kankars. The new alluvium is of light colour and poor in calcareous matter. The overall climate of the study area is sub-tropical continental monsoon type. Over 90 per cent rainfall is received during June to September from south-west monsoon. Pan evaporation exceeds the rainfall except for 2-3 months of rainy season. High evaporation rate just after rainy season in the range of 114 to 170 mm. per month is one of the factors which causes accumulation of salts on the surface.

Data Source and Methodology

The study is based on secondary data. Directorate of State Groundwater Board, Karnal provided data on groundwater level. Agricultural Department of Haryana, Karnal provided data on village-wise saline kallar areas. Data on number of tubewells was collected from Haryana State Minor Irrigation and Tubewells Corporation, Karnal. The smallest unit of observation is village as well as block.

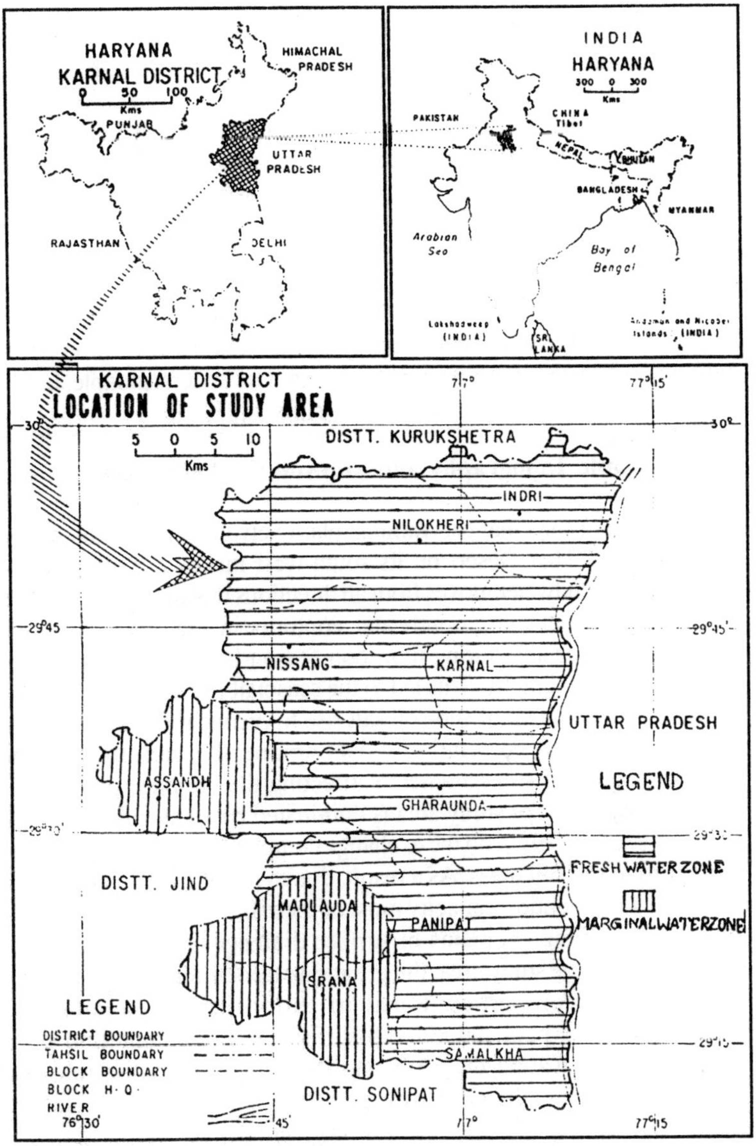

Fig. 11.1: Karnal District: Location of Study Area.

609 villages fall under the study area. Simple statistical technique has been used. Map analysis has been done to present the cartographic representation. Quantitative distribution maps, especially those employing choropleth technique, showing sub-soil water depth, extent of degraded soils have been used for clear understanding. Wherever required, graphs are used to show different aspects of data.

Findings

Soil quality and agricultural productivity are closely interrelated. Soil degradation in the study area is partly due to natural factors and partly man-made. Causes of soil degradation cannot be understandable, unless we study the whole Ghaggar-Yamuna plain.

Unfavourable Topography and Sub-soil Grohydrology

The location of the Ghaggar-Yamuna plain between the Siwalik on the north-east and Thar Desert on its south-west makes it a close basin. Therefore, groundwater moves from all directions towards the depression as is evident from the groundwater contours in Fig. 11.2. Drainage congestion, especially during monsoon season becomes accute due to bowl-shaped topography. This led to the seasonal rise in water table and the upward movement of sub-soil salt rise to secondary salinisation.

A fairly good correlation exists among the various climatic geohydrologic and soil parameters affecting the drainage problem of this area. Most part of study area is dominated by alkali soils. In alkali soils, the excess sodium on the exchange complex imparts poor physical properties to the soil mass. As a result, due to low infiltration rates and surface water retention capacity most of the rain water goes as run-off. As a result, serious problems of surface drainage exist. In the south-western part of the study area, the groundwater aquifers have relatively poor hydraulic properties. Canal irrigation, less exploitation of groundwater and poor natural sub-surface drainage condition have resulted in aggravating the problem. The high evaporation rate and saline water close to the surface evaporation more than the sub-surface flow resulting in the development of high salinity.

Sub-surface Lithology

The sub-surface lithology of the entire plain indicates that the thickness of the aluvium decreases from north to south and from east to west.

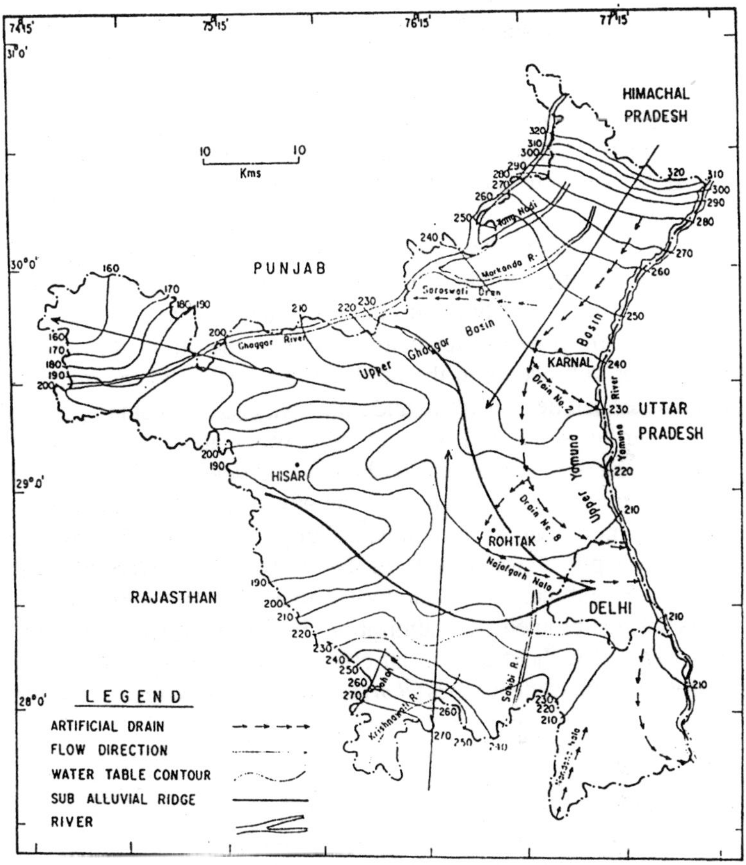

Fig. 11.2: Hydrogeoogical Map of Ghaggar-Yamuna Plain.

Transmissivity also decreases along with thickness of the alluvium. The distribution of alluvial sediment shows that the percentage of sand is more than clay and silt in fresh groundwater areas of north-east in the study area. The sand content decreases and the aquifers become thinner and isolated as we go towards south and west of Karnal. In the south-west part of the study area, extensive thick clay layer behaves like semipermeable membrance which produce osmatic pressure, electrical potential differences and salt sieving. The occurrence of marginal ground water in this part gives rise to the probability of close relationship between lithology and salinity.

Water Losses in Irrigated Systems

Introduction of irrigation in any area leads to seepage from irrigation canals, distributaries and fields. A study conducted by Jain (1961) revealed that 71 per cent water loss occurred when the entire system is unlined. When the whole system is lined, still, water loss is as high as 54 per cent. Irrigation water losses of such high order in semi-arid region of the study area where the geo-hydrology and topographical conditions are unfavourable for natural drainage gives rise to water table. But at the same time, groundwater quality plays a vital role. Three types of groundwater quality exist in the study area. Fresh groundwater belt occurs in the north-eastern part, whereas marginal groundwater belt lies westward and saline water exist in small pockets towards further south and west. In fresh water areas, loss of water through seepage is neutralised by shallow and deep tubewells. But in saline and marginal saline areas of the study area, the groundwater recharge is more than its withdrawal. Figure 11.3 shows the rise and fall in the depth of water table in the study area comprising fresh groundwater and marginal groundwater areas. The rise in water table creates serious problems of salinisation. In the initial stage, the crop yields reduced without any visual effects. The land may have to be abandoned when white patches (*Reh*) appears in otherwise fertile fields.

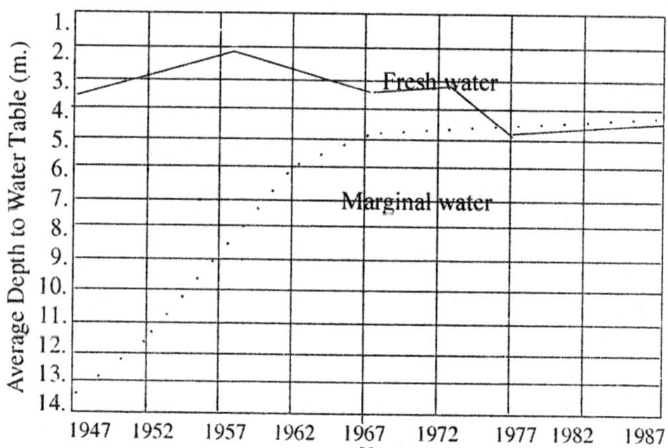

Legend

	Station	No. of Wells
———	Karnal	11
. . . .	Assandh	6

Fig. 11.3: Rise and Fall in Depth of Water Table.

Extent of Waterlogging

The analysis of groundwater depth was done to find out extent of waterlogged areas since 1960. For this purpose, data from 170 observation wells were taken. One observation well covered an area about 2200 ha. The study revealed that approximately 2,23,520 ha. of productive land was affected due to high water table in 1960. The groundwater contours show that the problem was more acute in fresh groundwater areas comprising Indri, Karnal, Gharaunda and Nassang blocks. The problem was at its peak during 1970 when about 2,79,360 ha. out of 3,71,684 ha. were affected due to high groundwater table. There was substantial decrease in waterlogged areas during 1991, when only 5008 ha. land were affected due to high water table. Though the problem diminished to a large extent, but in a large part of area (73.62%) the groundwater table depth was still between 3 metres to 10 metres from surface. Moreover, an upward movement of groundwater table was noticed in marginal to saline groundwater areas of Israna block, parts of Assand and Mandlauda blocks (Fig. 11.4).

Spatio-Temporal Extent of Alkaline and Saline Soils

While waterlogging alone can reduce crop productivity, the salinity and alkalinity is a great threat to the farmlands. Process of degradation of soil starts when the water comes close to the surface (within 3 meters), weathering of aluminosilicate minerals provides a steady supply of alkali bi-carbonates which accumulate in undrained basin. In the ensuing dry season, as a result of evaporation, the soil solution concentrates resulting in an increase of sodium absorption ratio (SAR) of the soil solution which, in turn, results in increased absorption of sodium ions on the exchange complex and also increases soil pH. The displaced calcium precipitates resulting in the formation of alkali soils (kallar land) over the years. Due to the adverse effect of excess exchangeable sodium and of accompanying high pH, plant growth is restricted due to transformation of some essential plant nutrients. The toxic effect of specific ions, i.e. sodium and boron is often an additional factor determining plant growth relationships in the alkali soils.

Expansion of Alkali (Kallar Land) Soils

Two types of degraded soils have been visible in the study area: saline and alkaline. Alkaline soils exist in fresh groundwater areas, whereas

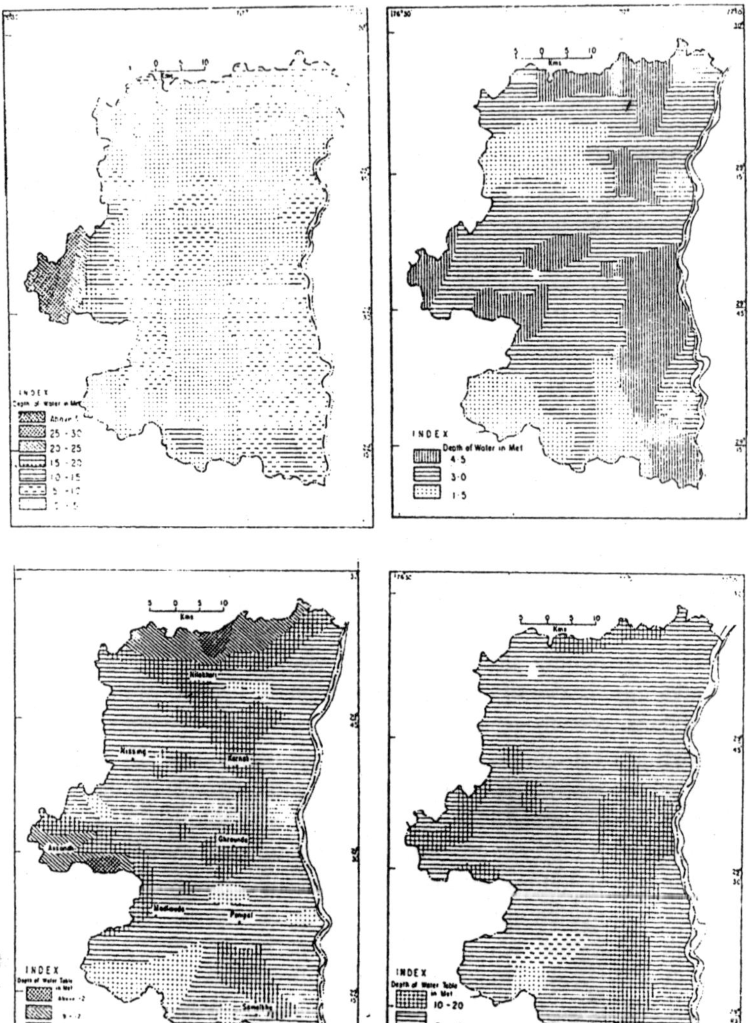

Fig. 11.4: Sub-Soil Water Depths, October 1960-1991.

saline soils exist in the areas falling under marginal to saline groundwater areas. The study reveals that the fresh groundwater areas have less kallar land in comparison to marginal groundwater areas. Figure 11.5 shows the village-wise fluctuation of kallar land in the study area since 1971. Table 11.1 shows the decadewise fluctuation of kallar land.

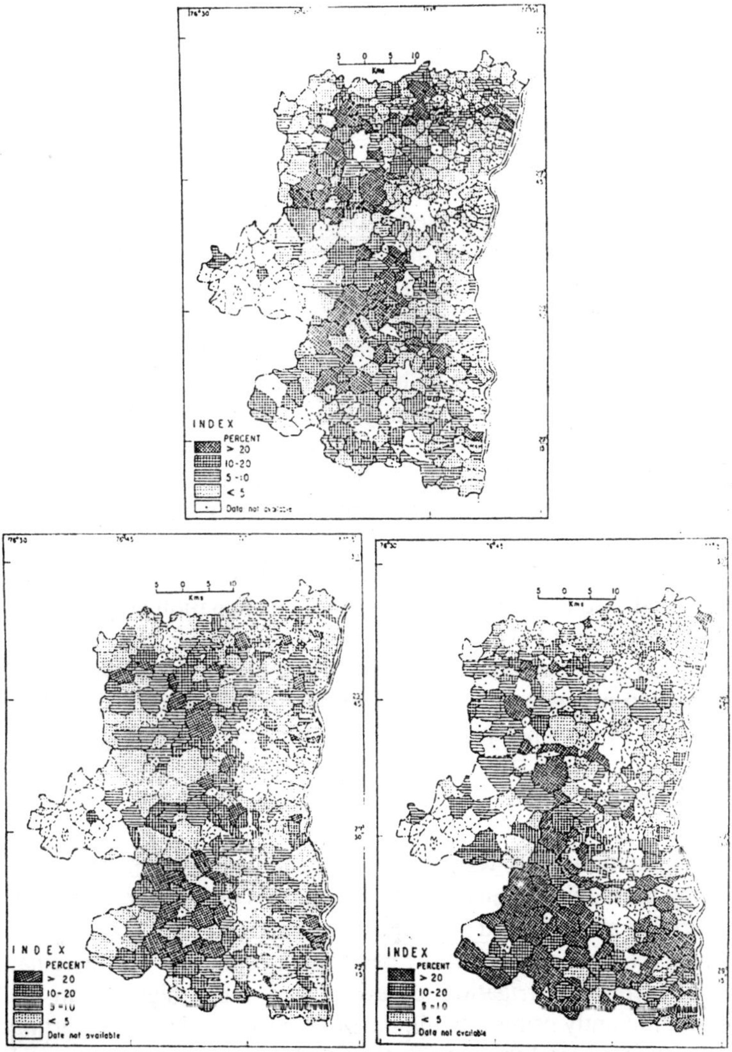

Fig. 11.5: Village-wise Area Under Kallar Land 1971-1985.

Table 11.1: Total Percentage of Kallar Land During Three Decades (1971-1991).

Year	*Percentage*
1971	9.81
1981	8.11
1985	10.24
1991	7.47

Reclamation Measures and Strategies

Vertical drainage has proved to be an effective measure for controlling groundwater table. The presence of favourable aquifer and good quality water has resulted in lowering of groundwater table in Gharaunda, Karnal, Nilokheri blocks. The number of shallow tubewells have increased 4 times since 1970. But the problem is more complicated in saline and marginal groundwater areas. Isarana, Samalkha, Nissang and Assandh blocks fall in this category. Several experiments show that poor quality water can be used for irrigation in conjunction with fresh water. Leaching in rainy season is another method. Sub-surface drainage can facilitate the leaching process for reclamation of degraded soils. In addition to this, gypsum has proved to be most effective 9696.65 ha. kallar land have been reclaimed out of 20726.11 ha. till 1991-92. Out of 609 villages in the study area, 15 villages utilized the government sponsored reclamation programmes. These 15 villages have reclaimed their entire kallar affected land. About 131 villages reclaimed their 50 per cent affected land. The reclamation programme in 429 villages is in its infancy stage while remaining 34 have not yet started reclaiming their land (Fig. 11.6).

Conclusion

Soil degradation due to waterlogging and salinity has been increasing in many developing countries. The seriousness of the problem lies in the fact that affected land is comparatively superior. Further, water is a scarce resource in most of the developing countries. The marginal cost of increasing irrigation to new areas is much higher than the cost of efficiently using water in the irrigated areas. Thus, converting fertile irrigated land to unproductive land is a major environmental risk. Though the problem is about to be solved in fresh groundwater areas of the study area, but, the problem of soil degradation is

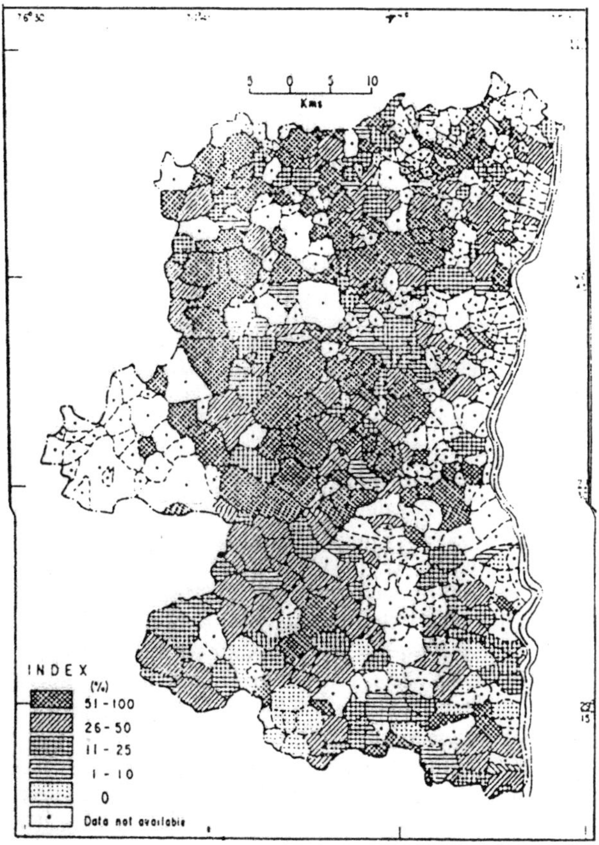

Fig. 11.6: Village-wise Reclaimed Area 1986.

increased alarmingly where the groundwater is marginal and completely saline. The remedial measures, if implemented in right earnest, could be helpful in keeping the productive land under permanent agriculture. Farmer's education and their participation is the only feasible long-term solution that one can find. It would be difficult to contain this problem by merely technical approaches since this is an environmental problem caused by series of factors, i.e. lack of education, stress on immediate economic benefits and, above all, physical environment of the area. If the problem is not tackled immediately, the farmers of Haryana are not far away from a major social catastrophe as almost half the area of the state is about to face the problem in coming days.

Acknowledgements

In writing this paper, we have been benefited immensely from discussions with Professor B. Thakur, Head, Department of Geography, Delhi University.

REFERENCES

Abrol, I.P. and Bhumbla, D.R. (1971), *World Soil Resources*, F.A.O. Report, pp. 41- 42.

Abrol, I.P., Chhabra, R. and Gupta, R.K. (1980) "A Fresh Look at the Diagnostic-Criteria for Sodic Soils", *International Symposium on Salt Affected Soils*, Feb. 18-21, Karnal, pp. 142-47.

Aggarwal, R.R. (1957), "Alkali Soil can be Reclaimed," *Indian Farming*, pp. 7-12.

Aggarwal, M.C. and Khanna, S.S. (1983), *Efficient Soil and Water Management in Haryana*, Haryana Agricultural University, Hisar Publication, p. 118.

Bowonder, B and Ravi, C. (1984), *Waterlogging from Irrigation Projects: An Environmental Management Problem*, Centre for Energy and Technology, Administrative Staff College, Hyderabad, pp. 27-45.

Dargan, K.S. (1982), *Crop Production in Salt—Affected Soils*, New Delhi; Oxford and I.B.H. Publication, pp. 276-91.

Diamant, B. (1980), "Environmental Repercussions of Irrigation Development in Hot Climates," *Environmental Conservation*, Vol. 7, pp. 53-58.

Jain, J.K. (1961), "Salinity Problems in the Arid Zone", *Proc. Symp.*, UNESCO, Tehran.

Kulkarni, D.B. (1960), "Reclamation of Waterlogged Lands in Bombay Deccan," Trans Fourth Congress Irrigation and Drainage, I.C.I.D., Vol. 3, p. 11.

Kumar, A. and Abrol, I.P. (1986), "Grasses in Alkali Soils", *Technical Bulletin No. 11*, C.S.S.R.I., Karnal, pp. 95-102.

Narayana, V.V.D., Pandey, R.N. and Gupta, S.K. (1977), "Drainage of Alkali Soils", *Indian Association of Hydrologists*, 1, pp. 21-28.

Rao, K.V.K. and Kumbhare, P.S. (1984), "Studies on Design and Performance of Sub-surface Drainage System", C.S.S.R.I., Karnal, Annual Report.

Singh, J. (1971), "Agricultural Colonization of Cultivable Wastelands in India", *The Geographer*, Vol. IX, pp. 135-49.

Singh, A. (1986), "Reclamation of Saline and Waterlogged Soils Through Sub-Surface Drainage", Report of Mundlana Project (Part 1), H.L.R.D.C., Chandigarh.

Singh, O.P. (1979), Hydrological Studies in A Decade of Research, C.S.S.R.I., Karnal, pp. 131-39.

Tyagi, N.K. and Narayana, V.V.D. (1983), "Evaluation of On-Farm Water Management Constraints in a Surface Irrigation System," *International Commission on Irrigation Drainage Bulletin*, 32(2), pp. 11-16.

Uppal, H.L. (1972), "Serious Waterlogging in Punjab and Haryana—How Cured and Measures to Prevent its Recurrence", C.B.I.P., *Symposium on Waterlogging Causes and Measures for its Prevention*, Vol. 11, pp. 41-53.

12

Impact of New Mining Technology on Land Degradation in Coalfields of Chhotanagpur

MAHESHWARI PRASAD

Introduction

Technology is the key of all the modern activities of human beings – social and cultural, economic and political. The adoption of ever developing new technology has become visible in all these spheres, but perhaps more clearly in the sphere of economic activities because our material demands are taking gigantic leaps everyday. In fact, technology has been successful in working wonders in all walks of material life. The technological impacts do not remain confined to productive activities but also tend to swallow in their vortex the fabric of environmental balance of the region concerned. Technology is no longer an obedient maid for fulfilling the materialistic demands but has become a tool of destruction of the very home of man – the Environment – which is the product of complex and dynamic interaction of physical, chemical, biological as well as social systems. This violent reaction or the consequence is the backlash of the modern technology.

Let us pause here for a moment to define the term technology. Unfortunately, it is elusive of an easy definition, though in our common parlance we are apt to use it frequently. The conceptual ambiguity partly results from the use of automation and technology as synonyms, the former meaning an automatic system of working. In fact, automation and mechanisation both are related terms – the difference between them being only one of degree.

The word technology has a very wide concept covering all aspects of material culture. It is the combination of the trio of science, art and invention. It is the means of increasing the productivity of human beings. For this purpose such methods and tools are developed which multiply their skills and energy. "It is at once a science and an art" (*Cowles Encyclopaedia*, 1967, p. 291).

Mining technology is a type of industrial technology embracing a wide range of science and engineering. It is, therefore, primarily based on mechanical technology which refers to different machinery and equipment. A basic feature of this technology is increased production by reduction or replacement of human labour. Its another important feature is the dynamic nature; it is ever changing and enhancing man's productive capacity as he is always keen to accept the new technological practices.

At this stage a pertinent question emerges in respect of the status of acceptance of new technology in Chhotanagpur. Technological acceptance is gradual and its level varies from area to area as well as sector to sector of human activities. Generally, its five stages are recognised, viz., elementary, animal traction, initial motorisation, motorisation and automation. The Chhotanagpur region is economically backward and poor, and technologically archaic. Here capital is scarce but labour abundant. Consequently, acceptance of new technology is not spontaneous. On an average, the region is in initial motorisation stage – the stage of partial mechanisation. However, the status of acceptance, as mentioned above, varies significantly from one economic sector to another. Without going into much detail on this point, it may be said that the region is on the threshold of the final stage in the steel industry, and in the third and fourth stages in the mining industry.

Thus, in the latter sector it ranges from almost primitive to modern depending upon the geological condition of the mineral deposits, their mode of occurrence, size, depth, profitability, etc. (Fig. 12.1). By and large, it is modern in case of coal mining which has undergone sea changes in application of new technology to accelerate the production during the last two decades. Technology works both ways; it helps in exploitation as well as repair of the damages-environmental or otherwise. Presently, coal mining has adopted new technology almost entirely for production purposes. Its environmental backlash is consequently severe, acute, persistent and painful. The purpose of this study is to highlight the more conspicuous aspects of the

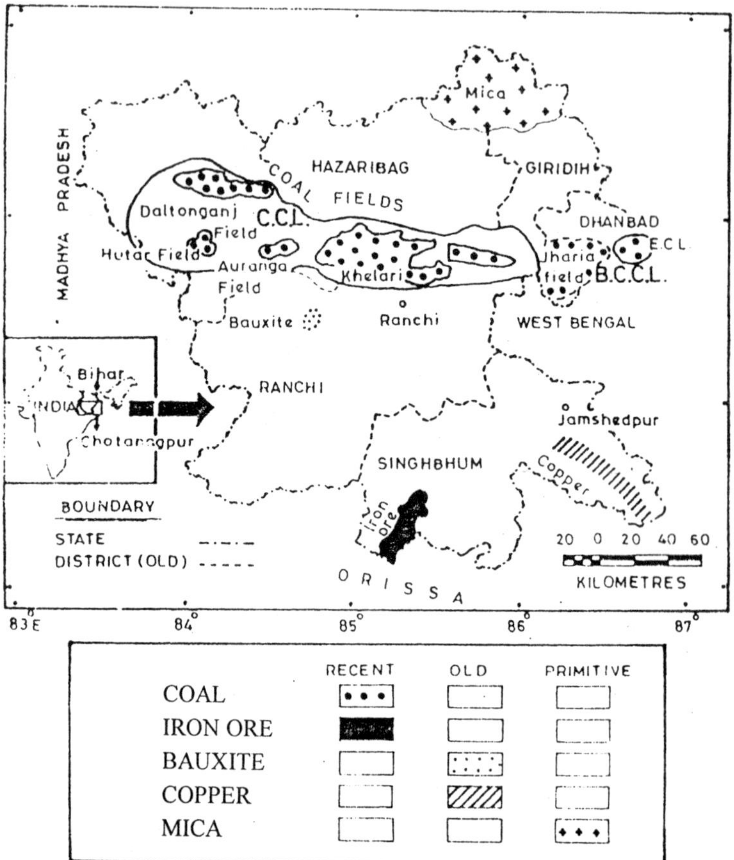

Fig. 12.1: Spatial Pattern of Mining Technology in Chhotanagpur.

environmental impacts of hi-tech coal production in Chhotanagpur region.

The region, the veritable treasure of minerals in Bihar, comprises the districts of Garhwa, Palamu, Hazaribagh, Giridih, Chatra, Bokaro, East and West Singhbhum, Ranchi, Gumla and Lohardanga. It covers about 55 per cent of the total area of Bihar. Several of the above districts are quite new. For the purpose of this paper, however, the boundaries of the old districts have been shown on the map (Fig. 12.1) because the discussion is not district-wise but focussed on the geological distribution of the coal deposits. It would perhaps also facilitate the readers to have an easy, clear grasp of the spatial pattern of coalfields without getting involved in the intricate political boundaries of the districts. Furthermore, the discussion is with particular reference to the C.C.L. and B.C.C.L. coalfields which account for most of non-coking and all the coking coalfields as well as constitute the principal scene of intensive mechanised mining in both old and new collieries of the region. The coalfields belonging to the E.C.L. and falling within the political territory of Chhotanagpur are quite small.

Significance of Coal Deposits

A variety of minerals occur in the Chhotanagpur making it the richest mineralised region of the country. The important minerals are coal, iron ore, bauxite, mica, limestone, copper, manganese, kyanite, uranium, fire-clay, china clay, asbestos, feldspar and quartz, etc. It contains about 90 per cent of the medium coking coal and 16 per cent of the power grade coal reserves of the country. North Karanpura coalfield alone has 13,578 million tonnes of coal reserves. The Jharia coalfield of the region is the only store house of prime coking coal reserves (5.3 billion tonnes) in the country. Presently, the region accounts for about 35 per cent of the national production of coal, 32 per cent of bauxite, 29 per cent of copper, 31 per cent of silver, 75 per cent of uranium, 65 per cent of kyanite and 17 per cent of iron. It is a pioneer in mica production in the world and produces about 50 per cent of its national output.

Of all the minerals, the deposit of coal is the largest, and quite naturally coal mining is the oldest, dating back to 1842, and the most dominant activity of mineral exploitation in the region. It has spearheaded industrial revolution not only in the region but in the whole country. It is also bound to play its crucial role in shaping its future industrial destiny of the country with extensive coal deposits, belonging to the Gondwana system, in the fields of Jharia (375 sq. km.), East Bokaro (202 sq. km.), West Bokaro (169 sq. km.), North Karanpura (1175 sq. km.), South Karanpura (195 sq. km.), Ramgarh (68 sq. km.), Giridih (7 sq. km.), and nearly 47 sq. kms. in the coalfields of Auranga, Hutar, Daltonganj, TISCO, ISCO and the D.V.C. The State of Bihar has the largest reserve of coal in India amounting to 64,371 million tonnes which is approximately 33.21 per cent of the total coal reserves of the country. The total coal reserve of Chhotanagpur is estimated to be 47,639 million tonnes in coal seams of 0.5 m. and above in thickness and down to a depth of 1200 metres which is approximately 74 per cent of the total reserves of the state and 24.58 per cent of the country as a whole (Table 12.1). According to V.K. Singh (1991).

Table 12.1: State-wise Coal Reserves in India (As on 01.01.1993).

State	*Reserves in Million Tonnes*	*Percentage of Reserves*
1. Bihar	64,371	33.21
(a) Outside Chhotanagpur	16,732	08.63
(b) Chhotanagpur	47,639	24.58
2. Orissa	46,218	23.84
3. Madhya Pradesh	39,022	20.13
4. West Bengal	25,123	12.95
5. Andhra Pradesh	10,837	5.58
6. Maharashtra	6,276	3.22
7. Uttar Pradesh	1,276	0.64
8. N.E. Region	864	0.43
Total	**193,777**	**100.00**

Source: Adapted from Coal India and You, Directorate of Personnel, Coal India Ltd., Calcutta, 1993.

General Manager, Environment and Head of Natural Resources, C.M.P.D.I., Ranchi, the Coalfields of Chhotanagpur account for about 27 per cent of country's coal resources of 192 billion tonnes.

The coal mining industry has undergone structural changes in the post-1970 period. The coking coal mining was nationalised in May 1972, and non-coking coal in May 1973. The Coal India Ltd., the apex body, came into existence in 1975. Presently, the coal deposits of the region are owned and operated by three subsidiaries of the C.I.L., viz., the Central Coalfields Ltd. having jurisdiction over the non-coking coalfields of the North Koel and Damodar Valleys and B.C.C.L. over the coking coal deposits of the Damodar Valley in Dhanbad district. The Barakar basin coalfields of Nirsa-Chirkunala area lying in the border zone with West Bengal as well as those of the Rajmahal area are owned and operated by the Eastern Coalfields Ltd. which has its major chunk in the State of West Bengal. There are 54 collieries on revenue and 72 ongoing projects – 47 mining and 25 non-mining, under the C.C.L. management. About 19,228 acres of land are under the leases of C.C.L. and B.C.C.L. areas.

Production Scenario

Coal is the most important natural fuel of the country. Although the share of crude oil has appreciably increased in the post-1970 period, coal has continued to have the lion's share; viz. 78.42 per cent in 1970-71, 78.23 per cent in 1980-81 and 67.88 per cent in 1990-91. The Chhotanagpur region with its large deposits of coal has persistently made a significant contribution to the accelerated production of coal in the country. In pre-nationalisation period, the growth rate of production was, however, not significant, but after nationalisation an average incremental increase of 3 Mt./yr. has been recorded, touching a level of 4.6 Mt./yr. by 1994-95 (Table 12.2).

Table 12.2: Trend of Coal Production in Chhotanagpur.

Year	*Coal Production (Mt.)*	*Annual Growth Rate (Mt./Yr.)*
1950	16.50	—
1955	18.50	0.4
1960	23.50	1.00
1965	29.50	1.20
1971-72	31.00	0.23
1981-82	41.00	1.00
1985-86	53.00	3.00
1989-90	67.00	3.50
1994-95	90.00	4.60

Source: Tewari, B. (1988) "Role of Coal in the Development of Chhotanagpur", C.M.P.D.I., Publication.

Induction of New Technology

The quantum jump in coal production has been basically achieved by inducting new technology of mining and resorting to more and more production by the Open-Cast method. Growing reliance on mechanised open-cast method of mining is a key feature of rapidly expanding coal mining industry in the region. Open-cast production shared about 82.8 per cent of the coal production of C.C.L. and 41.90 per cent of that of B.C.C.L. in 1990-91. The actual and projected production of coal in all the three coalfields of the region are presented in Table 12.3. However, the pattern of coal production in the E.C.L. is not exclusive of its area in West Bengal and, therefore, should be taken to give only a picture of the general trend of the two modes of production and not specific to its part in Chhotanagpur. It is quite evident that the reliance on mechanised open-cast production will continuously grow in the C.C.L. area from 32.90 million tonnes in 1995-96 to 64.50 million tonnes, almost doubling itself, in 2001-02. In contrast to it, open-cast production would only nominally grow in the other two coalfields, viz. the B.C.C.L. and the E.C.L. in the above period.

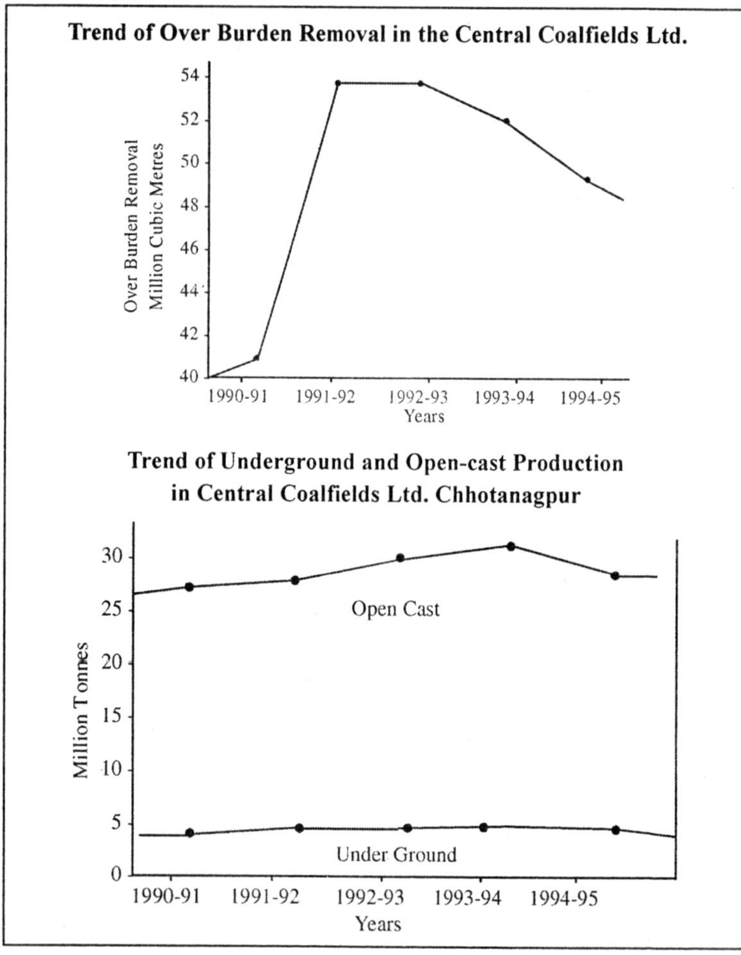

Fig. 12.2: Coalfields C.C.L. Ranchi.

Source of Data: Data Bank, C.C.L. Ranchi.

Open-cast production is environmentally very sensitive as it necessarily implies a higher level of mechanisation and application of new technology to production of coal. Indirectly, it means greater induction of new technology to the open-cast projects, mostly in the C.C.L. area, than in the underground mines. It does not, however, mean that new technology has not been adopted in case of these mines. But the important point to remember is that the new technology of mechanised production is more destructive to the environment of open-cast projects than that of the underground mines. In fact, its degrading impact is very nominal in the form of subsidence, pits, etc. as compared to its environmental havoc in case of the former.

Table 12.3: Actual and Projected Production of Coal.

(*in Million Tonnes*)

	Actual				*Projected*		
	1975-76	*1981-82*	*1991-92*	*1992-93*	*1995-96*	*1996-97*	*2001-02*
1. *B.C.C.L.*							
Underground	17.03	15.71	12.08	11.54	14.50	15.00	17.00
Open-cast	03.86	07.31	14.93	16.52	16.80	17.00	19.00
Sub-total :	**20.89**	**23.02**	**27.01**	**28.06**	**31.30**	**32.00**	**36.00**
2. *C.C.L.*							
Underground	06.85	07.19	04.43	04.22	05.10	05.20	05.50
Open cast	14.11	22.92	26.77	28.16	32.90	35.80	64.50
Sub-total :	**20.96**	**30.11**	**31.20**	**32.38**	**38.00**	**41.00**	**70.00**
3. *E.C.L.*							
Underground	23.56	17.82	14.65	16.00	17.50	18.50	22.54
Open-cast	02.62	06.73	09.87	10.50	14.50	16.90	19.06
Sub-total :	**26.18**	**24.56**	**24.52**	**26.50**	**32.00**	**35.40**	**41.90**

Source: Based on Coal India and You, Directorate of Personnel, Coal India Ltd., Calcutta, 1993.

The destructive teeth of new technology are principally the draglines (above 20 cum), rope shovels (20 cum), hydraulic shovels (4.5-12.5 cum), dumpers (170 T), dozers (410 HP & above) and drills (250 mm). The capacities of these equipments, collectively called HEMM, are constantly being upgraded and, consequently, their destructive powers getting boosted almost every day. The devastation continues unabated with enhanced vigour. The recent changes in the size of the equipments are brought out in Table 12.4.

Table 12.4: Change in Equipment Size.

Name of Equipment	*Unit*	*Earlier*	*Present*
1. Dragline	Cum	15	24
2. Electric Shovel	Cum	4.6-8	5-20
3. HYD. Shovel	Cum	-	3-12
4. Rear Dumper	T	25-50	35-170
5. RBH Drill	mm	100-250	100-311

Source: C.M.P.D.I., Ranchi.

Plant and machinery (PM) claim the largest share in capital investment of any modern, new-technology oriented open-cast mines. For example, the Piparwar Project being developed with Australian collaboration and having a life of 35 years, the capital investment upto the targeted six years (1990-1996) is Rs. 838.59 crores while PM (HEMM denoting "heavy earth moving machinery; CPH, i.e. Coal handling plant; and CPP, i.e. Coal Preparation Plant) investment is Rs. 519 crores or approximately 61.89 per cent of the above investment involving foreign exchange provision of A$148.88 million. The transfer of new technology has often involved import of both the expertise and equipments from the collaborating countries viz. Poland, former Czechoslovakia, Canada, Japan, etc. at least in the initial period of development of the projects, as in the above case from Australia.

The population of all types of basic open-cast equipments, viz. shovels, dumpers, dozers, drills and draglines has remarkably increased during 1976-1992 and 1995 in the coalfields of C.C.L., B.C.L. and E.C.L. (Fig. 12.2). But the increase in the number of dumpers and shovels is quite striking. Again, the population of all the types of equipments is higher in the C.C.L. than either in the B.C.C.L. or the E.C.L. area (Table 12.5) highlighting greater application of new mining technology in the area of the former.

Impact on Environment

Coal mining of Chhotanagpur is presently trapped in the vortex of two 'D's, i.e. demand of coal and degradation of environment. Environmental degradation is usually proportional to the extent of mining operations (Fig. 12.3) and the B.C.C.L. and C.C.L. sharing about 43 and 45 per cent of the total coal production of the region are the main problem areas of environmental degradation. Thus, the industrial and economic gains of coal mining cannot be regarded to provide enough compensation for the environmental damages caused by it. The region has significantly gained economically, but seriously suffered environmentally.

Table 12.5: Population of Basic Opencast Equipments (1976-1992)

Company	*Equipment*	*01-04 -1976*	*01-04 -1982*	*01-04 -1991*	*01-04 -1992*	*01-04 -1995*
1. C.C.L.	Shovels	48	112	137	156	165
	Dumpers	248	585	910	1050	1040
	Dozers	53	140	203	229	230
	Drills	41	97	161	179	191
	Draglines	—	—	—	—	—
2. B.C.C.L.	Shovels	04	48	143	159	—
	Dumpers	14	215	553	601	—
	Dozers	06	49	170	184	—
	Drills	04	33	124	139	—
	Draglines	—	—	02	02	—
3. E.C.L.	Shovels	06	43	77	85	—
	Dumpers	26	169	323	357	—
	Dozers	21	72	94	106	—
	Drills	11	39	50	57	—
	Draglines	—	—	—	—	—

Source: Based on Coal India and You, Directorate of Personnel, Coal India Ltd., Calcutta, 1993.

The new technology of mining and associated operations has adversely affected the entire eco-system of the coal fields as they directly impinge on it in a multiplicity of ways (Fig. 12.4). The dimensions of environmental degradation are shown in Fig. 12.3. Broadly speaking, these are:

(i) Devegetation, i.e. removal of grass, plants and trees.
(ii) Topographic Deformation, i.e. pits, cavities, heaps of overburden.
(iii) Edaphic Loss, i.e. loss of soil mantle, structure and nutrients.
(iv) Drainage, i.e. derangement, lowering of water table.
(v) Pollution, i.e. air, water, noise.
(vi) Dereliction, i.e. genesis of unliveable environment.

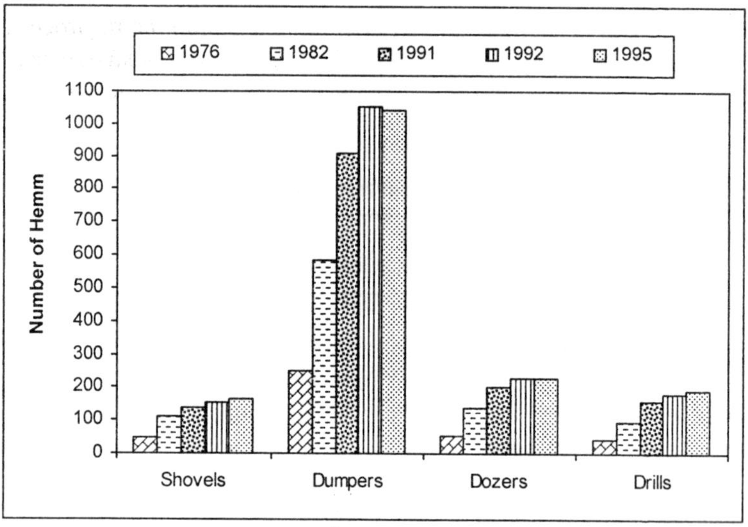

Fig. 12.3: Growth of Basic Open-cast Equipments in Central Coalfields Limited, Chhotanagpur, 1976-1995.

A brief discussion of these follows below:

Devegetation

Bio-diversity in the environment is an essential component which affects its sustainability, in the form of providing food for other biota. Formerly, coal bearing areas of Chhotanagpur were densely forested but now vegetation has become the first casualty of new technology of mining, especially the open-cast technology. Various kinds of vegetation ranging from grass to tall trees have to be removed first in the process of clearing the land for mining. The grasses and the small plants are valuable cover of the top soil, and are uprooted without much effort. The tall trees are ruthlessly felled and removed. Rapid depletion of forests gets initiated. The C.C.L., B.C.C.L. and E.C.L. have plans to deforest about 21 sq. km., 03 sq. km. and 1.1 sq. km. respectively of forests in Chhotanagpur in near future. The diversion of river Damodar will lead to destruction of forest in an area of about 0.57 sq. km. The vegetation cover lying outside the area of active mining suffers from reduced transpiration on account of heavy coating of dust and toxic materials; they slowly die out along with the associated fauna, while some species of the latter may migrate as micro-climatology changes drastically.

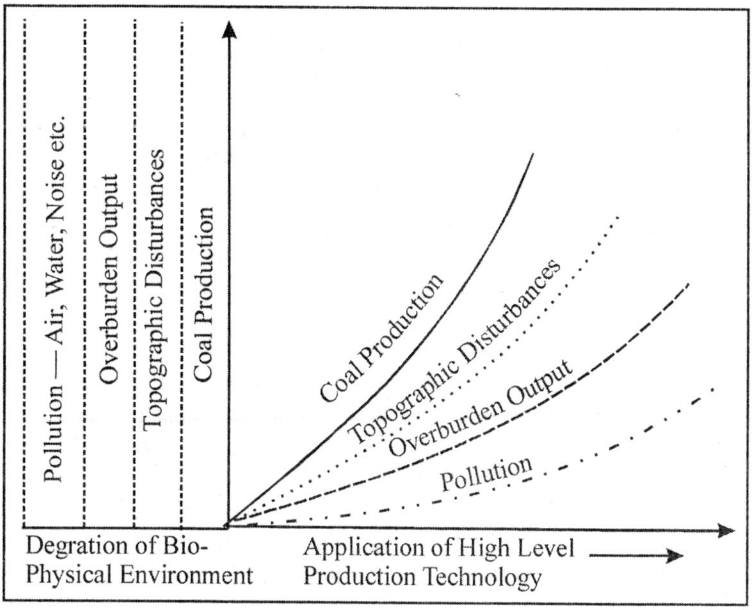

Fig. 12.4: Idealised Relationship between Modern Mining Technology and Environmental Degradation.

Topographic Deformation

The new technology of open-casting is a voracious eater of land. The area of active mining, overburden heaps, coal stock piles, infrastructure, associated industries, etc. claim far greater area than the actual lease area in the coalfields. On an average the open-cast projects require land at the rate of approximately 400 ha. per million tonne of coal production per year. Topography is very badly ravaged by the process of delayering; the surface is ruptured, strata after strata exposed and the monster of new technology devastates the whole area. The land resource suffers both destruction and degradation. On an average 60 per cent of the total land required for an open-cast project is damaged.

Coal mining is now not only extensive but also much deeper. Earlier the maximum depth was 100 m. At the present time, it is 140 m. and planned to be increased to 300 m. in future. The seams are generally multiple in nature necessitating the exposure of the bed-rock of the lowermost seam workable by open-cast method. The more extensive and deeper the mine, the greater the amount of overburden output. The input dumping of overburden has to be delayed until the deepest seam is worked out. Consequently, huge heaps of overburden having the appearance of artificial hills and ugly, deep scars nearby dominate the landscape of the new open-cast projects. Wind direction has changed in some localities due to decapping of hills and creation of huge dumps. In the C.C.L. region the external dumps occupy an area of 11.18 sq. kms. and twenty-six subsided spots another 5.25 sq. kms. In the B.C.L. region, the external dumps occupy an area of 6.31 sq. kms. abandoned opencast pits 4.34 sq. kms. and subsidence an extensive area of nearly 63 sq. km.

There are seven major open-cast projects in the C.C.L. and two in the B.C.C.L. area. The two major open-cast projects of the E.C.L. are outside the Chhotanagpur division and, hence, beyond the purview of this paper. Thus, there are nine major open-cast projects in the C.C.L. and B.C.C.L. areas having the production capacity of about 32.5 million tonnes per year and overburden removal capacity of 61.6 million cubic metres. Evidently, on an average, for production of one million tonnes of coal 1.89 or nearly million cubic metres of overburden have to be removed. (Table 12.6).

Table 12.6: Major Open-cast Projects.

Sl. No.	*Mine*	*Production Capacity (MTY)*	*OB Removal Capacity (Mcum)*
	Eastern Coalfields Ltd.		
1.	Rajmahal	10.5	19.4
2.	Sonepur Bazari	3.0	14.2
	Bharat Coking Coal Ltd.		
1.	Muraidih	2.5	4.1
2.	Block II	2.5	11.2
	Central Coalfields Ltd.		
1.	Ashok	6.5	7.4
2.	Piparwar	6.5	3.2
3.	K.D. Hesalong	4.5	5.7
4.	Rajarappa	3.0	9.5
5.	North Urimari	3.0	7.9
6.	Gidi – A	2.0	6.0
7.	Kedla	2.0	6.6
	Total	**46.00**	**95.2**

Source: **C.M.P.D.I., Ranchi, Report prepared on the Eve of the Visit of Chinese Team in April, 1995.**

In the Central Coalfields Ltd. which presently has the largest number of open-cast projects in Chhotanagpur, the overburden removal was between 2-6 Mcum per year but, presently, nearly 18 Mcum. It has been continuously mounting in all its thirteen areas during 1990-95. (Table 12.7).

Table 12.7: Trend of Overburden Removal in the Central Coalfield Ltd. 1990-91 to 1994-95 ('000 Cu.M.)

Name of Coalfield Area	*1990-91*	*1991-92*	*1992-93*	*1993-94*	*1994-95*
1. Barkakana	2382	2078	1272	1101	1056
2. Sayal	2002	2106	1980	2121	1682
3. Argada	5166	5296	5700	5492	4654
4. North Karanpura	5880	5865	6403	6015	6073
5. Rajhara	690	711	662	635	605
6. Piparwar	589	872	1986	2650	3171
7. Rajarappa	6335	6986	7001	6002	6035
8. Kuju	3786	3993	2901	2289	2138
9. Hazaribag	3028	2904	2795	3079	4216
10. Kargali	4650	5767	5809	5859	5087
11. Giridih	1005	1023	1170	1154	1163
12. Dhori	5338	5779	5793	5826	4955
13. Kathara	7573	8751	8616	8188	7123
Total C.C.L.:	**40444**	**52131**	**52088**	**50411**	**47958**

Source: Data Bank 1990-91 to 1994-95 C.C.L. Ranchi.

It is poised to further acceleration in the years ahead. The disturbances of subterranean strata in deep underground mines lead to subsidence, seepage of surface water, development of surface cracks, cavities and at times lead to tragedies like Chasnala (28 December, 1975) and Gajlitanr (26 September, 1995) disasters of mine-flooding. Again, the underground fire is further aggravated by passage of oxygen through the surface cracks.

Edaphic Loss

After vegetation, top soil is the second casualty. In almost all the collieries, where largescale open-cast production has been taken up, there is no scientific method of having separate stockpiles of either the top soil or the sub-soil. It is usually mixed up with the boulders and pebbles of overburden. The nutrients are washed out or destroyed by exposure to the sun. Wherever the ravages of underground or surface fire continue, both the top and sub-soils are baked and burnt, a typical example of which is the Jharia Coalfield still having nearly 70 active fire sites affecting an area of 15 sq. km. The leachate from the stockpiles of coal further barren the soil. Planting is, therefore, difficult and when attempted survival rate of the saplings is extremely low.

Drainage Problems

The drainage of the coalfields has become awfully complicated on account of obvious reasons. Surface drainage follows the cracks, pits and cavities. Derangement is conspicuous; small streams may be entirely lost and pools of water develope in quarries and hallow spaces.

The diversion of the courses of rivers may become unavoidable to extend the mining area under the river-bed. Such diversions are on the anvil in the region. For example, the D.V.C. has a plan to diver the course of river Damodar in a stretch of 4 kms. to extract about 5 million tonnes of coal locked in the bed of this river. Another diversion of the courses of Damodar and Konar is planned by the C.C.L. near Tenughat for extracting 90 million tonnes of coal. All these plans are feasible with the available new technology, though not yet attempted perhaps because presently abundant coal is available in other areas without diversion.

On account of more intense mining with new technology, there is a chronic shortage of water in the coalfields. The established flow lines of surface water are disturbed by cracks and openings, pits and cavities. Not only this the water table goes down because of removal of rock formations and dewatering of deep mines and large open pits. Thus, interruption of sub-surface aquifers and adverse impact on groundwater regime often create water-famine in the coalfields. It is a critical factor in recreation of green cover in the affected areas.

Pollution

Rampant pollution of all the vital elements of environment, viz. air, water, soil, etc. is the product of hi-tech coal mining in the region. Mechanical operations and ground movements of heavy equipment lead to intense vibration and noise pollution.

Suspended particulate matter (SPM) and respirable dust are quite obvious pollutants of air. The plying of HEMM and trucks, etc. releases fumes and noxious gases. The problem of mine-fire is as old as the coal mining itself; it occurs in the old coalfield of Jharia as well as relatively new fields of West and East Bokaro, and North Karanpura. On account of the presence of noxious gases, smoke and steam, air pollution is more intense in the fire-area than in the non-fire area. (Table 12.8).

Table 12.8: Ambient Pollution Level (Average figure).

Year	*Non-fire Area*			*Fire Area (Residential)*		
	SPM (Mg/m³)	*SO_2 (PPM)*	*NO_2 (PPM)*	*SPM (Mg/m³)*	*SO_2 (PPM)*	*NO_2 (PPM)*
1978	206	0.004	0.002	582	0.013	0.008
1979	217	0.006	0.005	640	0.014	0.009
1980	252	0.007	0.004	631	0.018	0.011

Source: N.D. Agarwal, 1991.

The quality of surface and sub-surface water has highly deteriorated in the coalfields. Leachate from run-off from overburden waste dumps is an important factor of water pollution. In fact, water suffers from both chemical and biological pollution. Soil wash-off and erosion are common pollutants of water. On account of intense mining and location of a number of washeries, the Damodar carries high suspended solid concentrates, i.e. nearly 3 to 4 grams/Lit. against the tolerance limits of 100 mg/l.

Both underground and opencast mining and related operations are prone to generate extreme noise pollution. In open-cast mines, most of the HEMM operations, viz. shovels, dozers, scrappers, dumpers as well as coal and overburden handling systems entail noise in the region of 100 to 107 dB(A). Ambient noise level in congested coalfields like Jharia is in the range of 55 to 67 dB(A) during night time. The level of noise pollution is quite high when we compare it with whisper (20 decibels), conversation (65-70 decibels) and riveter (90-100 decibels), CMRS (Coal Mining Research Station), Dhanbad has conducted a study and found that mechanical operations in the underground mines create noise pollution in the range of 90-100 dB(A) which is in excess of permissible noise level of 90 dB(A) for industrial area.

Large scale open-cast mining with new technology also calls for massive blasting operations. Since 1970's this mode of production has become a common feature in the region and, consequently, ground vibration a serious hazard. The immediate impact is vibration, noise, dust, toxic fumes and smoke, damage to land and associated ecology.

Dereliction: The Genesis of Unlivable Environment

The quantum jump in production capacity and adoption of new technology continues in the region without much concern for environment-laying barren the land, polluting water, defiling air, denuding forest and degrading the quality of life in the coalfields. The inevitable is dereliction – the end-product of environmental degradation (Fig. 12.5) in mining and industrial areas. The environmental quality gets so much deteriorated that healthy habitation becomes a day dream. It is not only bio-physical deterioration but psycho-social deterioration as well. Derelict mentality is no less harmful than derelict land. Unless the hiatus in the livable environment is removed by bio-physical restoration and economic regeneration, human habitation is not possible in a derelict area.

Where is the dereliction in the coalfields of Chhotanagpur? It is a question likely to be put by those who like to close their eyes from the ongoing degradation of the environments in the coalfields or by those who fail to realise that dereliction behaves like cancer; the former afflicts the environment, the latter the human body. The cancerous effects of dereliction is quite visible in the Jharia coalfields as well as in the environment of all the major open-cast coal projects of the region.

The Hope for Better Environment of the Future

One of the greatest evils of modern civilisation is its profligate use of natural resources without due concern for environment. The spirit in coal production in Chhotanagpur, based on application of new technology, might have helped the country and the region to tide over the economic crisis, but certainly it has destabilised the environment of the coalfields. By its very nature mining is an environment degrading activity.

Presently, in all the coal projects – an Environmental Management Plan (EMP)—is an integral part of mining feasibility report. The

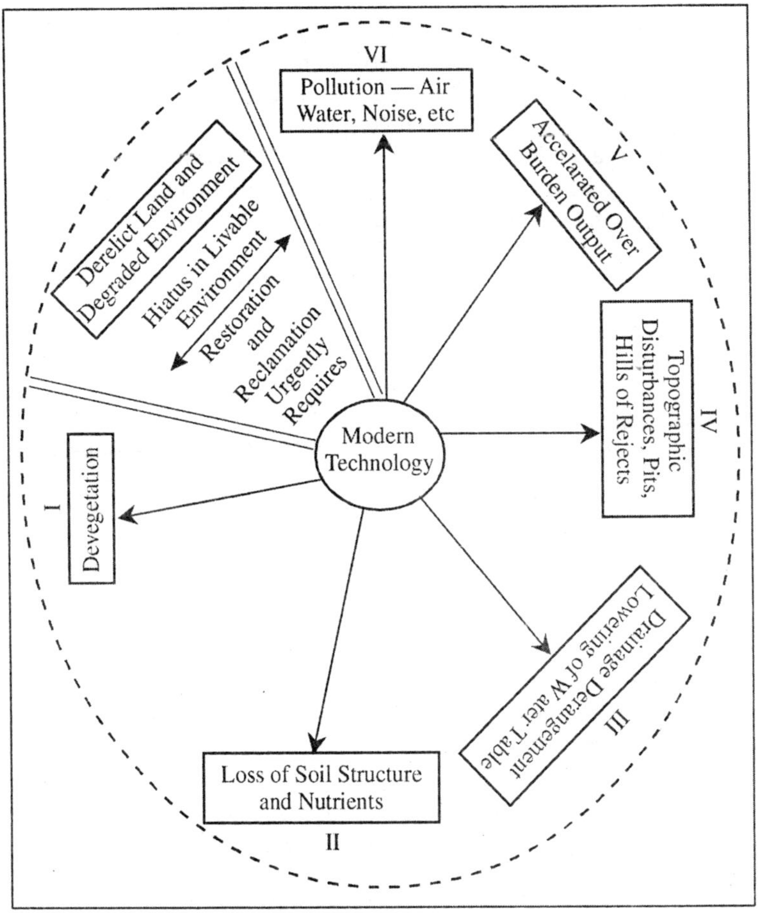

Fig. 12.5: Dimensions of Environmental Degradation by Induction of High Level Production Technology in Mining Activitie.

C.M.P.D.I.L., Ranchi has already prepared nearly 150 E.M.P.S. and during Eighth Plan additional 150 E.M.P.S. have been contemplated. But what is more important is not the plan, but its successful implementation.

Environmental control measures are quite expensive and the area to be annually covered by them is quite large. The Expert Committee set up by the Ministry of Environment and Forests (MOEF) on Restoration of Coal Mines has assessed that coal mining, mostly in the C.C.L. renders an area of 600 hec. of land per year biologically unproductive and the same would rise to 1400 ha. by the year 2000 A.D. To launch a plan for effective control of the environment in such a large area is financially a gigantic task.

It has been worked out that for pollution of one MTY (Million Tonne per year) of coal, the cost of E.C.M. (Environmental Control Measures) comes to Rs.7.92 crores which is about 11.28 per cent of the total project cost for about 6 MTY (Table 12.9).

Table 12.9: Capital Cost of Environmental Control Measures in Selected Open Cast Project of C.C.L. (Rs. Crores)

Open-cast Projects	*Capacity in MTY*	*Capital Cost*	*Capital Cost for Environmental Control*
1. Kadla	2.00	160.66	20.91
2. Paraj (East)	1.75	105.88	9.64
3. Topa (Reorganised)	1.20	81.56	8.52
4. Jharkhand	1.00	69.35	8.05

Source: V.K. Singh, 1991, p. 13.

A vital question is: where from the huge amount of capital investment will come? Even if there is specific provision of fund for E.C.M.; it is generally inadequate. For example, in case of ambitious O/C Piparwar Project, there is no clear policy for the purpose. The mineable reserve of coal in the project area is about 197 million tonnes and ultimately the raw coal output will be 6.5 MTY during 1995-96. The total outlay for EMP measures is Rs. 28.77 crores which works out to about 4.42 crores for one MTY production of coal. Since there are no dedicated equipments especially provided for *reclamation* only certain percentages of the cost of items of HEMM and CHP have been proposed to be set aside for EMP measures. For example, 50 per cent of the cost of 5 Nos. of 300 KW. Dozer and 25 per cent of the cost of 2 Nos. of Grader area earmarked for *restoration*. Likewise, meagre provision has been made for anti-pollution measures. Identical situation prevails in other new O/C projects. The region has already a huge backlog of environmentally degraded area as legacy of the past mining, and a backlog is being created for the coming generation.

It is heartening to notice a change in absolutely exploitative policy of the coal mining industry. Environmental consciousness has started growing and complete restoration of the coal areas devastated by coal mining may become a compulsion for the concerned mining companies in near future. The coming generation should certainly have a better, livable environment in the coalfields of Chhotanagpur.

REFERENCES

Agarwal, N.D. (1991), "Mine Fire : A Threat to Environment", Workshop on *Concern and Care for Environment in Chhotanagpur*, C.M.P.D.I., Ranchi, October 5-6.

Coal India and You (1993), *Directorate of Personnel*, Coal India Ltd., Calcutta.

Cowles Encyclopaedia of Science, Industry and Technology (1967), p. 291.

Data bank 1990-91 to 1994-95, C.C.L., Ranchi.

Final Capital Cost Estimate for Piparwar-Integrated-Mine-cum-Coal Beneficiation Project of C.C.Ltd. (1993), Updated Economics (Draft), C.M.P.D.I. Ltd., Regional Institute III, Ranchi.

Ganapathy, K.S., Mohan, Agarwal, A.K. (1991), "State of Environment of Jharia Coalfield", Workshop on Concern and Care for Environment in Chotanagpur, C.M.P.D.I., Ranchi, pp. 5-6.

Mukhopadhyay, M.K., Roy, J. and Saxena, A.K. (1991), "Effects of Mining on Land and Forests in Chhotanagpur" Workshop on Concern and Care for Environment in Chhotanagpur, Oct. 5-6, C.M.P.D.I., Ranchi.

Report prepared on the Eve of the Visit of Chinese Team (1995) April.

Singh, R.K. (1991), "Mineral Resources of Chotanagpur", Workshop on Concern and Care for Environment in Chhotanagpur, C.M.P.D.I., Ranchi, Oct. 5-6.

Singh, V.K. (1991), "Environmental Management vis-à-vis Mineral Development specially Coal in Chhotanagpur", Presented in the Workshop Concern and Care for Environment in Chhotanagpur, C.M.P.D.I., Ranchi.

Tiwari, B. (1988) "Role of Coal in the Development of Chhotanagpur", C.M.P.D.I. Publication.

13

Environmental Degradation due to Slate Mining
A Case Study of Khanyara Mines, Kangra District, Himachal Pradesh

BHUPINDER SINGH MARH and SANJAY S. PATHANIA

The economic potential of any region is founded primarily upon its natural resources, their availability being closely associated with the economic well being of the people. Natural resources have been used by man since times immemorial. The exploitation of natural resources has led to environmental degradation. Environmental degradation literally means "unfavourable change of the environment". Mining involves extraction of minerals or rocks from the earth's crust by removing the vegetation cover, the topsoil, rock and other materials that lie above required mineral. Extraction is done by open pit mining, strip mining, auger mining, dredging and hydraulic mining. Most non-metallic materials like sand and gravel, limestone, sandstone, slate and brick clays are extracted from open pit mines. Their impacts are more often noticed and criticised by the public. Their environmental impacts include disruption of vegetation, soil and groundwater, noise from both mining and the processing plants and dust especially from limestone and gypsum plants.

Slate mining generates very large amount of waste materials which is disproportionate to the surface area mined. The site of extraction, once identified, remains operational for long periods of time and the debris generated in the mining operations keeps on accumulating down the slope. The magnitude of these effects mainly depends upon the steepness of the terrain, amount of precipitation, characteristics of materials and the method of mining. Surface mining activity does not only influence the quality of land, but also have far-ranging impacts on air, water, vegetation and human and wildlife. Debris accumulation due to surface mining makes land unsuitable for agricultural purposes and settlement systems. Various pollutants generated in mining operations affect the quality of human life in the surrounding areas. The removal of the protective vegetation cover aggravates the problem of soil erosion. This also results in drastic reshaping of the surface and subsurface drainage patterns.

Some problems are peculiar to the mining activities in mountainous areas. Debris accumulated in the valleys leads to damming of streams. The subsequent breach of such dams can lead to flood hazards downstream. In rainy season, debris flow and associated hazards can occur mainly in surface mining regions. These can take the form of debris falls, rockfalls, slump, debris slides and rock avalanches. The situation in the Himalayas is complicated because of still continuing vertical uplift in the Himalayan region. Earthquakes associated with neotectonic and morphotectonic movements are rather frequent in this region. These also complicate the problem by leading to debris flows and avalanches of material accumulated by surface mining.

In the present study the effects of slate mining in Khanyara mines of Kangra district of Himachal Pradesh are analysed and evaluated. Slate is extracted by open-pit mining. This operation in Himachal Pradesh has been going on for a very long time and has affected the land surface considerably.

Review of Literature

Open-pit mines have received a lot of attention from environmentalists in recent years. Such mining involves removal of vegetation cover and the overburden to provide access to the mineral. The waste materials are often indiscriminately dumped, the washout from which affects the surrounding landscape. Both internationally and nationally many important studies have addressed to this problem. In 1967, the U.S. Department of Interior (quoted by Tank, 1976) concluded that the effects of mining upon the environment vary widely depending upon the steepness of terrain, amount of precipitation, temperature, chemical characteristics of minerals and methods of mining. UNESCO, as reported in Archer *ct al.,* (1987) identified different kinds of impacts of unwise mineral extraction. Firstly, it was suggested that the exploitation of minerals is an irreversible drain on the supply of naturally occurring raw-materials. Secondly, in the process of mineral extraction, toxic solids, liquids and gases are injected into the biosphere faster than its natural cleansing capacity. Thirdly, the mismanagement of mining activity leads to reduced productivity and adverse effects on human health. Mac. Andrews and Sien (1979) have reported that in Malaysia, tin is extracted by open cast methods. Problems of erosion and silting remain largely uncontrolled and their effects are visible near Kuala, Kuba, Selangor etc. Siltation has its effects on aquatic life. Mining has resulted in the removal of forest vegetation and washing of soils for ore has left sterile tailings on ground.

A study of the effects of the Sandur iron and manganese mines has been reported in Nag (1991). Mining operations have adversely affected the topography of the area, vegetation and fauna. The process of deforestation has led to soil erosion which, in turn, resulted in silting of Narihalla reservoir. The liberation of fine particles during mining has caused siderosis, silicosis-tuberculosis, asthama and other diseases which affect the respiratory tract. Jaitli (1993) has studied an area of Jodhpur district of Rajasthan. This area is today witnessing a major ecological setback because of unscientific and unplanned mining of gypsum, limestone and sandstone. Menezes and Monte (1994) have reported from the west coast that massive quarrying for stone and sand is turning the ecological oasis of Dahanu into a wasteland. Nearer the study are Dehradun and Mussoorie belt of Uttar Pradesh is famous for limestone mining. In this area limestone mining is haphazard and unscientific. It has caused considerable landscape degradation, environmental pollution and ecological imbalance (Awasthi, 1983, 1984). Detached and haphazard mining has exposed the already fragile geosystem to further damage (Nath and Nath, 1990).

Geographical Aspects of Study Area

The Khanyara slate mines, the present study area, are located on the southern slopes of the Dhauladhar north and east of the town of Dharmshala in Kangra district of Himachal Pradesh (Fig. 13.1). The study area forms a part of the Beas river basin. The Khanyara slate mines are situated 11 to 15 kms. north-east of Dharmshala, at altitudes varying from 1525 to 2135 m. Slates are extracted from the southern slopes of Dhauladhar just above the Khanyara village. Mines are spread over a band which is over 5 kms. long and 200 to 1500 metres wide extending between Thathari and Thatt.

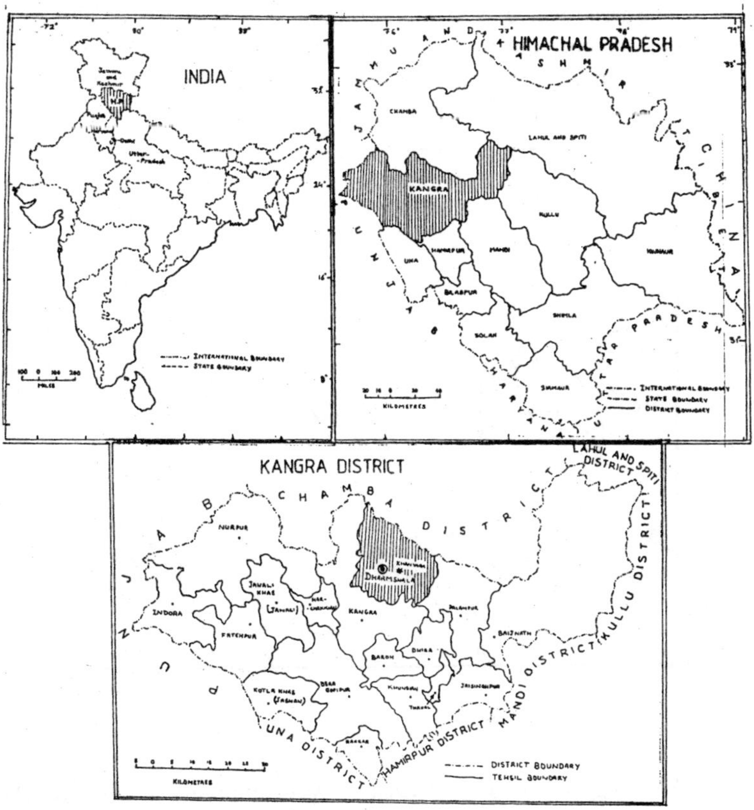

Fig. 13.1: Location Map of Study Area.

The word Khanyara is derived from *Khan* meaning mine or quarry. Thus, this place name denotes "a village of quarries". In the ancient times the Raja used to be the proprietor of this land. But the British Settlement Officer Mr. Barnes made this land a property of the village proprietory body. Two perennial streams, Manjhi and Manuni drain the area. The drainage pattern is dendritic (Fig. 13.2). The slate bearing horizon is wedged between Dhauladhar Granite and Shali Group (Dharmkot limestone) of rocks having thrust contacts. The slates belong to Chail Formation (Chandpurs) (Fig. 13.3) and are bluish-gray to blackish-gray in colour. They are both thinly and thickly bedded having even to slightly undulating surfaces.

The area receives very large amount of rainfall. About 70 per cent of rainfall comes during July to September, the maximum being about 1055 mm. in July. The annual range of temperature of this area is about 17°C, maximum and minimum being 22.8°C and 6.5°C respectively.

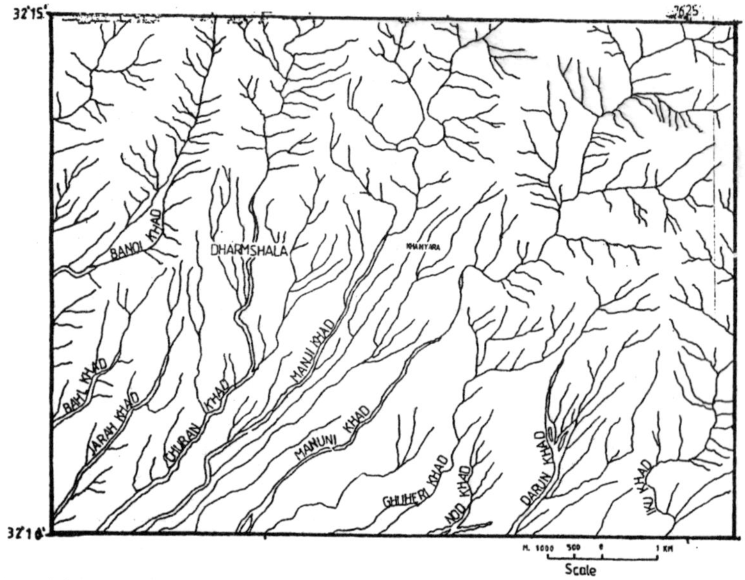

Fig. 13.2: Drainage Pattern of Study Area.

The vegetation of this area is wet-temperature and includes various temperate species. The main varieties are pine and oak. Rododendron trees are also common. Initially, about 366 acre area divided into two zones around Manjhi and Manuni streams was affected by mining. But, now, it has been extended to about 625 hectares. At present, the following are the main slate mining zones in Khanyara area (Fig. 13.4).

1. *Manuni Area*: The slate bearing horizon on both the sides of Manuni stream has been designated "Zone A". The mining pits are situated 2-4 kms. north-east of Khanyara village. There are 213 working quarries of various dimensions in this zone. There are 108 quarries along the right side and 105 along the left side of Manuni stream. The main areas under mining are Phatt, Mandali Tawa Kongari, Tappalli, Choo, Tibbu, Chakki, Baharunu, Salotu and Konda.

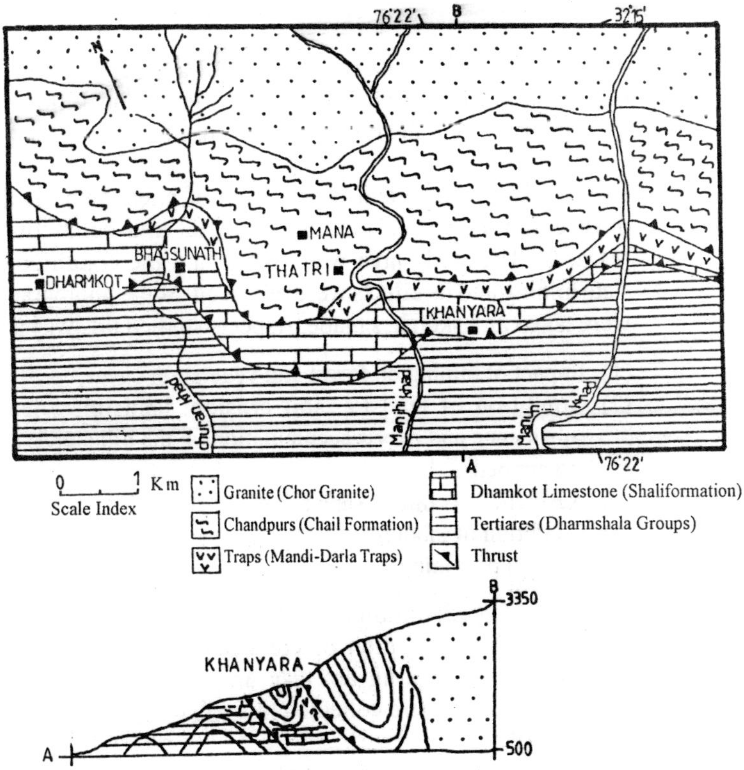

Probable Geological Section Along A-B

Fig. 13.3: Geological Map of Study Area.

2. *Manjhi Area*: In this area slate mining is located on both sides of Manjhi stream and is labelled as "Zone B". These quarries are situated 3-5 kms. north-west of Khanyara village. There are 115 working quarries located within a distance of 2 km. between Thathari and Kuthali Bagali. The main areas under mining are Mana-da-Nal, Thathari, Kala Choo and Kuthali Bagali.
3. *Thatharna Area*: Quarrying in this area started in 1975. This includes Nilighar, Londa, Gurlu Nal and Ukhal Nal in the area between Manjhi stream in north-west and Manuni stream in the north-east. It occupies 1.5 km. long stretch on the southern slope of Khanyara area. There are officially 75 slate mines in operation in this zone.

Data

For this study both primary and secondary data have been collected. The primary data are collected fresh in the field. The following methods of primary data collection have been used in this study.

1. *Field Observation*: Field observation of phenomena is the most commonly used method in social and behavioural research. The study area was visited and the condition of slate mining and its effects were observed. Some photographs were taken and sketches made. Aspects studied this way are:
   (1) Impact of mining on forest and vegetation cover.
   (2) Condition of land i.e. soil erosion, landslides, land-subsidence and debris accumulation on the hill slopes and stream beds.
   (3) Aesthetic loss due to slate mines.
   (4) Condition of labourers.
   (5) Method of mining.
2. *Interview Method*: Information regarding the socio-economic aspects of the population of the study area, their way of thinking about mining activity and its impact and the historical aspects were collected through personal interview method.
3. *Questionnaire*: In the present study information collected through questionnaire are associated with environmental degradation e.g. effect on local environment, flood frequency, siltation in rivers and reservoirs. Information about problems due to blasting, health problems and government policies were also collected by this method.

Secondary data were collected from government offices and local authorities. Some of the data about production and value of slate in Himachal Pradesh were collected from Geological Wing of Department of Industries, Shimla. Data about historical and ownership aspects, production and value of slate in Khanyara mines and its royalty were collected from the Khanyara Gabli Dhar Panchayats and Khanyara Gabli Dhar slate Quarries Board.

Land use distribution was mapped from Survey of India Topographical sheet No. 52 D/8 (1962-63). An assessment of the extent of land degradation due to mining was made through visual interpretation of IRS 1-A LISS-2 Imagery of the area of June 1989 (Fig. 13.4).

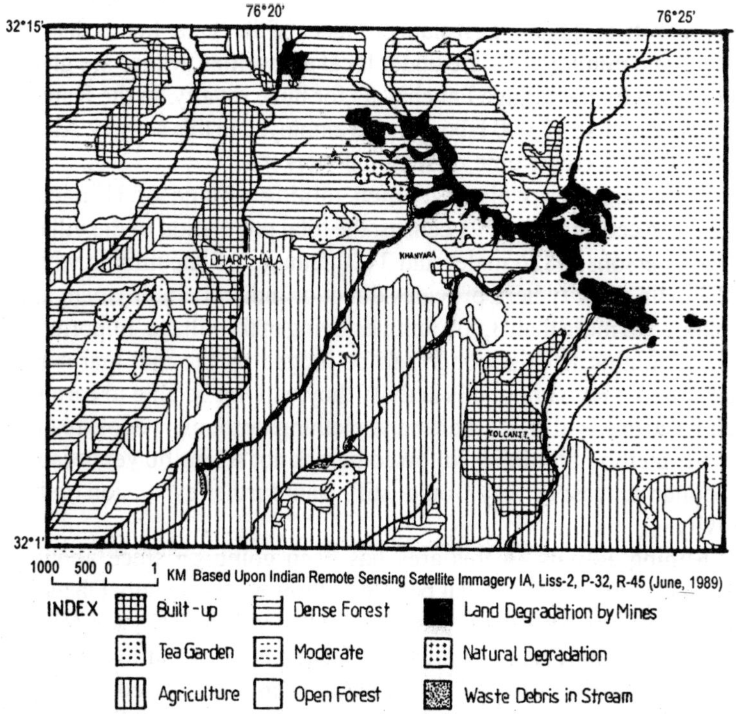

Fig. 13.4: Khanyara and its Surrounding Areas (Land use Pattern).

Interpretation and Analysis

Slate is a non-metallic material extracted by surface mining. It is a fine-grained metamorphic rock characterised by perfect cleavage. Mineralogically, slate is metamorphosed shale and is composed by very fine grains of micas, chlorite, felsphar, quartz, oxides of iron and many other minerals. The usual colours of slate are black, blue black, gray and light green. They are used mostly for roofing and also for flooring and wall panelling. Slate flour is used in paints and rubber products. In the interior parts of the state of Himachal Pradesh slate mining and related activities provide the only kind of employment to skilled and unskilled workers. The state receives large amount of revenue from this activity.

Himachal Pradesh is one of the leading states in the country as far as slate mining is concerned. Good quality slates are found in Chamba, Kangra, Kullu, Mandi and Shimla districts of Himachal Pradesh. There are a total of about 550 slate mines in the State. The Kangra district is the largest slate producer of the State. In this district all the state mining areas are under local authorities. Mines are given on contract by the village Panchayats for huge royalty. In the Kangra district more than 12,000 labourers are employed in this activity. About half of them are part-time or seasonal workers.

Khanyara is a well-developed mining village of Himachal Pradesh. According to 1991 census, the total population of the village is 7246 persons (3826 males and 3420 females). Slate mining which is the main activity here provides direct or indirect employment and livelihood to more than 3,000 families living in the village. Importance of slate mining to this village can be judged by the fact that many locals think that without this activity they will not be able to earn for their basic needs of food, shelter and clothing. Although people realise that the method of mining followed here is not scientific yet because of economic reasons they would, still, want further expansion of this activity.

Mining activity in this area has been going on since about a thousand years. It is said that people of three castes namely Dhaugri, Hali and Scippy were the first settlers of this village. They are said to have come from Chanauta area of Chamba district and lower areas of Kangra district. The people of Dhaugri community used to quarry slates, carry them on their backs to lower areas of Kangra district and sell them. It was only in 1867, that Mr. R.W. Shaw took the quarrying right in this area from *Zamindar* of Khanyra village for the Kangra Valley Slate Company. In 1954, the quarrying rights were transferred to village Panchayats. At present, the Panchayats of Khanyara and Dari villages through Khanyara Gabli Dhar Slate Quarries Board, which was constituted in 1975, lease out slate mines in the village Shamlat (commons) land to the contractors.

The total forest land in Khanyara under "Khanyara Cooperative Forest Society" is about 12,699 acres. But the forest area has been decreasing due to removal of trees for mining activity, increasing pressure of population and consequent increasing demand for timber, fuelwood and fodder. The effect of quarrying related deforestation is felt both upstream and downstream of the mining area.

Landslides bring down considerable amount of debris from the upper slopes and accumulate it in the lower areas. Mainly it has been attributed to unscientific mining practices. Both these phenomena affect forest cover either by uprooting of vegetation or by their burial. Considerable amount of good agricultural land has become useless due to accumulation of debris. Many streams of the study area now experience more frequent floods of large magnitude. Debris removal by floods and soil erosion on the hill slopes which has been accelerated by degradation also leads to the problem of excessive siltation of Pong Dam Reservoir located downstream on Beas River.

Water for all domestic purposes comes from natural springs and surrounding streams. It has been reported that considerable amount of deterioration in the water quality does occur during rainy season. Slate mining and its attendent activities throw up considerable amount of dust in the atmosphere. This is blown toward the village and deposited on everything there.

Scars left by quarrying and landslides and accumulating heaps of debris lead to the loss of Aesthetic beauty of the area. This probably is the most visible and accepted effect of mining activity in the study area. Mining activity and consequent destruction of trees has disturbed or completely destroyed the natural habitat of many species. Wild animals have virtually run away from the area due to the land blasting activities.

Most of the waste materials in the study area are lying at a very steep angle, very near the angle of repose. A slight tremor can lead to violent sliding of this material. It is now well known that the study area is located in an earthquake-prone zone. Thus, the danger of earthquake-induced landslides is very real in the study area. Slate mining in the study area is done using outdated and unscientific techniques which are labour-intensive and involve large-scale land blasting. As a result, there are many mining accidents in the study area. It was reported during the field work that two to six persons get killed almost every year due to mining accidents and between two to five persons are injured every month. Some persons have lost such vital organs or eyes in mine blast accidents.

The health related problems faced by the people of Khanyara village include malnutrition, tuberculosis, asthama, epilepsy, jaundice, anaemia, diarrhoea and gastric problems. Although the local people tend to believe that there is no relationship between the mining activity and these health problems. Yet, it seems probable that some of these health problems might in some way related to the mining activity.

Illiteracy, hard labour, malnutrition, impure water and widespread use of local countrywine lead to many health problems faced by the local people.

It was observed in the field that the mining activity in the study area faces some labour-related problems, such as bonded and child labour. A system of bonded labour has evolved in this area. The contractor pays each labourer a sum of Rs. 5,000 to 10,000 as advance and in return the labourer works for him throughout his life. Children in the age group of 10-15 years also work in the slate mines. The people are not able to protest these practices as the economy of the village is totally dependent upon the slate mining activities.

The whole area of slate mines is under Khanyara Gabli Dhar Slate Quarries Board. In 1988-89, production was 1,00,000 m. tons valued at Rs. 5,00,000. After this period, production as well as total revenue have decreased. In 1992-93, production of slate was 50,000 m. tons which fetched Rs. 2,6,50,000.

At present, the Panchayats of Khanyara and Dari villages through Khanyara Gabli Dhar Slate Quarries Board, lease out slate mines to the contractors and collect royalty according to the size of slate extracted from contractors. The amount of royalty collected is divided among the two panchayats. The Khanyara panchayat gets 45 per cent and Dari panchayat gets 55 per cent of the total collection. In the year 1992-93, the largest amount of royalty of Rs. 9,80,402 was collected.

During field survey it was observed that the slate mines at Kuthali Bagali and Kala choo localities in Manjhi area and Thatharna area are located on very steep ground (slope angle varies between 60° and 80°). Such steep slopes are unsuitable for mining activity. It was also noticed that the slate mines of Phatt, Choboo and Barathi on the right bank of Manuni stream and Mandali and Thatt on the left bank of Manuni stream are infested with landslides, debris falls and land subsidence, affecting the forest and vegetation cover of the area. It was also observed that in the Thatharna area slate mining has left behind precarious overhangs which on collapsing cause great loss. The accumulation of waste material on lower slopes is not only harming the forest wealth but also is a menace to Khanyara village which is located at the foot of this slope. The Geologists of the Himachal Pradesh Department of Industries (Geological Wing), have recommended that out of 409 slate mines in Khanyara regions, 161 are unsuitable for mining and should be closed.

An analysis of the IRS imagery reveals that in Khanyara area, forest land has been badly degraded. Total forest land under degradation in Khanyara area is 3.020 sq. km. The waste material from mines is deposited in the valleys of Manuni and Manjhi streams. It is observed that Thatharna area was densely forested but during last three decades this forest land has been badly degraded. The scars due to slate mining removal of forest and landslides have reduced the aesthetic beauty of the area.

Conclusions

The Khanyara slate mines although are mainstay of the economy of the study area and, thus, provide the only means of employment to the local population and an important source of revenue to the local village Panchayats and the government, yet are also leading to extensive environmental problems in the study area. Mining by outdated and unscientific methods and indiscriminate dumping of debris and consequent damage to local ecology are major problems encountered by the area. Other related problems are those of bonded and child labour, health and aesthetic loss.

REFERENCES

Archer, A.A., Lullig, G.W. and Snezhko, I.I. (eds.) (1987), *Man's Dependence on the Earth: The Role of Geo-Sciences in the Environment*, Paris: UNESCO.

Awasthi, S.C. (1990), "Environmental Appraisal of the Mining Area in Dehradun-Mussoorie Belt of U.P. Himalaya, in Nath and Nath (eds.), *Environmental Pollution: Conservation and Planning,* Vol. 2, Allahabad: Chugh Publications, pp. 182-96.

Jaitli, H., Gupta, P.K., Motiani, D.D. and Kaushal, S.S. (1993), "Sandstone Mines Eating into Their Lungs", *The Hindu Survey of the Environment 1993 (Annual),* Madras: M/s. Kasturi & Sons Ltd.

Mac Andrew, Collin and Sien, Chialin (eds.) (1979), *Developing Economies and the Environment – The South-East Asian Experience*, McGraw Hill.

Menezes, J.P. and De Monte, L. (1994), "Quarrying Poses Threat to Dahanu". *The Times of India,* April 13, 1994.

Nagchaudhuri, B.D. (1983), *Introduction to Environmental Management*, New Delhi: Interprint Publications.

Nath, P. and Nath, S. (eds.) (1990), *Environmental Pollution: Conservation and Planning,* Vol. 2, Allahabad: Chugh Publications.

V.S. Department of the Interior (1976), "Surface Mining : Its Nature, Extent, and Significance", in Ronald Tank (ed.) : *Focus on Environmental Geology*, New York: Oxford University Press, p. 328.

14

Environmental Degradation and Possible Solutions for Restoring the Land

A Case Study of Magnesite Mining in the Indian Central Himalayas

AFROZ AHMAD

Global Situation

Throughout the world, mining activities generally have a marked environmental impact and cause a number of changes in key characteristics and processes of the ecosystem. For example, mining schemes may spark off changes in the structure, fertility and stability of soils, the quality of air, the quantity and seasonality of water supplies, the abundance or scarcity of biota and the areal extent of particular types of ecosystem, etc.

UNESCO's Man and Biosphere Programme, project 10, has undertaken to evaluate the undesirable effects of major engineering works of global concern, including mining, which have resulted in serious environmental problems (UNESCO, 1976). For example, the problem of soil deterioration due to gold mining in the lake country of California is of serious concern (Smith and Kay, 1986). The amount of dust containing MgO that is generated from a magnesite factory in Austria has had a serious effect on the wax tube of plant leaves: the crystalline structure of the wax tube fuse is lost and results in damage to the vegetation cover (Bermadinger, Grill and Golob, 1988). Mining has also seriously affected the Kawerong-Taba river system in Papua New Guinea due to excessive sedimentation and clearance of 220 hectares of tropical forests (Petre, 1978). Saline water intrusion, vibration and dusts from the Neyveli Lignite Mining Project of India have also contributed to degradation to the ecosystem (Banerjee and Natarajan, 1985).

Mining Scenario in India

In India, the total land area under mining is estimated at about 7.854 km^2, which constitutes less than 0.24 per cent of the land area of the country. In 1982, there were about 4,052 working mines in India, excluding those producing oil, gas, atomic energy, minerals and minor minerals. Of these, 2,854 were mining non-metallic minerals, 720 were mining metallic minerals and the remaining 478 were mining coal and lignite. Among the non-metallic minerals, magnesite mining plays an important role because of its major use in India's steel industry.

Mining must be one of the most destructive of all of man's activities on the environment. It destroys the original eco-system and replaces it with only an empty pit, a sterile wasteland, or both. Consequently, mined land represents a fascinating challenge to scientists who are attempting to find ways of restoring the original ecosystem.

This present study was undertaken to rehabilitate a damaged, mined ecosystem using an ecological approach for sustainable development.

Project Background

The magnesite industry in the environmentally sensitive Himalayas was initiated in 1971 with the immediate objective of mining magnesite by the open-cast method and processing it in shaft kilns to produce "dead burnt magnesite" or magnesite sinster. The mine is located at Jhiroli village in Almora district, Uttar Pradesh State, Central Himalaya. It lies between latitude: 29°45′30″ N and 29°47′30″ N and longitude: 79°44′E and 79°46′ E (Fig. 14.1). The area has a proved magnesite reserve of 3.9 million tonnes and this rich deposit stretches between the river Kali in the East and Alaknanda valley in the West (Valdiya, 1968). The deposit is located on the western mid-slopes of Jhiroli ridge which is 1,888 metres above mean sea level and extends over 2.9 km.

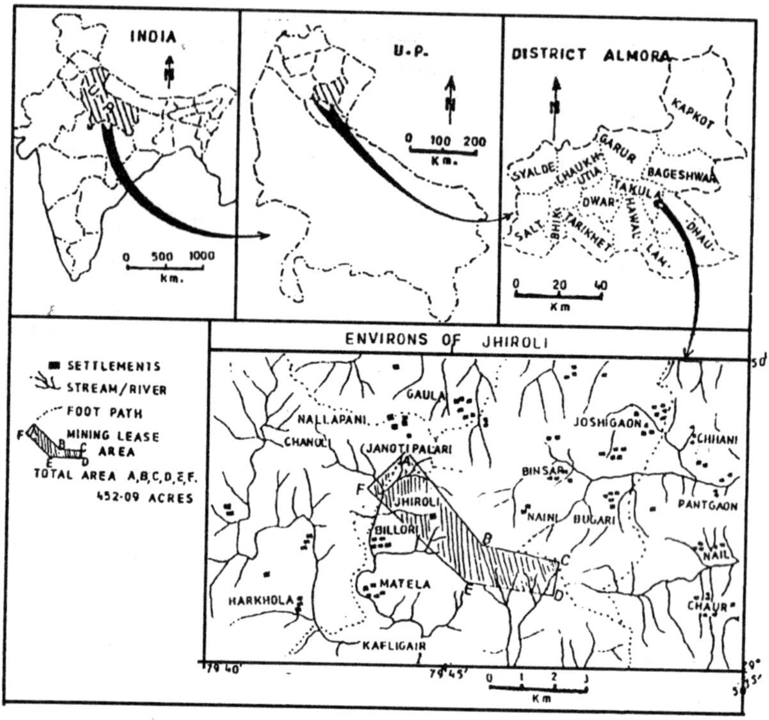

Fig. 14.1: Map Showing Location of Jhiroli Magnesite Mining Project in Indian Central Himalaya.

Terrain

The magnesite project is located in a hilly region where the elevation is 1,350-1,750 metres above mean sea level. The hill slope of the area is generally concave to convex with a general slope inclination range between 34° to 70°.

Socio-cultural setting

The land is presently used for agriculture grazing forests human settlement and as fallow ground. The main agricultural crops grown are rice, followed by wheat and coarse grains. There are 29 villages adjoining the project site and the area is fairly remote and unsophisticated.

Flora and Fauna

The main mining area is dominated by reserve forests, particularly pine (*Pinus oxburghii*): the hill top is mostly covered with xerophytic plant species. Wild fauna were once rich in the area but, after the mining industry was set up, ruthless denudation and changes in habitat forced the fauna to migrate to the opposite hill side adjacent to the mined area.

Climate

The area falls under what UNESCO would define as a temperate zone. Annual rainfall varies from 100 cm. to 330 cm. and generally occurs during the monsoon period from mid June to September. Winter sets in immediately after the monsoon and continues up to the end of February. In summer, nights are always cool and day temperatures ranges between 7.9°C to 30.2°C. In winter, wind speeds range between 1 km/h. and 20 km/h. with an average of 8 km/h. During the summer, wind speeds vary between 2.5 km/h. and 12 km/h. with an average of 6 km/h. The wind direction is from north-west to south-east in winter and from west to east in summer. The prevailing direction is from west to east.

Geo-hydrology

Presently, there is a general shortage of water apart from in the rainfed drainage channels, although even most of these remain dry except during rainy periods. The only aquifer present in the area is one natural perennial stream (Billori Nala).

The magnesite deposits in the project site are confined to the carbonate rocks. Magnesite dolomite are usually massive crypto-crystalline rocks and very dark to pale grey in colour. Magnesite itself is white to dirty white a specific gravity of 3 and a heterogenous texture. Chemically, magnesite is 40 per cent Mgo; 2.0 per cent SiO_2; 1.77 to 2.82 per cent Fe_2O_3 and 1.01 to 4.09 per cent CaO.

Environmental Impact Assessment of Mining

An environmental impact assessment (EIA) of magnesite mining operations in the environmentally sensitive Himalayas has been made.

The main environmental impacts are on the physical and chemical environments, i.e., earth's atmosphere, water and processes; biological components, i.e. flora and fauna, and socio-economic and cultural environments, i.e. landuse, recreation, aesthetic and human use value, and cultural status (Fig. 14.2). One of the most serious environmental impacts in the Himalayas is soils impoverishment and loss which leads to biological desert since the environment already has a very low biological potential, i.e. the plant and animal production and overall situation exceeds the natural capacity of the ecological conditions.

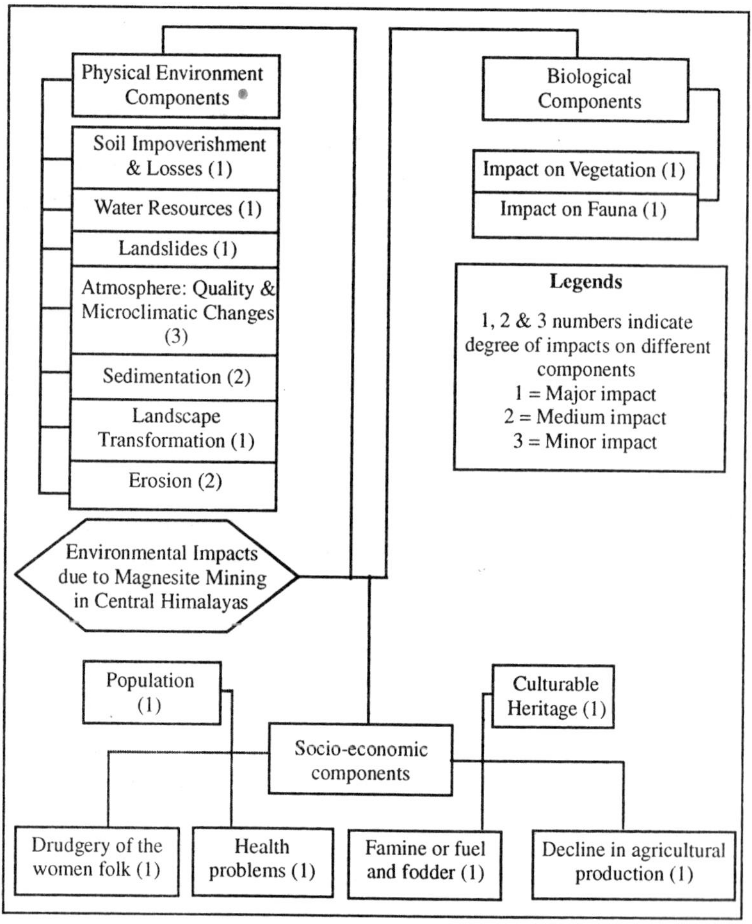

Fig. 14.2: Analysis of Environmental Impacts Associated with Magnesite Mining Project.

Soil Impoverishment and Loss

Soil quality, landuse and the beautiful landscape have all been radically affected by open-cast magnesite mining using power shovels and bulldozers. The first 25 cm. of top soils have been ruined by the operation of bulldozers and by getting mixed up with dolomite over-burden and rejects. Because of the steep slopes, the excavated scree and fine silt easily roll down and remove the soil cover. Ultimately it is deposited on the agricultural fields at the bottom, which leads to a reduction in plant biomass, in the land's carrying capacity for livestock, in crop yields and in human well being and, eventually, to the intensification of sterile biological desert. The rate of erosion (4 cm/d to 16 cm/d) and sedimentation (9 cm/d to 26 cm/d) are based on measurements made by installing two erosion pins (Table 14.1). The data represent mainly the month of July when significant rainfall occurred: rainfall in other periods of the year was scanty and insignificant. This means that data taken at four day intervals for both erosion pins have been used to draw the following conclusions. The data show clearly that the rate of sedimentation is higher than the rate of erosion at both recording stations. The main reason for this higher rate of sedimentation than erosion is the addition of soil that is carried from other, unrecorded, sub-drains in the area during the rains and finally gets deposited as silt in the main drainage channel. The development of a fan-shaped structure of eroded materials following the destruction of highly fertile agricultural land (terraces) clearly indicates the severity of the erosion. On the other hand, the heavy amount – 1,000 to 2,000 m'/d – of solid wastes such as over-burden, waste rocks, tailings, etc., mainly generated during mining, is leading to the siltation of

Table 14.1: Erosion and Sedimentation Rate of a Drainage Channel in Mined Area.

Erosion pin no.	*Date*	*Width of cross section (m)*	*Erosion (cm)*	*Sedimentation (cm)*
01	3.7.90	2.28	16	26
	7.7.90		06	11
02	3.7.90	43.50	04	09
	7.7.90		04	18

perennial water resources and agricultural lands in Billori and Matela villages, leading to a decline in agricultural production.

The project operation has also utilised 62.31 hectares of land from several villages located in the vicinity (Table 14.2). The local people were mainly dependent on agricultural practices for their livelihood and the utilisation of this land has decreased the total crop production and increased food problems among the local inhabitants.

Table 14.2: Land Acquired by Mining Industry from Villages During Project Execution.

Village	*Land (Acres)**	
	Acquired	*Proposed*
Jhiroli and Naini	102.45	0.63
Billori	1.20	11.83
Matela	1.55	1.18
Sinduri (Kafligair)	19.65	11.86
Baskhola	21.30	NIL
Karasibunga	7.14	1.36
Check Sakar Pathan	1.75	NIL
Total	**153.91**	**26.62**

* 2.47 Acres = 01 Hectare.

Source: Almora Magnesite Limited, Land Records (1986).

Restoration of Damaged Mined Ecosystems

The growth of the world's population especially in India, and the associated problems this brings, such as nutrition and supply of other essential goods, makes it necessary to cultivate all available areas to give high yields and simultaneously maintain soil fertility. This means it is necessary to rehabilitate all ecosystems that have become degraded as a result of mining activities.

In view of this, strategies have been developed and launched to restore damaged mined ecosystems in the Central Himalayas using an ecological approach which takes into account the inter-relationship between water, land, air and all living things including people; and involves all user-groups in the management and use of locally designed, suitable and acceptable strategies for resolving environmental problems to ensure ecologically sustainable development.

Execution of Environmental Impact Assessment

A detailed Environmental Impact Assessment (EIA) was carried out and an Environmental Impact Statement (EIS) prepared as an independent project of the GB Pant Institute of Himalayan Environment and Development. On the basis of the EIS, it was inferred that environmental degradation is of serious concern in the area and, therefore, rehabilitation of the critically degraded ecosystem is urgently needed. In view of this, a joint project between the GB Pant Institute and Almora Magnesite Limited was launched in 1989, to be equally funded by both organisations. A Mining Rehabilitation Review Committee (MRRC), consisting of experts from various fields, was also constituted in consultation with both parties to give advice on successful rehabilitation programmes, monitoring of the work and also for resolving conflict (if any arises) between the two participating parties.

The restoration programme was divided into two sections:

1. Selection of a site for restoration.
2. Restoration approach.

Selection of a Site for Restoration

On the basis of topography and degree of degradation, the mined area was categorised as follows:

(a) Very highly damaged land or very highly sterile biological desert: this includes the landscapes created after over-burden and ore-rejects were dumped there, which are devoid of plant cover.

(b) Highly degraded land or highly sterile biological desert: this includes land which has been disturbed by toxic emissions generated from the dead burnt magnesite plant where the raw magnesite is processed. These processes have serious affected vegetation cover as well as the top soil layer.

(c) Much less degraded land or living biological desert: the land under this category is the least degraded from mining although vegetation cover is still poor.

Site Selection

A spoil dump of 2 hectares in area was selected for restoration out of land categorised as very or highly damaged. The salient features of this land are:

- Rainfall is scanty, undependable and confined to a few days during July and August.
- Poor soil moisture due to the presence of boulders and coarse solid particles which are very influential in creasing the infiltration rate.
- Drainage is excellent about 90 per cent of the rain water is drained into the Billori stream water table by deep percolation.
- The soils are poor in nitrogen and organic matter and alkaline in nature due to the high percentage of magnesium present.
- Irrigation facilities are poor.
- Temperatures are high during the summer months.
- Uncontrolled grazing and lopping by the villagers are common features.

Restoration Approach

On the basis of the Environmental Impact Assessment (EIA), it became evident that the landscapes affected by mining may not be developed directly for agricultural purposes due to poor soil cover, poor organic matter, the presence of big boulders, the alkaline nature of the soil and coarse soil particles – not less than 10 mm in size. Therefore, the first priority was to afforestate and to use the site for, agriculture only after a few years' rotation of forestry and improvement in soil physico-chemical properties. The whole restoration programme has been divided into the following sub-categories:

1. Preparation of land
2. Afforestation of spoil dump
   - 2.1 Nursery development
   - 2.2 Stabilisation of soil by grasses
   - 2.3 Plantation of suitable free species
   - 2.4 Manuring and fertilization

   2.5 Irrigation
   2.6 Biological fencing
3. Reviving fauna
4. Control of erosion and sedimentation
5. Control of air pollution
6. Environmental awareness campaign.

The costs involved in launching the different restoration measures, i.e. preparation of land (terracing, adding farmyard manure and 5 cm. of top soil), nursery irrigation, installing a rain water harvesting tank, part time manpower expenses and other miscellaneous expenses are shown in Table 14.3.

Table 14.3: Amount spent (July 1989 – 31 May, 1990) by GB Pant Institute and Almora Magnesite Limited for Different Rehabilitation Activities

	Amount (Rupees)	*Spent/contributed*
Activities	*GB Pant Institute*	*Almora Magnesite Ltd.*
Preparation of land (terracing levelling, contour)	NIL	60,000.00
Manuring	25,990.00	NIL
Nursery	1,950.00	NIL
Water management	13,578.75	NIL
Miscellaneous expenses (video, filming, photography, etc.)	6,819.00	NIL
Board and lodging	NIL	5,000.00
Part-time man power expenses (plantation, land husbandry, etc.)	61,655.90	32,307.50
Mining Rehabilitation Review Committee	NIL	10,000.00
Total Rupees	**109,993.65**	**107,307.50**

Preparation of land: Approximately 2 hectares of disturbed mined land was prepared by terracing and maintaining slopes with a 20° to 30° steep gradient. As top soil cover was completely absent, a 5 cm. thick layer of top soil and a uniform 4 cm. thick layer of farmyard manure was applied over the entire area before starting the afforestation programme. The whole process of land preparation was carried out by local people from villages adjoining the mining project. They were paid Rs. 24 per day although some villages, including some NGOs, participated free of charge once they were made aware and understood the direct benefits for them of protecting their land and water resources from further degradation.

Afforestation on spoil dump: The rehabilitation area is under heavy pressure from grazing, browsing and environmental degradation. It was therefore decided to use only those tree/vegetation species that can grow under such harsh environmental conditions. Fast growing tree species were given priority, taking into account local climate and their value in meeting the growing demand for fuel and fodder, and in improving soil physico-chemical properties. Plantations of the more valuable species, especially in agro-forestry configurations (possibly with food crops), play a significant role in rehabilitating degraded land. Under this afforestation programme, emphasis has been put on the objectives of keeping external inputs low and sustainable forestry. The whole afforestation programme has been divided into the following categories:

Nursery: A nursery has been developed near the rehabilitation site with good irrigation facilities. The main aim is to provide saplings for the plantation site and to monitor the impact of dust generated from the dead burnt magnesite plant.

Soil stabilising grasses: The soil of the plantation site is highly inhospitable due to poor soil cover, poor moisture and the high content of MgO (40 %) which makes the soil naturally very high in alkaline. Because of this, afforestation was first preceded by grass planting, especially on slopes, to stabilise uncompacted soil and to prevent erosion and landslides. As the rainfall in the area is undependable and mainly concentrated during July-August, 18 tonnes of grass germules representing 3 species were transplanted during July 1989. The main species tried and found successful include: Love grass (*Erogrostis curvula*), Kudji (*Puraria nirsuta*) and Kikui (*Pennisetum clandesitnum*). Besides these Rumex (*Rumex hastatus*) regenerated after one month of starting the plantation programme. The performance of grasses was very promising and the first harvest was taken by the villagers after 6 months of planting. In this time the soils, including the unstable spoil dump, were stabilised.

Plantation of suitable tree species: Trees were planted on a site of approximately 2 hectares in area including slopes, on which over-burden, including big boulders and very poor alkaline soil was dumped. The choice of suitable species was dependent on those that can survive under such harsh environmental conditions.

Approximately 10,000 saplings were planted one month after the grasses were planted. The planting was carried out immediately after a good shower of rain (i.e. in July-August). Much emphasis was given to mixed forests rather than monoculture to avoid complete destruction of the forests in the event of tree-specific pests. Mixed tree species are also better able to meet growing demands for fuelwood, fodder and for improving soil fertility. About 5,000 saplings are now successfully showing good growth. The major tree species tried and found successful for plantation on degraded mined land include: Banj (*Quercus leucotrichophora*), Kareel (*Acacia nilotica*) and Pine (*Pinus roxburghii*) (Table 14.4). Planting fast growing species of trees and grasses will solve the villagers' problems of fuel, fodder and timber, besides improving the environmental conditions (Table 14.5).

Table 14.4: Tree Species Tried and their Survival Rate on Mined Spoil Dump.

Vegetation		*Species under plantation*	*Survival (%)*
Trees	1.	Banj (*Quercus leucotrichophora*)	10
	2.	Kareel (*Acacia nilotica*)	80
	3.	Pine (*Pinus roxburghii*)	90
	4.	Poplar (*Populus ciliata*)	50
	5.	Panger (*Aesculus indica*)	30
	6.	Bakain (*Melia azedarach*)	70
	7.	Bhimal (*Grevia optiva*)	30
	8.	Silver Oak (*Grevelia robusta*)	75
	9.	Timil (*Ficus auriculata*)	30
Grasses	1.	Love grass (*Erogrostis curvula*)	95
	2.	Kudju (*Puraria hirsuta*)	25
	3.	Kikui (*Pennisetun clandesitnum*)	90
	4.	Rumex (*Rumex hastatus*)	90
Shrubs	1.	Rambansh (*Agave sissiliosa*)	90
	2.	Kilmora (*Berberis asiatica*)	80

Use of transplantation techniques: A new transplanting technique was used in the area for the tree species that have coppicing properties and grow naturally in the vicinity, i.e. Pine (*Pinus roxburghii*) and Banj (*Quercus leucotrichophora*). This technique involved transplanting small trees from where they grew naturally and from the mining site where they were under threat from blasting and denudation, and replanting them at the rehabilitation site, with care taken not to damage their root systems. Special care was also taken to prepare a suitable pit and to irrigate well after transplanting.

Manuring and fertilisation: The spoil dump is very poor in organic matter so one application of farmyard manure was made during the land preparation to help increase soil moisture capacity and to improve the soil's physico-chemical properties. A second application of farmyard manure was made after 6 months of raising the plantation. The farmyard manure was bought from nearby villages where the farmers use it as fuel. Consequently, alternative sources of energy, such as LPG (Light Petroleum Gas) were given to them at a subsidized rate by the government to ensure that manure would always be available for future use in the rehabilitation programme. Inorganic fertilisers were not applied, in line with the principles of keeping external inputs low and sustainable forestry.

Table 14.5: Anticipated Economical/Environmental Uses of Planted Trees in Magnesite Mine Spoil Dump Study Area.

Vegetation		*Species under plantation*	*Environmental/Economic Significance*
Trees	1.	Banj (*Quercus leucotrichophora*)	fodder, medicine, soil fertility, soft word
	2.	Kareel (*Acacia nilotica*)	timber, fuelwood, poles, tannin, soil stabilisation
	3.	Pine (*Pinus roxburghii*)	timber, soft wood
	4.	Poplar (*Populus ciliata*)	soft wood, fodder, shelterbelts
	5.	Panger (*Aesculus indica*)	soft wood, fodder
	6.	Bakain (*Melia azedarach*)	timber, fodder, soft wood, soil fertility
	7.	Bhimal (*Grevia optiva*)	fodder, fibre
	8.	Silver Oak (*Grevelia robusta*)	timber, shelterbelts
	9.	Timil (*Ficus auriculata*)	softwood, fodder
Grasses	1.	Love grass (*Erogrostis curvula*)	fodder, soil fertility
	2.	Kudju (*Puraria hirsute*)	fodder, soil fertility
	3.	Kikui (*Pennisetun clandesitnum*)	fodder
	4.	Rumex (*Rumex hastatus*)	soil stabilisation
Shrubs	1.	Rambansh (*Agave sissiliosa*)	fibre, biological fertility
	2.	Kilmora (*Berberis asiatica*)	medicine, biological fencing

Irrigation

It was observed that scantly and undependable rains coupled with poor soil quality, excess run-off and poor infiltration rates make the area poor in soil moisture. Irrigation under such conditions plays an important role in successful plant growth. Rain water harvesting was found to be the only solution for irrigating the plantations. One rain water harvesting tank (8 metres × 4 metres × 1 metre in size) was constructed to harvest rain water during the monsoon period.

However, the monsoon contributes only 27 per cent of irrigation to the plantation. Mechanical irrigation of the plantation area by drip or sprinkler methods is not possible due to the lack of available water. Water had to be applied manually which is labour intensive but the only solution under such harsh environmental conditions. The area was hand watered twice: first after the initial planting and again 6 months later.

Biological Fencing

The rehabilitation site is heavily grazed and lopped due to poor vegetation cover in the area and the tradition of having as many goats and cows as possible as a source of income. Most animals depend on grazing and lopping and are seldom stall-fed. Under such conditions, protection of the plantation area from domestic animals, especially goats and cows, was essential, particularly for the first few years. A living fence was planted along 95 per cent of the periphery of the rehabilitation site using shrubs such as Kilmora (*Berber is asiatica*) and Rambansh (*Agave sissiliosa*) together with one metre high, locally available stones for checking the entrance of animals and for promoting natural regeneration of the vegetation.

All these rehabilitation programmes were jointly implemented by Almora Magnesite Limited (Mining Industry) and the GB Pant Institute of Himalayan Environment and Development. Other environmental safeguard measures such as reviving fauna, control of erosion, landslides and sedimentation and control of air pollution have been implemented by the factory itself. All these measures had a direct effect on the economic performance of the factory but not significantly because the government subsidised the environmental safeguard measures. The government also played an important role by implementing environmental legislation.

An inspectorate known as the Pollution Control Board was set up at state level to monitor the Environmental Management Plan (EMP) of the factory. The factory has to prepare an EMP each year and to have it approved by the Ministry of Environment and Forests before further land can be granted. Therefore a sound EMP for the factory is a pre-requisite for further leases to start mining.

Reviving Fauna

To revive fauna in the rehabilitation area, the following measures were implemented by the mining industry:

- Artificial nesting places were introduced and berries, other fruit trees and shrubs were planted to make a suitable habitat to attract birds and wildlife from the adjacent forests.
- To revive fish fauna in Billori nala and other water resources available downstream, a stabilisation pond was constructed to minimise the toxic effect of effluent discharged into the rivulets.

Control of Erosion, Landslides and Sedimentation

The erosion, sedimentation and landslides problems associated with mining have seriously affected the productivity of agricultural land downstream of the project. The following measures were implemented by the factory to combat these problems:

- A series of check dams and retaining walls were constructed at certain places out of locally available rubble masonary to arrest rolling debris and prevent sedimentation of perennial water resources and agricultural land down-stream.
- Garland drains are currently being constructed to prevent the uncontrolled flow of precipitation, to prevent village livestock from entering and to promote natural regeneration of vegetation.
- The scale of open-cast mining and deep hole mining has been minimised.
- Substantial support is being provided to check the large scale movement of the over-burden dump slopes.

Control of Air Pollution

The amount of dust generated during crushing and grinding of raw magnesite in kilns accounts for 6 tonnes/day. It is mainly composed of MgO, Fe, A1, SiO_2 and Ca, besides Toxic emissions such as SO_2 and NO_2 which cause widespread air pollution and resulting degradation of agricultural land and vegetation cover. The following measures have been undertaken to check the menace of air pollution:

— The dust is being collected by fixing filter bags/blower cyclones for final collection.
— The dust containing trace elements such as Mn, Fe, A1 or Si is being used to make bricks for building and for levelling the pits.
— The slack height of dead burnt magnesite plant has been increased to reduce the concentration of suspended particulate matter.
— To reduce the dust menace, wet scrubblers have been introduced at the magnesite hand loading site and water is sprayed on the road and dump yard. Water is collected in tankers from a nearby, glacier-fed perennial river which is about 13 km. from the site.
— A shelterbelt of fast growing Poplar (*Populus ciliata*) and Silver Oak (*Grevelia robusta*) was planted near to the magnesite processing plant to reduce the spread of toxic emissions generated during the processing of raw magnesite and also to act as a sound-proofing barrier.
— An absorption muffler is being used to reduce the noise caused by the blower and pump.
— The noise and dust generated during blasting has been minimised by restricting the number of charges per day, keeping the delay intervals down to 17 milli-seconds, reducing the number of charges per hole, minimising the blast size, limiting the number and size of blast holes and the frequency of blasting, and maximising the charge depth.

Environmental Awareness Campaign

An environmental awareness campaign was launched by the Institute which invited non-governmental organisations to inform the local population about the management of resources in the hill area. Two campaigns were organised in Matela village and the factory compound to which children, adults, farmers, pastoralists and mining personnel were invited to learn about the sustainable use of resources and environmental conservation, with particular emphasis on the participation of local people in the rehabilitation programme.

REFERENCES

Banerjee, S.K. and Natarajan, A. (1985), Environmental Impact Assessment and control measures adopted in Neyveli Lignite Project in *Proc. 2nd World Congress on Engineering and Environment*, Delhi, India, November 7-9, pp. 1-25.

Bermadinger, E., Grill, D. and Golob, P. (1988), Influences of different air pollutants on the structure of needle wax of spruce *Picea abies* (1) Karsten. *Geo Journal* 17.2 pp. 289-93.

Petre, T. (1978), Environmental Aspects of River Basin Management in a Humid Tropical Country: Papua New Gumea In: *Proc. Water Techn.* Vol. 10, No. 3-4, pp. 335-38.

Smith, R.F. and Kay, B.I. (1986), Revegetation of serpentine soils: difficult but not impossible, *California Agriculture,* Jan.-Feb. Issue, pp. 18-19.

Stiles, D. (1988), Arid land plants for economic development and desertification control. *Desertification Control Bulletin* No. 17, pp. 18-21

UNESCO, MAB. (1976), *Expert consultation on Project 10: Effects on man and his environment of major engineering works*, MAB, Report Series No. 37, p. 46.

Valdiya, K.S. (1968), Origin of the Magnesite deposits of Pithoragarh district. Kumaon Himalaya, *Eco. Geol.*, No. 63, pp. 924-34..

15

Brick Kiln Industry and Related Land Degradation in Rural-Urban Fringe of Delhi

R.B. SINGH and JAG MOHAN

Introduction

Land, whether it is inherited, allotted, bought or captured is the most basic resource of agriculture production. Moreover, in a developing agrarian economy, this fact assumes even greater significance as there is a need to bring the maximum possible area under some sort of vegetation to the maximum potential of that particular land area. One of the most serious threats to food security is the increasing degradation of agricultural land from natural or human-induced changes due to the physical, chemical, and biological processes. The word 'degradation', from its Latin derivation, means reduction to a lower rank. The term rank relates to the present and possible uses, while reduction implies a problem for those who use the land. When land becomes degraded its productivity declines, unless steps are taken to restore productivity and check further losses (Blaike and Brookfield, 1987).

The rural-urban fringe is "the zone which lies immediately outside the city area and has strong interaction with the city proper in terms of daily communication, exchange of goods and services and bears an urban reflection on the physical, occupational and demographic structure. It is rural compared to the typical urban of the city area and urban to the typical rural land-scape" (Nangia, 1976). This encircling belt around a city also serves as a hinterland to the urban core catering to its daily essential needs like milk, poultry, fruits, vegetables, cereals and grains. Besides these essentialities, many heavy goods like bricks, construction materials etc. are also a part of the fringe's endowments to its urban core. Bricks, an essential feature of urbanisation in the developing countries, is a heavy commodity and cannot be carried over long distances without an increase in costs. Thus, the industry is forced to locate itself near the market, that is the city. The fertile agricultural land having good amount of clay and loan is considered the best for bricks. The irony of the situation is that the soil which could have provided a fairly large amount of agricultural produce for many more years to come is quarried in a span of 3–4 years for about 1.2 mts. – 1.5 mts. in depth in order to cater to the developmental requirements of urbanisation. Swelling urban population has necessitated vast lands to be quarried, so as to cater to the corresponding escalation in demand for brick and construction materials. Extensive areas are quarried for clay that is needed to mould the bricks. The kiln itself occupies approximately 0.8 hectares of land. The consequent landuse change has a telling impact on the land quality and land status. The quarried land often becomes degraded to such an extent that its restoration to a cultivable order is quite difficult. The issue of land degradation in Delhi's rural areas is one of the burning topics today as large stretches of agricultural land are being lost continuously to this irreversible developmental process.

Study Area

The National Capital Territory of Delhi comprises an area of 1,483 sq. kms. It is situated between the Himalayas and Aravalli ranges in the heart of the Indian subcontinent. It lies between 28°25 and 28°53 north latitudes and 76°50 and 77°22 east longitudes. It is surrounded by the State of Haryana on three sides and to the east, across the river Yamuna, by the State of Uttar Pradesh. Rural Delhi is divided into five Development blocks, i.e. Shahdara, Alipur, Kanjhawala, Najafgarh and Mehrauli (Fig. 15.1). The headquarters of the development blocks are located at the place after which they are named. The Kanjhawala block is one with the maximum concentration of brick kilns.

The Kanjhawala Development block is an important part of rural Delhi. It was earlier known as Nangloi and after delimitation in 1971, it became a separate block by itself. It is situated in the north-west part of the city. The whole block consists of 39 villages of which 27 come under Delhi tehsil and 12 under Mehrauli tehsil. The block has a total area of 17,420.79 hectares. The Kanjhawala block comes under

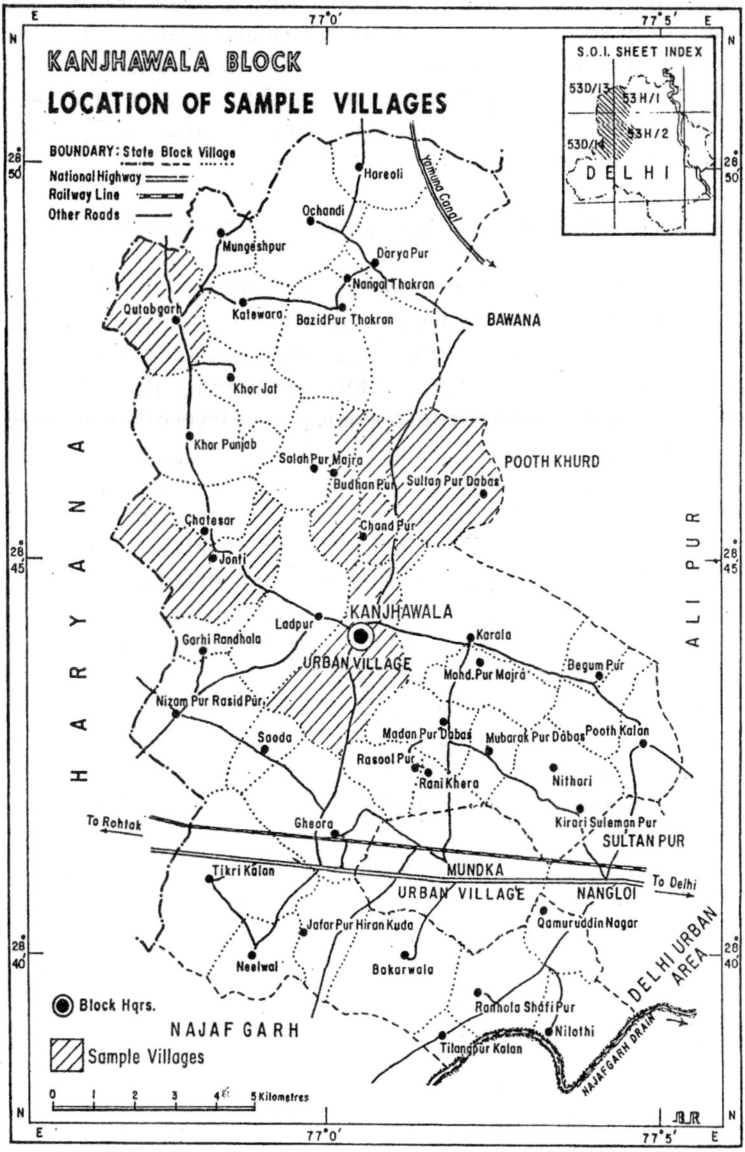

Fig. 15.1: Kanjhawala Block Location of Sample Villages.

Source: Compiled from 1:50,000 topographical Sheets Village Boundary from Delhi Census Map.

the Bangar or the old alluvium tract consisting of sandy loam and fine loamy soil. The soil is fertile with high moisture. The total population of all the villages is 1,77,666 of which males are 97,592 and females are 80,074. The sex-ratio is 820 females per thousand males. On the whole, the Kanjhawala block comprises 18.7 per cent population of the total rural population in Delhi.

Urban Development and Metropolitanisation Process

"The history of urbanisation in India reveals, broadly, four processes of urbanisation at work. These include the socio-cultural process, politico-administrative process, economic and geographical processes." (Ramachandran, 1989). Throughout the historical period, these processes were in operation. "The level of urbanisation was remarkably stable till about 1900 A.D., throughout the Indian history. It was never higher than 12 per cent or lower than 5 per cent of the total population. Since 1900, however, there has been a slow but steady upward trend in the level of urbanisation. By 1981, India's urban population of 159 million, constituted about 24 per cent of total population. Currently, every year about 3 million rural people migrate to the cities, and this flow is also showing a steady increase" (Ramachandran, 1989).

The chief reason for the rise in urbanisation is the rural-urban migration. While non-availability of jobs and food acts a "push" facto for the rural population, job availability in the urban area acts as the main "pull" factor. India has witnessed a new form of urban growth, namely, metropolitanisation. The 219 Class I cities with 12 million Class cities (in 1991) have accounted for more than 60 per cent of urban population. Along with rural-urban migration, migration from small towns to big cities has also contributed much to the metropolitan development (Table 15.1).

Delhi, the capital city of India, has not been left devoid of the urban and metropolitan processes. Delhi, which in the pre-Independence period, was limited within the walled city, spread to the boundaries of Uttar Pradesh and Haryana in the post-Independence period. The Delhi Master Plan of 1990, plans to include the neighbouring districts of Uttar Pradesh which are Ghaziabad, Loni and NOIDA in the Delhi Metropolitan Area (Ministry of Urban Development, 1990). This is the consequence of the increasing population pressure on the scarce resources of Delhi.

Table 15.1: Delhi at a Glance.

	Years		
Geographical Area (Km²)	*1980*	*1995*	
Total	1485	1483	
Rural	1039	798	
Urban	446	685	
		Years	
Population (In Millions)	*1981*	*1991*	*1996*
Total Population	6.22	9.42	11.224
Rural	.45	.95	1.13
Urban	5.77	8.47	10.09

Source: Delhi at a Glance, Directorate of Economics and Statistics, Government of National Capital Territory of Delhi, 1996.

Urban Delhi in 1981, accommodated about 1.15 million households. The next two decades would add another 1.3 million households. At present also, there is a shortage of 0.3 million houses. About 1.62 million new housing units would be required during the period 1981-2001. (Table 15.2).

Table 15.2: Housing Requirements in Delhi.

Years	*New Housing Required ('000)*	*Average per year ('000)*
1981-1886	323	65
1986-1991	379	76
1991-1996	434	87
1996-2001	483	97

Source: Ministry of Urban Development, Master Plan for Delhi, 1990, DDA.

In order to cater to the increasing housing demands in the past two decades, the Delhi fringe area saw the emergence of a number of brick kilns so as to fulfil the demands of construction materials and bricks. The bordering villages, which were earlier engaged only in agricultural activity, underwent a radical landuse change catering to the urban as well as rural demand for bricks, in order to house the thousand immigrants into the city. These immigrants live mainly in slums and form 40 per cent of Delhi's total population and require housing facilities in the cities. The process has, thus, begun for the metropolitanisation of the rural villages.

Contribution of Rural-Urban Fringe Towards Urban Requirements

Urbanisation, to manifest itself, requires intrinsic transformation in rural landuse and occupational structure. Changing agricultural land-use in the village is the first sign of impact of urban demands on village economy. "The city offers a market for products that a village is in a position to supply, such as milk, vegetables, flowers and fruits. A few enterprising farmers in the village may perceive and take advantage of this opportunity, leading eventually to daily contact with the city" (Ramachandran, 1989). Besides milk, vegetables and fruits, and other products necessitating the landuse change is the demand for eggs and meat from the urban areas. The rural landscape is now experiencing the emergence of large poultry and cattle farms. These products are also perishable and hygienic. Cleaning and packaging of meat and meat products is still in the developing stage in India. For this reason, these products are also being supplied by agriculturists of the fringe areas. The chief reason for such a change in the agricultural landuse around a city is the small gestation period and quick earning for the farmers. Thus, with ready market, more money and that too in a short time, a farmer opts for vegetables, fruits, poultry and meat over cereals and grains. Besides the eatables, other products supplied by the area are the heavy construction materials like bricks, stone, timber, sand and gravel etc. Being heavy, these cannot be carried over long distances, without adding to its price. The industries processing these goods are located just outside the city limits in the villages. Brick industry is one of the most prominent of these.

Acquisition of village land by the Delhi Development Authority (DDA) for development of urban residential areas increases the land-use transformation. Farm houses, godowns for industrial goods, foodgrains and small industries also mark the change in land utilisation from rural to urban. The agricultural land, being low in costs and accessible to urban centres, is ideal for location of farm houses, etc.

The urban requirements of skilled, semi-skilled and unskilled labour are also being fulfilled from the surrounding villages, bringing a change in the occupational pattern of the fringe zone. A lot of this migrated population from the other states also work in the brick kilns located in the fringe area.

The whole of Kanjhawala block is dotted with brick kilns which are the most integral part of these rural villages. The fertile land of villages provides a good resource base to brick kilns. Though agriculture is also dominating in several villages, it is hard to distinguish between the brick kiln dominating villages and the agriculture dominating villages. Agriculture is the basic activity in which most of the villagers are participating. As the social norms keep on changing, the joint family system is disintegrating leading it to a change in economic activities. The young generation is unable to continue the agricultural activities and they lease out their good fertile and productive land to the brick kilns on a 4-5 year contract basis at the rates varying from Rs. 0.2 million to 0.4 million, all paid at the beginning of the lease in cash. Once the lease is over they are left with barren, unlevelled and unproductive land devoid of all top soil and humus. On the other hand, the older generation is still continuing with agriculture to produce good quality of wheat, rice, other foodgrains and vegetables. With the commercialisation of agriculture most of the villagers are interested in wheat farming, vegetables and fruits. Local grain markets provide a good market base for their products. Marigold and rose gardening and poultry are some other activities which are gaining grounds.

Further to escape from the urban development controls and zoning regulations, many industries are now being set up within the '*Lal Dora*' (habitation) area of the village, eroding rural character of the village and adding to the haphazard settlement growth and environmental pollution. All these industries are responsible for the changing occupational structure of the villages. Small and ancillary industries are mushrooming in the surrounding villages of the Census towns. Under the various rural development programmes, rural youths are getting trained for various job-oriented courses.

Status of Brick Kiln Industry

At present, there are 150 brick kilns located in 20 villages of the block. The maximum growth period was of 1980-85 in which 55 brick kilns came up (Table 15.3). This maximum growth of brick kiln was due to the Rohini Residential Development Project, one of the biggest planned settlements in Asia. The Rohini Development

Project gave a boom to the brick industry raising the demand for bricks to a new high. This attracted a lot of brick manufacturers from the neighbouring states of Uttar Pradesh, Haryana and Punjab. During the same time, political disturbances forced, brick manufacturers from the State of Punjab migrate to Delhi. The industry saw a decline in the 90's due to the non-renewal of licencing to brick manufacturers, as the industry has been classified as one of the air polluting industries, releasing toxic effluents and smoke into the atmosphere by burning wood, coal and any other inflammable material as burning fuel.

In order to fulfill the growing demand of bricks new sites are captured. But due to rapid urbanisation and high demand of land for residential purposes, brick kiln spread has been abruptly halted. From 1990-95 only 14 brick kilns came up and there is no hope for the brick kiln owners to get new land sites for soil quarrying. This stagnation of brick kilns industry in the Kanjhawala block can end only by shifting this industry further towards the Haryana border.

Table 15.3: Growth of Brick Kilns in Kanjhawala Development Block.

Year	*No. of Brick Kilns Established*	*Total*
1970-75	16	16
1975-80	40	56
1980-85	55	111
1985-90	25	136
1990 and above	14	150

Source: Delhi Brick Kiln Association.

Burnt clay bricks are the most important building units of construction in India. As per the latest estimate of National Buildings Organisation (NBO), annual production of bricks in India is 51,200 million (1990) from 45,000 small/cottage scale units distributed throughout the country. Over 2.5 million workers are employed in this industry and most of the units are operated manually (Aslam, 1993). In Delhi alone, there are 340 brick kilns and on average each brick kiln produces 4.0 million bricks annually. With the expansion of urban city limits, brick kilns have been shifting toward the peripheral areas in the outskirts.

In Delhi the first commercial brick kiln was set up in the mid-1960s. As the demand for bricks increases, the number of brick kilns also increases. Delhi provides job opportunity and residential security to migrants from the poverty ridden areas of the whole of north India. This ultimately forced Delhi Development Authority (DDA) to set up new colonies and also regularised the illegal colonies set up by private developers. The 1980s decade saw the brick industry reach its zenith as hundred of brick kilns mushroomed in all the five development blocks. In case of Kanjhawala block, first brick kiln was set up in 1967 in the Bawana village named as Rana Brick company. At present, the whole of block is dotted with 150 medium and small kilns.

Impact of Brick Kiln on Land and Landuse Change

To know the impact of clay quarrying on land and land use change, six villages were selected in the Kanjhawala Development block, namely, Kanjhawala, Chandpur, Sultanpur Dabas, Jonti, Qutabgarh, and Budhanpur. These villages were selected on the basis of a detailed pilot survey and data were collected from various sources. Only those villages were selected in which 5 or more brick kilns were operating. By choosing this criteria it was easy to identify the exact condition of the land degradation (Table 15.4).

Table 15.4: Consumption of Land in the Study Area.

Number of Brick Kilns	*Total Land Consumed Till Date (in Hectares)*	*Land Under Quarrying (in Hectares)*
63	513.2	333.2

Source: Personal Survey.

There are at present 63 brick kilns in the six selected villages. Each brick kiln, on an average, requires at least 2.2 hectares of land at any given time; of this .8 hectares are taken up by the kiln itself and the rest is required for clay quarrying. Thus, every kiln, in a span of four years, removes top soil from at least 1.4 hectares of land. Till date, 513.2 hectares of land have already been quarried, while 333.2 hectares are at present leased by brick kilns for quarrying purposes and are under a severe threat of soil erosion and land degradation. With Hon'ble Supreme Court ordering the closure or shifting of brick kilns from the city limits, the question of areas most likely to be the potential sites of re-location has also arisen. The States of Haryana (Bahadurgarh, Gurgaon, Sonipat) and Uttar Pradesh (Ghaziabad,

Loni, Meerut) located in the National Capital Region are the areas bearing an immense potential for setting up of brick kilns from the point of view of distance from the market as well as availability of raw materials.

The most visible economic impact of land degradation is on agricultural productivity. Any reduction in productivity of land resources has an enormous effect on the socio-economic conditions. The immediate effect of land degradation includes reduced crop yields, increasing need for agricultural inputs and decreasing profits and reduction in the value of land.

Standard of living, which depends upon the farmer's income greatly declines with decrease in the per capita availability of land. Increasingly, less remunerative subsistence farming system force the peasants to leave the land uncultivated which then tends to become barren. All the above said reasons ultimately result in shifting the land resources from agriculture to other uses including industry which results in environmental degradation. This is more common in Kanjhawala block where there were 150 brick kilns operating till September 1997 (Table 15.5).

The decrease in agricultural land is complimented by an increase in cultivable wasteland in the decades 1971-91. The decrease in agricultural land was more than 13 per cent while increase in cultivable wasteland, which includes common land and pastures, was 63.08 per cent.

Table 15.5: Landuse Pattern in the Study Area 1971-1991.

(*in Hectares*)

Year	*Agricultural Land*	*Cultivable Waste*	*Area not Available for Cultivation*
1971	2706.6	460.52	426.11
1981	2788.8	500.8	327.2
1991	2350.59	751.03	494.78

Source: District Census Handbook, Census of India, 1971, 81, 91.

Area not available for cultivation decreased by 23.21 per cent from 1971 to 1981 and after a decade it increased by 49.38 per cent in 1991. After mid-70s various agricultural development programmes initiated by the government brought down barren and fallow land. But due to spread of urbanisation process, the area not available for cultivation increased by approximately 50 per cent. This area includes settlements as well as brick kilns and land under quarrying. Table 15.6 clearly depicts the relationship between growth of brick kilns and the simultaneous decrease in agricultural area.

Table 15. 6: Agricultural Landuse Status in the Study Area.

(in Percentage of Households Surveyed).

Households Leasing did Lease	*Land-use did not Lease*	*Agricul-tural Sell*	*Reason for Leave*	*Out their Agri.*	*Lease is Over Comm.*	*Type Leaving Subsis.*	*Agriculture Lands*		
							Eco	*Inability to Continue*	*Others*
72.9	27.1	30.9	33.33	38.9	52.43	47.56	40.8	28.77	30.4

Source: Personal Survey.

Nearly 73 per cent of the households surveyed had leased out their land. Such a high percentage is due to the effect of man's greed, inability to provide workforce for agriculture as well as levelling of fields. After the land is reclaimed a large percentage of land-owners again practice agriculture using a very high dose of fertilisers etc. An equal percentage prefers to sell it off, or leave this land barren for a few years. More than half (52.43%) of the households surveyed in the study area practise commercial agriculture, while the rest use the produce for their own subsistence purposes. Nearly forty per cent of the households whose traditional occupation was agriculture, have left it to earn hard cash instantly and a lot more from agricultural returns by giving it for soil quarrying. Another 30 per cent have given land for brick kilns as there was no one to continue agriculture in the family and the head person being too old and weak to do the work himself, the rest quoted reasons other than the said two.

Land Management and Reclamation Relating to Brick Kiln Areas

Like all the other big development issues, brick industry also remains in controversy and poses a big question mark on the sustainability of its surrounding physical and cultural environment. The traditional method of moulding bricks by hand requires good quality clay. This initiates the process of land degradation. To minimise the pace of land quality deterioration and to reclaim the already degraded land, various Governmental and Non-Governmental Organisations have devised various alternatives. However, with the lack of awareness and financial constraints, these alternatives are not in regular operation in the study area. In India, good quality clays are also the sites of brick industry. Extensive research work carried out at Central Building Research Institute (CBRI), Roorkee and elsewhere during the last few decades has established that bricks of acceptable quality can also be prepared by using usar, red and highly silt soils by adopting special techniques.

Utilisation of Flyash/Flyash Bricks as Sustainable Initiative

Fly ash utilisation is a national problem and has attained serious proportion. The thermal power generating capacity in India is around 40,000 MW disposing of about 35 million tonnes of flyash per annum. For every tonne of coal consumed 0.2-0.4 tonne of flyash is produced from thermal power stations in India. If the thermal expansion programme in the next decade is also taken into consideration the production of coal ash is likely to be about 130 million per year which is near about 3.5 lakh tonnes of ash per day. This indicates the enormity of the problem which the country has to face in disposal of the coal ash which is an environmental hazard. This problem has to be brought on war footing and flyash put to proper use to reduce surface and air pollution (Roy, 1993). Developed countries in the West utilise as much as 30 per cent of the flyash they generate by converting it into usable building products. Although much has been talked about utilisation of flyash in this country, not even 3 per cent of the total ash generated is utilised today. So it becomes necessary to utilise this ash to produce some useful materials. One of the established ways is to produce new and alternative building materials like flyash. Sand lime bricks consume about 70-80 per cent flyash in their manufacturing process.

Clay Flyash Bricks

In places where suitable clay, that is, agricultural soil, is degrading fast, good quality of bricks can be made by admixing flyash with inferior clay. Admixing 20-30 per cent flyash reduces drying and firing shrinkage and improves compressive strength of the bricks. There is an energy saving to 10-15 per cent in firing the bricks also. Quality building bricks of compressive strength 70-100 kg/cm^2 and moisture absorption 15-20 per cent can be made.

Lime Sand Flyash Bricks

Lime sand flyash bricks can be made near the thermal power plant sites to reduce the transportation cost of flyash. The finished product can be transported to nearby places as in the case of clay building bricks. Hot water/steam, available from power plants, can be utilised for autoclaving the bricks. Suitable moulding machines are now available in India for casting the bricks. In this process 8-15 per cent of lime is admixed with flyash along with chemical additives and bricks are cast in bricks moulding machines and steam cured (autoclaved). The flyash bricks are slightly lighter, gives better insulation and does not require plaster. Quality building bricks of 75-200 kg/cm^2 and low moisture absorption can be made.

Reclamation of Excavated Land

Till today, thousands of hectares of good quality fertile land have been excavated 1.2 – 1.5 metres deep for the purpose of taking out clay for making bricks. At the end of the expiry of lease of the land, this excavated land is left as such after levelling. Land owner may use this field for raising agricultural crops. In such cases the yield is poor because of absence of top soil, which is richer in humus and minerals and large quantities of compost and fertilisers may not be within the reach of the farmer. Many a times, these excavated lands get flooded during rainy season destroying the standing crops. These low lying lands may also be a source of environmental hazard, such as erosion of adjacent fields, breeding of insects etc.

To preserve the quality of agricultural land and to stop erosion from the adjacent fields, there is a strong need to reclaim these excavated pieces of land by refilling them; whether possible, with permissible waste materials, especially those excavated lands which are in close proximity to thermal power plants could fill up their land with flyash which has been shown to support the growth of many crops. Alternatively, the top of the flyash refills may be covered with 30-35 cm. thick layer to good quality soil which would take care of crop raising. "This process will not only help in regeneration of lost fertile land for agricultural purposes but also in removal of flyash from thermal power plants which have become a big disposal problem and environmental hazard" (Aslam, 1993).

This process of reclaimation by filling up excavated land with flyash involves substantial cost and this may be the reason why it has not been given a practical shape so far. Since it is an extremely vital step from the point of view of environmental conservation for which the Government and public are equally committed, there is an urgent need that all concerned parties such as the Ministry of Environment and Forests, (MoEF), National Thermal Power Corporation (NTPC), Central and State Pollution Control Boards (CSPCB) and All India Bricks and Tiles Manufacturers Federation (AIBTMF) should jointly evolve a formula in which the cost of land filling with flyash is shared on the basis of natural justice and social obligations. There is no way to escape this responsibility and sooner the modalities are finalised and put into practice with strict controls, the better it would be in the national interest.

Other Raw Materials

Another way of preserving good quality fertile soil is to replace a part of it with other permissible materials, such as certain industrial/ agricultural wastes or other locally available materials. Though many of these materials are sometimes added to compensate for some weaknesses in soils for the purpose of brick manufacture, many a times these wastes could be added to partly replace clay so as to conserve good quality soil. This replacement should ensure that is does not adversely affect the quality of bricks. In fact, in many cases admixture may be designed so as to produce improved quality of final product.

Carbonaceous wastes are added as a minor constituent (5-10%) to the alluvial soils and a major constituent (30-40%) in black and red soils to take advantages of both their mineral matter as well as fuel value of carbonaceous matter. In the manufacture of bricks from clay waste admixture, it has been observed that maximum strength in bricks is obtained when wastes are mixed to the soil in such a proportion that residual carbon in the admixture is in the range of 1.2-1.8 per cent. Maximum saving of fuel is also achieved in this range of residual carbon content.

Law as an Instrument of Land Management

Delhi faces serious environmental problems that have been exacerbated by the stress of its growing population and by its efforts to industrialise and modernise. Delhi may also see continued degradation of her rich agricultural land unless environmental controls are enforced. From the very beginning brick kiln industry is considered as a sensitive industry. The following laws clarify the concern about the brick kiln industry:

Main Law: Delhi Land Reforms Act, 1954 : Section 81 (related to denouncing quarrying as a landuse).

Related Laws : Delhi Brick Field Rules, 1941. The East Punjab Control of Brick Supplies Act, 1949. The Delhi Bricks (distribution, sales movement and price) Control Order, Supplies Act, 1949). The Delhi Coal Control Order, 1963.

Conclusion

In this study an effort has been made to study the present status of brick industry, its role in development as well as degradation of the environment and capability to cope with the future scenario in a sustainable manner. Urbanisation and metropolisation have played a major role in the development of brick kilns around Delhi. Although this was necessary element, the unplanned and haphazard exploitation of the finite land resources have played havoc with the landuse pattern and land quality. In order to meet the future requirements of housing without jeopardising the environmental quality, it is necessary that the flyash brick technology is adopted and policies are formulated so as to introduce eco-marked for brick and subsidising the cost of bricks made through this eco-friendly material and technology. Utilisation of flyash to fill up the excavated land is another measure which can be put into use, so as to minimise the erosional forces.

The fringe which was meant to be area of mixed landuse outside the city is slowly loosing its rural character, while catering to the urban requirements. Thus, there is urgent need to look into these urban needs and their fulfilment as an issue of morality and responsibility towards the environment, of which we are a part. Brick industry has played significant role in altering the landuse pattern of the fringe area, from agriculture and related activities to quarrying and consequent renouncing of the land as barren and uncultivable.

REFERENCES

Blaike, P. and Brookefield, N. (eds.) (1987), *Land Degradation and Society*, Methuen Publication, London.

Government of India, (1971, 1981, 1991), *District Census Handbook: Delhi.*

Government of India, Ministry of Urban Development, (1990), *Master Plan for Delhi*, pp. 115-122.

Government of NTC of Delhi, (1996), *Delhi at a Glance*, Directorate of Economics and Statistics.

Hajela, R.B. and Gupta, R.C. (*et. al.*) (1991), "Manufacture of Improved Brick Utilising Inferior Saline/alkali Soils, *Brick and Tile News*, New Delhi, pp. 38-40.

Jain, A.K. (1988), "Urban Planning and Growth: A Case Study of Delhi", in R.C. Sharma (ed.), *India's Urban Land Policy and Development Finance*, Seema Publication, New Delhi, pp. 143-161.

Nangia, S. (1976), *Delhi Metropolitan Region: A Study in Settlement Geography*, K.B. Publication, New Delhi.

Nusrat, A. and Mahadev, P.D. (1991), "Environmental Impact of Brick Loam Quarrying an Agricultural Soil", *The Indian Geographical Journal*, Vol. pp. 83-88.

Ramachandran, R. (1989), *Urbanisation and Urban System in India*, Oxford University Press, New Delhi.

Roy, S. (1991), "Utilisation of Flyash in the Brick Manufacturing Industry", *Brick and Tile News*, New Delhi, pp. 25-29.

16

A Geographical Analysis of Land Acquisition for Urban Development
A Study of Delhi

ATUL KAPOOR and B. THAKUR

Introduction

In India the areal limits of settlements from hamlet to towns and cities are set by the government. The limits of the built-up area, as different from the administrative area are known by such terms as 'abadi', 'lal dora' etc., in the case of rural settlements any addition to existing built-up areas would require the sanction of municipal corporation or the town planning department. Such increase in built-up areas of towns and cities is an ongoing process which involves acquisition (purchase) of land by private land developers and its conversion into urban uses with the consent of the proper authorities. This process of land acquisition and its development is more or less spontaneous, but the focus of the study is on land acquisition and development by the government or its authorised agencies. In Delhi Union Territory, the first was important upto 1962 and, thereafter, land acquisition and its development by private agencies has been prohibited by law and the government has stepped in. This work focuses on the latter aspect of land acquisition in Delhi Union Territory and relates it to the process of urban growth and urban development.

The earlier studies in urban geography focussed mainly upon the origins and characteristics of individual towns. Later, the theme changed to study the rural-urban fringe and identification of the inter-village and intra-village differences in the impact of metropolitan expansion and its consequences. Other studies emphasised the value of primary data through personal interviews. Land acquisition, its disposal, later development, are all relevant from a geographical point of view. Therefore, the purpose of study is to highlight the role of government agency in the process of development after acquiring the land for public interest.

Review of Studies

Geographical studies on land acquisition are non-existent. However, this problem has been studied by non-geographers, particularly those interested in town planning and administration. Geographers have, however, studied the problems of land use in the rural-urban fringe and in the city itself. While the problems of planning these landuses have been touched upon, there are no specific references to the question of land acquisition. Geographical studies in the area of the rural-urban fringe are many and even around Delhi, more than a dozen studies have been completed. Research scholars in Delhi area have been particularly interested in landuse in and around Delhi. But there are no serious studies on land acquisition as yet.

Economists, on the other hand, have taken more interest in this field. Bijlani (1978) in his paper on 'Law and Urban Land', described the acquisition of land as foremost requirement for planning. This requirement can only be fulfilled by the legal rights. That is because of limited boundaries, expansion is obvious need for all Indian metropolitan cities. He has suggested a revision of land Acquisition Act and a smaller process of acquisition denying the right to the owner to be heard. Luther (1978) analysed urban land as most advantageous depending upon its location and situation. He briefly traced out some suggestions for the improvement in Land Acquisition Act. Gupta (1978) undertook a detailed study of Bombay Metropolitan Region Development Authority Act and mentioned about the necessity of area development through schemes by local authority. He emphasised that the land should be acquired in time and made available for needy sections of population. Land acquisition is only possible when the acquiring authority has the necessary legal powers and rights. These rights should be backed by courts and law as it must follow the procedure laid down in law. Eminent legal scholars have related the acquisition of land with expansion of urbanisation. Shukla (1969) has dealt with the concept of public purpose for acquiring the land. For him the concept of public purpose has never been static. It is, therefore, always expected that the legislatures would review in the body of the Act, the meaning of the public purpose of land acquisition. Mukerjee and Willcox (1969) in their study focussed upon acquisition as an important tool for plan implementation. They further described the nexus of development for public purpose and not for individual gain. They proposed that the compensation paid should not be in cash, but, may be in bonds or land. But the compensation given should be reasonable and close to market value. The brief review of literature shows that geographers should draw attention to the study of the acquiring of land and its consequences.

The Need for Geographical Studies

Land acquisition, despite its drawbacks, is necessary and will continue. Land is a central theme in geography and geographers have a duty to study the use of land in urban areas. Further, a geographer's prime concern is in the study of man and his relationship with the environment. However, in an urban environment, the land is precious even more so as compared to rural areas, as the pressure of land is more in urban areas due to the high density of population. There is an inequitable man-to-land ratio in India and in case of urban land, this inequality deepens even more. The total urban land in India has intense pressure and as the pace of urbanisation increases, the problem of land gets further accentuated. Urban geography focuses on the problems emerging from this inequitable situation in individual cities or towns. The acute shortages in housing have created heavy demands for residential land in urban areas. This has resulted in the development of slums, congestion and in over-crowding in the well established residential areas. In addition, land is a place for living, working and enjoyment. But what is most essential is healthy and livable environment and this can be achieved through proper landuse planning. It is the principal tool for securing spatial development. At present, in a large number of urban areas facilities have been strained. The planning of landuse and a healthy environment in a town or city is a major theme in urban geography now-a-days.

Study Area

Although all Indian cities are facing a rapid influx of low income population, Delhi seems to be attracting a large number of high income and also low income families. This can be seen in the emergence of numerous posh colonies in Delhi as well as multitude of slums, where poverty is highly visible. According to provisional population tables, Delhi has had unexpected population increase in the decade 1981-91. The population of Delhi in 1991 census was 9.3 million and the projected population was less than this. Whereas the growth rate had declined from the previous census. However, increase of population in absolute terms is more important than rate of increases, since the planners deal with the problem of providing land. Since more and more people are coming to Delhi more and more land will have to be acquired, development and sold to the people. Failure to do so, will result in uncontrolled urban growth without urban amenities.

The process of land acquisition which took place for development work was rather slow and difficult. The Second Master Plan for Delhi, 1982–2001, proposed to acquire 20,000 hectares of land. The problems of challenging the purpose of acquisition and delays in possession of land lead to obstruct the process of acquisition of land. The role played by courts while acquiring land is the main hurdles, irrespective of legal sanction given by Land Acquisition Act.

Rural-Urban Fringe

The increase in population of Delhi Union Territory was the phenomena of post-partition era, which witnessed migration of about two lakhs of muslims from and influx of about five lakhs of displaced persons into the city. Interestingly, in 1951 due to the influx of refugees following partition, the population of Delhi increased by 90 per cent. The population roughly went up from 1961-71 by 14 lakhs and from 1971-81 by 21 lakhs, but from 1981-91 by as much as 30 lakhs, while earlier it was being said that the population of Delhi increased by 4 lakhs, two lakhs from an increase in the residents and two lakhs by migration, every year. This would hold true for the period 1961-71 an 1971-81, but not for the period 1981-91.

Migration into Delhi is a major factor, but this cannot be regulated by government. In a democratic country the freedom to migrate cannot be denied to people. It is a continuous process unless some strong steps be taken with planning for growth of medium and small towns. In the words of Jagmohan (Former Lieutenant Governor, Delhi) "It is neither feasible nor desirable to stop the process of migration, only an excessive rush is to be avoided. In fact, the cities are not prepared to absorb the migrants. The more planned and purposive the process of absorption, the better it would be for the migrants and the city".

However, it is the 'pull' factor that accounts for migration from villages to the urban centres. Thus, it is the peculiar combination of 'push' and 'pull' factors that is accelerating the migration in Union Territory of Delhi and determining its characteristics. And, as is evident from 1981 census, the Union Territory of Delhi had the growth by 5.8 per cent.

As per 1981 census, the major share of Delhi migrants were from Uttar Pradesh (48.2%), Haryana (15.5%), Rajasthan (7.6%), Punjab (9.8%), Madhya Pradesh (2.4%) and other states as shown in map (Fig. 16.1). Irrespective of these percentages the migratory people do influence the social/regional composition of Delhi Union Territory. In some places the people migrate only because there they have similar caste of people. This kind of tendency is more prevalent in whole of urban Delhi. In case of poor people, they do not have any choice while selecting the residential area. They have to live in low quality areas like slums. However, the planners only take care of rich migrants and the poor migrants are neglected.

The 1981 census remarked that "rural-urban migration has increased significantly over the period 1961-81. This has certainly led to the acceleration of urban growth". Prime reasons stated for rural to urban migrations are employment and education or for family reasons. Of the total migration, 11.1 per cent is from urban to urban areas which means people are moving out of small towns. This flow has increased after 1961.

The physical expansion of Delhi between 1911–1991 led to the decrease in rural villages towards the structural change in urban growth (Diwakar and Qureshi, 1993). At present, Delhi is adding each year about 2 lakh persons and the trend could lead to an increase of about 3 lakh persons per year by 2001. If the present growth rate of population is maintained, Delhi would have staggering population of 12 million in its urban areas and only 6 lakh in the rural areas. Even the area of rural population is decreasing. But except resettling of migratory people there is no other alternative left for the planners.

The Scale of Land Acquisition – An Overview

The Master Plan of Delhi, 1962-81, envisaged acquisition of 30,000 hectares. The purpose was to increase the population holding capacity

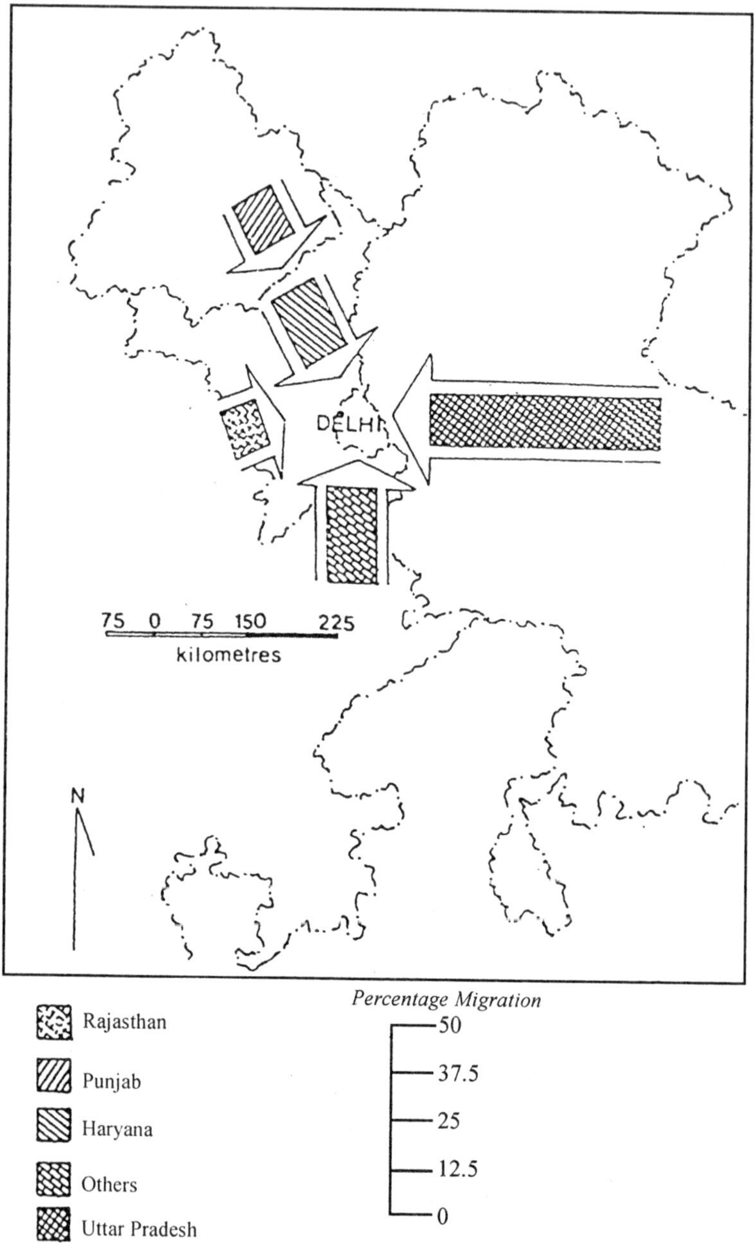

Fig. 16.1: Influx in Delhi, 1981.

of Delhi Urban Area by merging of rural villages after acquisition of land. The Master Plan 1962-81, directed the balance landuse pattern for the acquired land, so that the improved living condition can be obtained. Table 16.1 shows the planned landuse for the acquired land during 1962–81. The implementation of planned development paved the way for acquisition of land in Delhi Union Territory. The land was acquired from the villages around Delhi Urban Area for the development purposes. The analysis of this study is based on the acquisition done during last 40 years, since 1951.

Table 16.1: Proposed Planned Landuse by the End of First Master Plan, 1962-81.

Landuse	*Covered Area (hectares)*	*% coverage*
Residential	19000	43.5
Major Commercial	1000	2
Industrial	2500	6
Government	3000	7
Recreation	10000	23
Public and Semi-Public Facilities	3500	8
Agriculture (Nurseries)	120	0.5
Transportation		
Roads and Railways	4500	10
Total	**43620**	**100**

Characteristics of Land Acquisition

Data Base

The data on land acquisition in Delhi Union Territory is available with the Department of Acquisition, Delhi Administration. The Delhi Union Territory has been divided into six land acquisition zones. Each zone of land acquisition have a separate cell for maintaining records of land acquisition. The data available with the Land Acquisition Department is the only authentic source and the same is analysed in this study. The nature of data available with the Land Acquisition Department, has several information regarding the award number, name of village from where the land was acquired, year of final acquisition and those lands which are pending in the court, because of writ petitions of owners of land, are not available. The level of the data is village and unit is hectare in which the land is acquired in different years starting from 1951. The land acquired in each of the 224 villages in eight periods were computed from the raw data. The aggregate land acquisition in five year period for each village forms the basis of this work.

Land Acquisition Process

As there is increase in the population of Delhi Union Territory, the need for urban land uses also increased. The land has been acquired to meet the needs of the increased population. The Land Acquisition Act, 1894, still governs the acquisition of land for public purpose.

The proposal of residential schemes for accommodating the increased population has been formulated by the Delhi Development Authority. The proposed scheme envisaged to acquire the amount of land for various landuses. So the Delhi Administration serves the notices to the owners of land under Section 4 of Land Acquisition Act. There is a provision under Section 5A for objections by land owners, whose land is going to be acquired.

Further, Section 6 of the Land Acquisition Act highlights the intention of acquiring a particular piece of land and in Section 7, the Land Acquisition Collector makes an order for final possession of the notified land. However, the compensation paid in lieu of acquired land is at the prevailing market rate. The date of the first notification under Section 4 was taken into account while settling the compensation.

When the final possession takes place the process of development can start. The whole process of acquisition or development is slow and time consuming.

Acquisition in Different Periods

Delhi has about 1483 square kilometres of area of which, 591 square kilometres are urban, the rest being agricultural which includes 195 rural villages. Besides the already built-up areas and rural areas in villages, there are areas coming within villages in the urbanisable limits. These lands can be acquired by Delhi Development Authority. There are nearly 89,100 hectares of area left which are still to be acquired. To date, the Delhi Development Authority has acquired 23,000 hectares and notification has been issued for 46,935 hectares. The Delhi Development Authority is the legal authority, which is involved in all kinds of acquisition and development work in Delhi Union Territory.

The planned development of land was started after 1962, when the First Master Plan for Delhi, 1962–81, came into force. The large chunk of land was acquired between 1966–70 (Fig. 16.2), for the construction of houses in Union Territory of Delhi. The land acquisition after First Master Plan for Delhi, 1962–81, came into force. The land acquisition after First Master Plan, that is, 1981, was high, because Delhi Development Authority's ambitious projects, namely, Rohini, Papankala and Narela etectra, were started during this period.

'000 Hectares

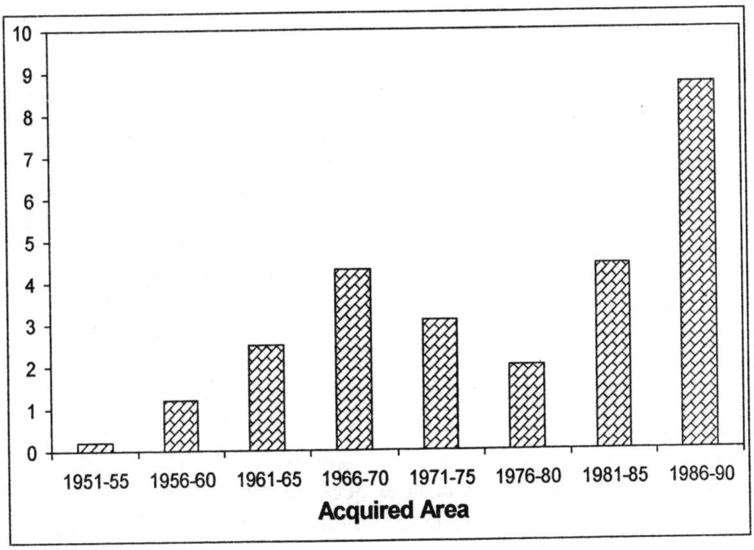

Fig. 16.2: Acquisition of Land in Time Periods Delhi Union Territory, 1951-89.

Source: Department of Land Acquisition, Delhi Administration.

The acquisition in various lot sizes during different periods is shown in Figure 16.3. Acquisition in large lot sizes (200 hectares or more) began from 1966–70 period and continues to the present, 1986–89. This indicates a preference for bulk acquisition as against acquisition in bits and pieces. After 1980's, Delhi Development Authority started locating big residential colonies with capacity of around 1 million population. This shows, creation of a city in a city itself. The large lot size of 200 hectares is about a size of a city and the accommodation capacity can exceed to 100,000 population. Although, the large lot of acquisition involves large sums of money, the scale of planning and development is also high in the large lot size. There was no acquisition between 1951-55. During this period a lot of land was available in the form of refugee property, that is, *refugee* muslims, who left Delhi for Pakistan. Some of them were given out/occupied by the *refugees* (Hindus) from Pakistan; others were acquired for development, that is, for selling to *refugees*.

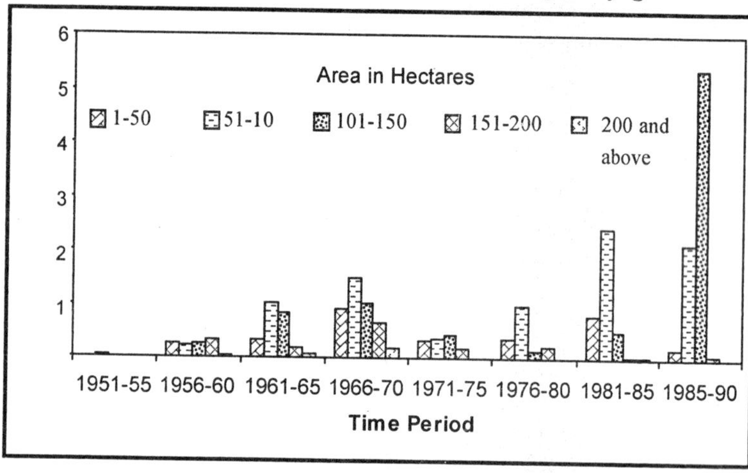

Source: Department of Land Acquisition, Delhi Administration

Fig. 16.3: Acquisition of Land with Lot Sizes Delhi Union Territory, 1951-89.

Between 1955–65, thc land acquisition was on a small scale, that is, in small lot size (1-50 hectares). This shows thc financial constraints of Delhi Development Authority and its small administrative size and its inability to cope with large acquisition. This is natural, but as it gained experience, the scale of acquisition became largcr.

Stages of Land Acquisition of Villages

The scale of land acquisition may also be looked at from the point of view of relative proportion or per cent of total land that has been actually acquired. Thus, we may classify villages as: (a) Initial acquired, 1–20 per cent, (b) Partial acquired, 21–40 per cent, (c) Moderate acquired, 41-60 per cent, (d) Substantive acquired, 61–80 per cent, (e) Complete (full) acquired, 81–100 per cent.

In 1951, the total inhabited number of villages in Delhi Union Territory were 304. During 1951–61, the number of villages declined to 276, which was the result of full acquisition of villages merged into Delhi Urban Area. Between 1961–71, more villages were incorporated into the Delhi Urban Area and the number of villages in Delhi Union Territory were reduced to 243. So in 20 years period, the number of villages which were fully acquired was 61. During 1971-81, the number of villages further reduced to 214. This resulted in the expansion of Delhi Urban Area. However, the number of villages which were fully acquired (Fig. 16.4) and yet to be included in Delhi Urban Area are 38. These villages were fully acquired in one lot for the two major residential schemes, that is, Rohini and Papankala.

The villages where land has only been initial and partially acquired, are the future prospects for expansion of Delhi Urban Area. The moderate and substantive acquired villages are going to be acquired fully in coming future for urban uses. However, the acquisition is a continuous process and will require several decades while Delhi Union Territory is fully urbanised.

Progress of Land Acquisition

The expansion of Delhi Urban Area could not take place considering the continuous extension of urban area. But the towns in Delhi Union Territory have emerged even far away from the Delhi Urban Area. So the progress of land acquisition in Delhi Union Territory has been divided into (a) time gap acquisition, after a gap of 5-10 years, (b) starting years acquisition, 1951–60, (c) middle years acquisition, 1961–1975, (d) end years acquisition, 1976-89, (e) continuous acquisition, land acquired in every period (Fig. 16.5).

The villages which were acquired continuously, were closely located to census towns and Delhi Urban Area. This is because, the land was not acquired in large lot from the villages and only acquired in consecutive periods in pieces. The villages from where land has been acquired during 1961-75 were treated as middle years acquired villages. Figure 16.5 shows cluster of villages from where land was acquired in middle years. This highlighted that the land was acquired in large lots and not in bits or pieces.

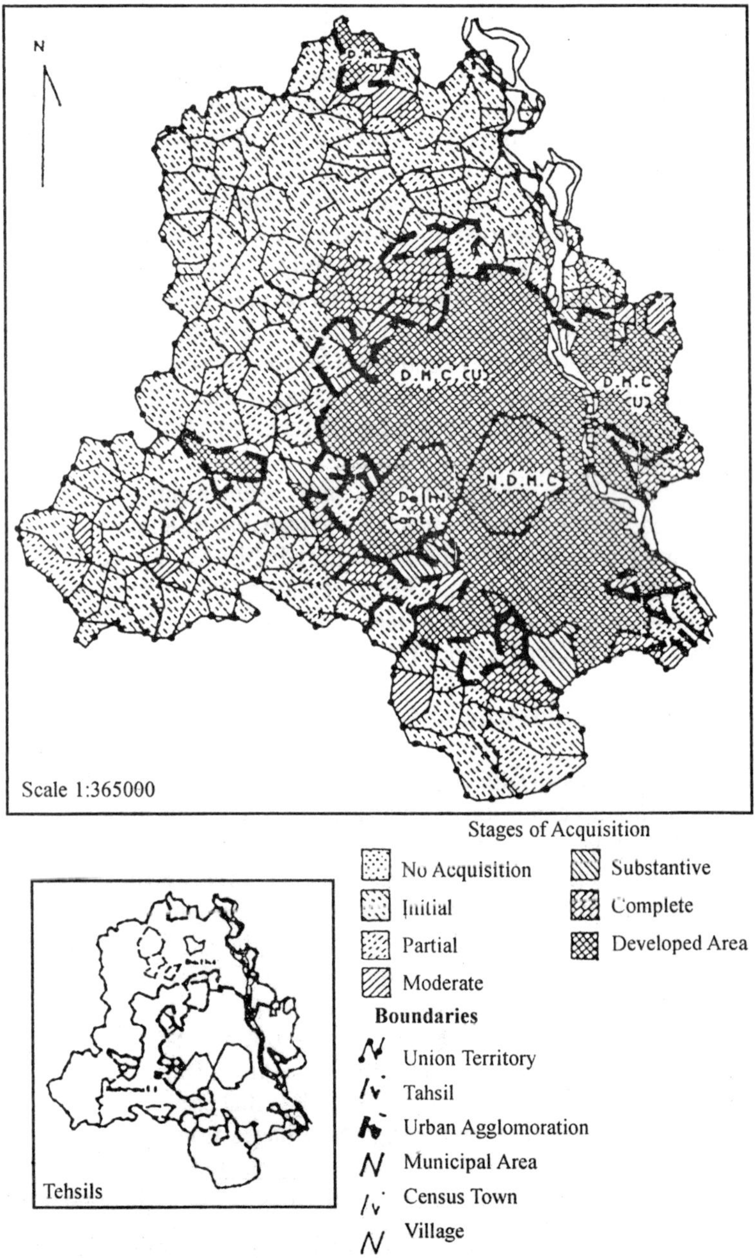

Fig. 16.4: Stages of Land Acquisition of Villages (Delhi).

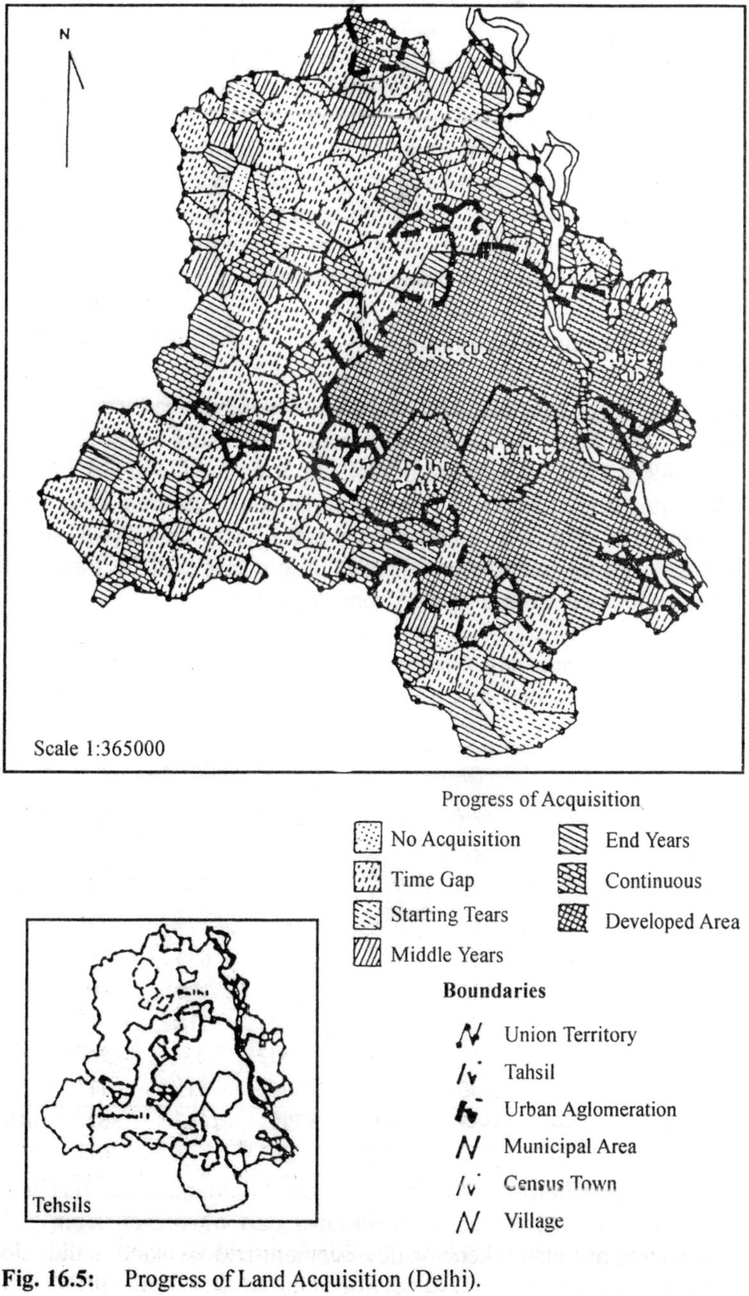

Fig. 16.5: Progress of Land Acquisition (Delhi).

There were few villages from where land has been acquired before 1960's. But there were numerous villages which fell in time gap years. Which actually represent acquisition held in after long time gap, that is, first acquisition held in after one time gap, that is, first acquisition was held in 1957 and then in 1975. The acquisition held during 1976-89 was mainly from villages which were lying in the outer part of Delhi Union Territory. These villages were taken for acquisition because the urban area is expanding more rapidly and, accordingly, the process of acquisition gained momentum in recent past.

Distance Factor in Acquisition

The land around the city center, that is, Shahjahanabad was taken as the distance factor for the measurement of total land acquired during 1951-89. As thee is increase in distance from the Delhi Urban Area the acquisition of land was less in comparison to the areas near Delhi Urban Area. The 1-10 kilometers were already developed, so the villages falling under more than 11 kilometers will be treated as villages located around Delhi Urban Area.

Table 16.2: Distance-wise Distribution of Land Acquired in Union Territory of Delhi, 1951-89

(*in hectares*)

	Distance from City Centre in Kilometers						
Period	*Below 5*	*5-10*	*10-15*	*15-20*	*20-25*	*25-30*	*Above*
1951-55	—	30	33	9	12	10	—
1956-60	—	253	210	252	338	42	37
1961-65	—	341	1025	851	178	77	60
1966-70	—	925	1482	1023	639	189	260
1971-75	—	335	353	436	180	10	22
1976-80	—	376	977	130	215	—	2
1981-85	—	815	2443	500	43-	24	183
1986-89	—	182	2154	5391	611	11	53
Total	—	**3358**	**9040**	**8592**	**2613**	**362**	**618**

The area below 5 kilometers from Shahjahanabad has been developed for urban uses. The major part of stretch from 5-10 kilometers are also taken for development and the land acquisition was on full swing in this stretch after First Master Plan, 1962-81, till

1985. Thereafter, the acquisition was less compared to the previous periods, because the land was already developed. Whereas, the land acquired between 10-15 kilometers were high and as we move further away from the centre the acquisition, subsequently, increased due to the urban expansion of Delhi Union Territory. But after 20 kilometers from the city centre (Table 16.2) the acquisition was declining and the future prospects for urban expansion are bright.

Transformation of Villages

The Delhi Union Territory comprised of three major towns and several minor towns, as per 1981 census. The towns are the resulted feature of development over the villages in Delhi Union Territory. Still the villages in Delhi Union Territory are the future scope for expansion of Delhi Urban Area. This is because, in 1921, there were 357 villages in Delhi Union Territory, which have been reduced to 231 in 1981. The effective transformation of villages into urban area began since 1931, when New Delhi was established. In last 60-70 years of development, the number of villages merged in Delhi Urban Area were 126. The magnitude of transformation was high during 1961-71 and 1971-81. Between 1971-81, 27 villages were given the status of census towns.

The future development of Delhi Union Territory and growth of Delhi Urban Area is inevitable, because the census of 1991 shows, more population than projected. The land required for the increased population can be fulfilled by acquiring and developing more land for urban uses.

Findings

The cities of India are growing fast. The land required to accommodate the growing population also increased. The legal authorities are desperate to solve this problem. The increase in population is due to two main factors, i.e., (i) immigration and (ii) natural increase in population. The process of growth is continuous and free from all restrictions. The share of urban population has also increased in India during 1981-91. So the resulted picture emerged from this large increase is to provide more land for urban uses. The only way to get through is acquisition of land adjoining to urban areas and develop it for urban uses.

The position of Delhi Union Territory is the same, moreover, related to other cities of India. Being capital of India, the ratio of increase of population in Delhi Union Territory is more as compared to other cities. The disturbed states, namely, Jammu and Kashmir and Punjab also are the main factors in increasing population of Delhi Union Territory. The people who migrate from these disturbed states preferred to resettle in Delhi Union Territory than in any other city.

The growth of population in Delhi Union Territory is increasing, inspite of the decrease in the population growth rate during 1981-91. The need of more land due to the sharp increase of population can resolve by taking more land for urban uses. The urban land in Delhi Union Territory has been increasing rapidly since independence. The increase in urban area of Delhi Union Territory has been mainly due to the status given to it after 1912, i.e., Capital City of India. Since 1951, the land has been acquired according to the requirement of that time. But after 1962, the acquisition of land has been increased because the planned development of Delhi Union Territory has started with inception of First Master Plan of Delhi 1962–81.

The main fact is that large areas of land were acquired by the Delhi Administration for the planned development of urban Delhi, beginning from 1957 and that this land was made available to the Delhi Development Authority and to other government agencies, like Delhi Transport Authority etc. for the eventual development of Delhi according to the Master Plan 1962–81. Before planned development came into effect, all construction including houses were mostly in private hands. This meant, that there were more number of persons who bought plots for building purposes, when the freehold system was prevalent in Delhi Union Territory.

The concept of planned development of Delhi is very clear and unambiguous. The land, which is acquired, will be developed and disposed off in a systematic and proper manner. There were certain difficulties that arose out of the scheme called Large Scale Acquisition, Development and Disposal of Land in Delhi, 1961. These difficulties were: (i) the stay order by the Court in favour of land-owner, (ii) the political pressure for dropping the land acquisition idea, (iii) encroachment on the acquired land by poor people. The administration has again to think over by drawing up schemes in which the alternative land was to be given to the public. The main method visualised by the government was either to use the land for housing the weaker sections of the society by providing cheap houses constructed through a government agency or under the control of government agency or alternatively to allot the plots by auction or to allot the plots through cooperative societies. The above scheme was framed with this object in view.

Progress of development of Delhi is not according to Master Plan, 1962. Delhi Development Authority is far behind to achieve its objectives because of several reasons as below: (i) process of acquisition is slow and time consuming, (ii) problem of compensation due to change in market price of land, and (iii) overlapping of various administrative organisations slows down the pace of development.

The implementation of Land Acquisition Act and Master Plans lacks only due to the above-mentioned reasons. The process of land acquisition takes several years for final acquisition which result into slow development of land also. The land acquired by paying the compensation and the price of land for compensation fixed on the basis of first notification issued by authority. So the land-owner is not satisfied with the compensation as the market rate of land increases between date of issuing first notice for acquisition and final notification. This resulted in litigation and stay orders, which causes delay in the Plan and its progress. The Land Acquisition Act, though useful and effective, had failed. The Act, on the one hand deprived the owner from its real value and, on the other hand, did not benefit the economically weaker section. Some changes are necessary to settle compensation payable to real owner, so as to avoid long litigations before the Court for payment of compensation.

The analysis did show that all those villages which were contiguous to Delhi Urban Area were fully acquired first and the process is a continuous one, until, there is no more land available for acquisition. The acquisition of land from the surrounding villages created a problem of change of occupation for fringe people and decline in agricultural land as well as agriculture production.

The slowness of development, can be seen in the case of Rohini. After 15 years of the Scheme (beginning from the time of its execution of the scheme), only half of the area has been developed. The other half of the area in this scheme is under the process of development. Eventhough, several other schemes also emerged on the map of Union Territory of Delhi at the same time, each is progressing at a slow pace.

The role played by Master Plan of Delhi 1962–81, has no match in India. This is because Delhi is the only city in India which has completed its First Master Plan and framed the second for the period of 1982–2001. During the First Master Plan period, the land acquisition was not in large lots. However, the acquisition made during Second Master Plan (1982-2001), was in large–scale. This clearly differentiates the strategy of two Master Plans, since the Second Master Plan is still in progress.

The land acquired during the First Master Plan of Delhi 1962–81 was 23,000 hectares and the land developed was 20,000 hectares. The main achievements of the First Master Plan was the improvement in infrastructure services and substantial part of Delhi Union Territory has been developed. However, the haphazard growth was not controlled during this period. This resulted in the increase in ugliness of the historical city.

The credit for city planning of Delhi Union Territory goes to the Delhi Development Authority, which is the sole authority for acquisition and development. But, the Delhi Development Authority's role is somehow controversial because, on the one hand, it is the only legal developer of land in Delhi Union Territory, and, on the other hand, it has the monopoly on all fronts of development. So it is not possible to take a stand on the lapses and or on the achievements of Delhi Development Authority.

The two residential projects of Delhi Development Authority (Rohini and Papankala) were aimed at to reduce the pressure of land by providing more land for urban uses. Therefore, large areas of land have been acquired for the development purposes. Some of these land have already been developed and rest are in the process of development. The land for phase I and II of Rohini residential scheme has been developed and the phase III has been started. In case of Papankala residential scheme the acquisition has been completed and the development is yet to start.

Although Delhi Development Authority has several plans in hand or on papers but has failed on all the fronts of land acquisition. There is a long list of registrants waiting for plots/flats. The prices of land and flats are increasing day by day. It is beyond the reach of Janata/ Economically Weaker Section. The Delhi Development Authority has restricted for allotment of only one plot/flat but one can have as many plots/flats from property dealers on big premium. There is no check on resale or power of attorney.

In land acquisition there are so many problems. Neither it has benefited the people of Delhi nor the land owner is satisfied with the compensation paid. Hence, the middle man, property dealers are benefited due to land acquisition and its sale after development.

Conclusion

The geographical view of the study concludes that the growth and development are reflection of many processes and land acquisition is one of them. The acquisition of land is essential to cope with the pace of change. The utilisation of land resources for urban purpose is inevitable, therefore, the role of agricultural land is reducing. The land coming under urban use can be selected on the basis of quality of land not suitable for agriculture. The better way of planned growth can be by using the Master Plans effectively.

REFERENCES

Bijlani, H.U. (1978), 'Law and Urban Land', in H.U. Bijlani and M.K. Balachandran (eds.), *Law and Urban Land*, IIPA, New Delhi, pp. 59-82.

Diwakar, A. and Qureshi M.H. (1993), 'Physical Processes of Urbanisation in Delhi', *Urban India*, pp. 94-106.

Gupta, L.C. (1978), 'Some Thoughts on Planning and Development of Urban Area'. H.U. Bijlani and M.K. Balachandran (eds.), *Law and Urban Land*, IIPA, New Delhi, pp. 101-06.

Mukerjee, B.N. and Willcox D.L. (1969), 'Development Planning and the Power of Acquisition of Property', The Indian Law Institute, *Law and Urbanisation in India*, New Delhi, pp. 103-34.

Shukla, V.N. (1969), 'Concept of Public Purpose and Land Reuse Planning, The Indian Law Institute, *Law and Urbanisation in India*, New Delhi, pp. 93-102.

17

Urban Encroachment on Rural Land in West Delhi

VANEETA CHANDNA

Introduction

Land is the basic fundamental resource which allows the synthesis of a wide variety of different factors in addressing the societies wants and needs. The increasing concentration of population in different parts of Delhi and multiple demands made by this population on the nearby rural areas has led the land in the city fringe to fall prey to the process of urban encroachment. The term encroachment as stated by Webster's International Dictionary is 'to enter or intrude by gradual steps or by stealth into the possessions or rights of another' or 'to advance beyond desirable or normal limits. At times, the word 'sprawl' is also used. The term sprawl as applied to urban growth in Webster's International Dictionary is 'to spread or to develop irregularly or gracefully.

The process of encroachment on rural lands is most common in the area of rural-urban fringe. The term fringe is used for 'the area of transition between well recognised urban land uses and the area devoted to agriculture'. (Wehrwein, 1959) It is also characterised by 'extending as a belt around metropolitan city where full city services cease to be available' (Anderson, 1952). Urban fringe is a highly fluid area. It is symbolically integrated with the city, but as a part of city jurisdiction may or may not be within its limits. (Harris, 1956).

Delhi's population has increased from 0.4 million in 1901 to 9.4 million in 1991 with a progressive growth rate of 2,209 over the period 1901 to 1991 (Census of India, 1991). Keeping in view this ever increasing population, there is a growing realisation that good fertile land is a finite resource and it should be protected in order to feed these teeming millions. The urban development of Delhi has increased nearly 13-fold since 1900, eating into surrounding rural and agricultural areas and absorbing more than 100 villages (Hardoy *et al.*, 1992). Hence, land is most valuable for agriculture from both physical and economic point of view, is frequently the most desirable for urban development as well. Consequently, competing demands on land can arise which can be met by the conversion of good farmland to urban use. This, in turn, reduces the ability of that area to provide food for domestic needs and also in response to future shortages elsewhere in the world. Hence, the agricultural lands deserve to be preserved on two counts : First, they are a valuable natural resources; and Second, they provide the breathing space which contributes towards keeping the land a pleasant place to live in.

Review of Literature

The physical supply of land is limited. With increasing population growth its demand is increasing in direct proportion. The fringe zone is facing increasing industrialisation, commercialisation and expansion of settlements resulting in the loss of prime agricultural land which has begun to be major concern for planners.

The work by Thakur (1991) is an attempt to delimit and examine the salient features of culture and landuse of the rural-urban fringe of Darbhanga city in Bihar. A statistical-ecological-physio-gnomic approach has been adopted for identification of aspects of interactions, its measurements and mapping. Since (1967) has aptly remarked that the emergence of the fringe zone with its complex problem of adjustments between rural and urban ways of life has assumed great importance and has drawn the attention of planners and social scientists.

Russwurm (1973) has worked on the development of cities on the prime agricultural land with reference to Ontario cities and has attempted at monitoring landuse change in Canada. Hart (1976) studied the urban encroachment on rural lands in the United States and found that urban expansion is a major cause for the permanent loss of cleared farmland despite its importance on the fringes of metropolitan areas. Best (1981) studied the various forces and processes, the problems and complexities of linking agriculture with urban development in the United Kingdom. A concrete study with objective methodology on the subject was made by Rikkinen (1972) wherein with an emphasis to how ownership boundaries affect urban sprawl and expansion and changes in landuse. Harvey and Clark (1965) took note of the nature and economics of the urban expansion in general.

Many attempts have been made in India on trends and patterns of encroachments. One such work is by Siddiqui and Hussain (1975) who estimated the loss of agricultural land for the city of Saharanpur. They made similar attempts for the towns of Modinagar and Ghaziabad. Aziz (1968) studied the impact of urbanisation on rural land and the way depletion of agricultural land takes place.

An insight into the economic aspect on the theme of conversion of rural land for urban use, change in land price has been attempted by many scholars. Shonkwiler and Reynalds (1980) suggested an appropriate method for analysing land sales data in circumstances where alternative uses for the land in an urbanising area appear likely. The study by Dunford, Martin and Mittel Hammer (1985) contributes to the empirical land value literature by incorporating subjective expectations into a bid price model for rural land in an urban fringe market. A conceptual model for rural land is presented in which the role of subjective expectations is explicitly acknowledged. This study is a modest beginning in quantifying the forces involved in the determination of rural land prices.

Lockertz (1989) explained how metropolitan expansion has affected farming, not only in terms of landloss for farms but also with change in the intensiveness in farming in the North Central Region-using two distinct measures – ANOVA and Multiple Regression Analysis. Sinclair (1967) developed the idea that the prospect of urban development can have a negative influence in agricultural investment and landuse. This became a popular conceptual device against which to interpret farm landuse change in metropolitan regions. He, however, failed to emphasise on the extent to which farmers vary in their responses to the threat of urban expansion. This gap has been filled up by Bryant (1981) who developed a conceptual framework linking variation in farmers evaluation and strategies to the strength of potential urbanisation forces of the Paris region.

Purpose of Study

Thus, land problem is multi-faceted, multi-sectoral and multi-disciplinary. The losing of agricultural land to urban uses has fostered strong public interest in the landuse along the rural-urban margins. Studies completed claim the loss of good productive land for non-agricultural uses. To this effect, the following questions are addressed for examination:

How much agricultural land is being converted? i.e. what is the loss of agricultural land for urban uses? In what direction is the urban encroachment taking place over time and what has been the change in landuse pattern? and, finally, what are the management policies of the government to control the rapid encroachment?

The present study intends to analyse the loss of agricultural land for urban uses for 1961-1991 and the type of land quality being consumed for the purpose. The main objectives are:

(i) To measure the rates and describe spatial and temporal patterns of urban encroachment.
(ii) To determine the conversion of type of land quality.

The Hypothesis to be tested is as follows:

The rate of conversion varies with the agricultural land quality in the fringe of Delhi or the better quality land puts more resistance to development. Similarly, land associated with more intensive agricultural production offers greater resistance. To test the above hypothesis, western part of West Delhi has been chosen to conduct the survey.

The Study Area

The Delhi Development Authority has divided the urban area of Union Territory of Delhi into different zones for planning purposes. There were eight zones in the First Master Plan in 1961 which increased to 15 in 1991. The boundaries of each planning zone were so selected that each of such division reflected its own character of development, pattern of landusers, physical character and other social and cultural values. The study area chosen for this study is a part of West Delhi comprising of G and K Planning Zones. The G division falls under the urban area and the K division in the rural area (Fig. 17.1).

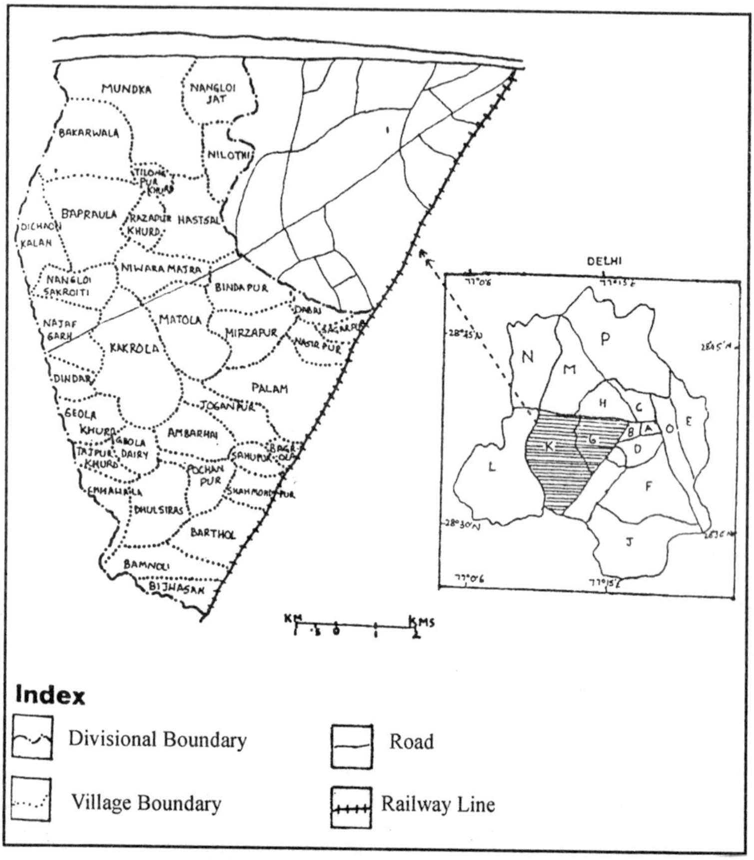

Fig. 17.1: West Delhi: Locational Setting.

Delhi is administered under two tehsils – Delhi and Mehrauli. In the study area there are 38 villages falling in Mehrauli tehsil, out of which six are census towns, while in Delhi tehsil the number of villages is four. (census of India, 1981). The DDA has acquired 20,000 hectares of land in West Delhi to accommodate population of more than one million, called the Dwarka (Papankala) Project. The entire area falls under the K division of rural Delhi. Hence, this area is very significant for studying the pattern and the rate of encroachment and the quality of land being converted.

Data Source and Methods

Village is treated as the basic unit for this study for which both primary and secondary data is collected. The primary data collection involved the village *patwari* and villagers. The data on village area and crops grown, accessibility, waterlogged area as well as the '*khasra map*' was procured from *patwari* which helped in locating the plot no., whether acquired or not. The notification data to be acquired for 1979-80 was collected from *Kanungo* at Tis Hazari.

The secondary data is obtained from various census and planning reports from School of Planning and Architecture and Town and Country Planning Organisation. Data on loss of land was generated from the Planning Wing of DDA. It also provided data on land acquired by the DDA for different urban uses, number of villages urbanised from 1961 to 91. The census Block of Delhi Administration provided the latest data on population of the village.

The macro level approach is adopted to find out the pattern of conversion of rural land for urban uses for the period 1961–91. For this purpose, K division is chosen which forms the most dynamic zone along the rural-urban fringe in West Delhi. It contains 38 villages for which notification record of the plot numbers for each village is obtained and mapped. The number of plots in each village varies from 60 to a few hundred. This shows the area acquired by the government.

Since it was not possible to study all 38 villages in the K division for quality of land consumed under the process of encroachment, a micro approach is adopted. A group of 5 villages comprising Baprala, Razapur Khurd, Hastsal, Tilongpur Kotla and Nilothi located in the northern part of Najafgarh Road adjacent to Pankha Road was selected. These villages were partially acquired by the government, hence, selected for the study (Fig. 17.2). At the micro level, unstructured and informal household and shop owners survey was conducted in the villages.

The detailed analysis of land quality was made for 1991-92. The variables selected to determine the land quality are : crops grown, nearness to road and drain, waterlogged area, ruggedness of the land, compensation provided by the DDA and proportion of irrigated land.

Data has been analysed statistically and presented through a number of cross tables, frequency tables and maps.

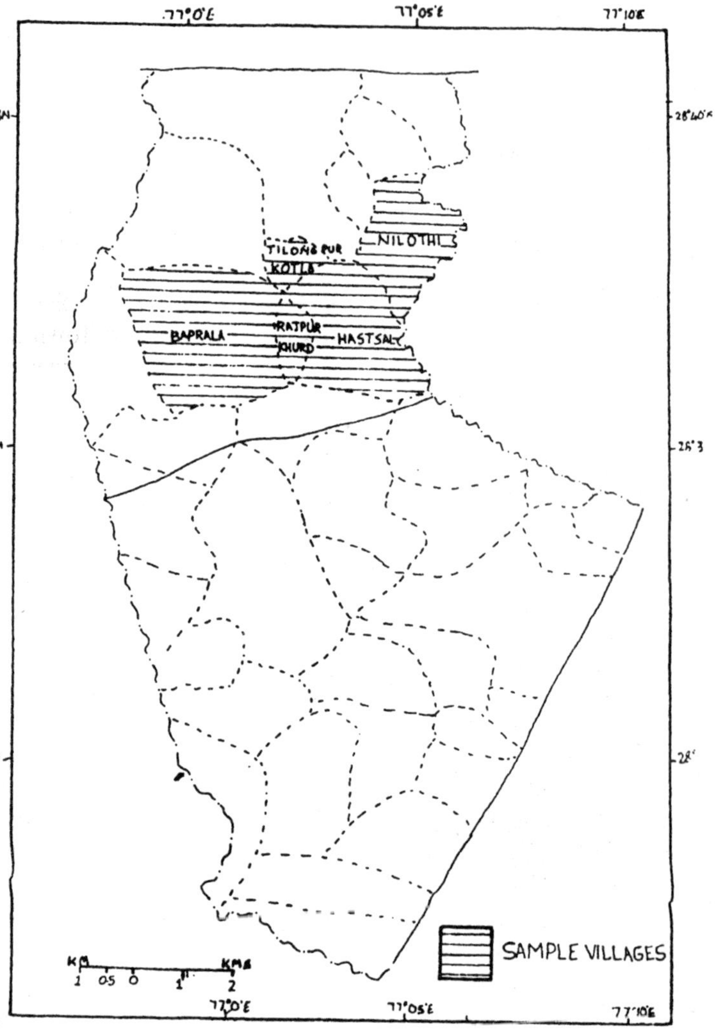

Fig. 17.2: Sample Villages.

Findings

The explosive expansion of urban population over the period of last 30 years resulted in swallowing of agricultural land. The resultant pattern, therefore, has become quite dynamic which has been fluctuating in space over time. This phenomenon is understood through scientific analysis of its spatial and temporal variations. Since 1947, three distinct periods of land development are discernible in Delhi. These periods are rooted in history and growth of economic base of the city (Bose, 1970). The initial factors accelerating encroachment on rural land were essentially population growth, both by immigration and net increase and urbanisation. This was further encouraged by development of residential and industrial areas.

The resultant pattern brought about by the conversion of agricultural land and exaggerated by ribbon developments along the highways and main roads. This has been followed by residential and industrial buildings haphazardly filling up the areas at the back or the front building lines. Much of the suburban growth has been unplanned and uncoordinated. This phenomenon has been studied through scientific analysis of its spatial and temporal variations.

An observation of the patterns of urban encroachment at an aggregate level, we find that the acquisition of the village has taken place in the extreme west of Delhi till Dhansa village adjoining the boundary of Haryana state. This indicates that the entire rural area has been acquired since 1961 but the acquisition at that time was in very small quantities (less than 25 hectares).

Between 1948 and 1961, the Punjabi refugees were housed in West Delhi in Moti Nagar, Ramesh Nagar, Subhash Nagar and Tilak Nagar following the partition of the country.

These residential localities were built on the agricultural land of the urban village of Basai Darapur, Tatarpur and Khampur Raza. Besides Khyala, Chowkhandi, Tihar were also engulfed within the ever expanding limits of the sprawling city. This process continued at a very fast rate with the passage of time and is still in process.

Temporal Acquisition

A decade wise acquisition within the last 30 years (1961–91) was analysed in order to see their patterns and trends in different time periods.

On an aggregate level, acquisition took place in all the villages since 1961, but, maximum acquisition took place along the rural-urban fringe. A detailed decade wise analysis has been made for the time period 1961–71, 1971–81 and 1981–91 (Table 17.1). In the decade of 1961-71, maximum acquisition took place in the villages adjoining the urban limit of Delhi. The villages acquired were Dabri, Hastsal and Nangloi

Table 17.1: Acquisition of Villages (1961-91).

	(Time Period)		
Land Acquired (In Ha.)	*1961-71*	*1971-81* (*% of villages acquired*)	*1981-91*
<25	44.73	42.10	34.21
25-50	10.52	05.26	18.40
50-75	07.81	02.63	13.25
75-100	—	—	07.91
>100	13.15	05.26	18.42
Not acquired	23.69	44.73	07.81
Total	**100.00**	**100.00**	**100.00**

Source: Land Acquisition Records, DDA.

Jat. DDA and private colonisers played an active part in it. Some villages like Chhawala, Dhul Siras were acquired under the Najafgarh Drain and Kakrola Drain to control the floods during the monsoon season. During 1971-81, very few villages were acquired with very little amount. 42.10 per cent of the villages were acquired in the category of less than 25 hectares of land. A scheme for the development of seven dairy colonies in rural Delhi was approved by the Ministry of Agriculture for the implementation by the Delhi Municipal Corporation during this period. During the period of 1981-91 maximum land was acquired in terms of number of villages and quantity of land. Rapid expansion of urban use was the reason for so much acquisition. In the north of Najafgarh Road maximum acquisition was in the village Mundka which has now been declared a Census Town. It is followed by Nilothi and Dichaon Kalan. South of the Najafgarh Road, twelve villages were acquired by DDA under the Dwarka Project to accommodate more than one million people. For this, 20,000 hectares of land have been acquired to be developed in two phases.

Spatial Acquisition

For finding out the direction of urban encroachment or its spatial patterns, plotwise notification of the village was analysed falling in the K division. 1981 has been taken as the transition year because after this year the rate of encroachment took place at an alarming rate. Before 1981, land only along the Najafgarh Drain was acquired

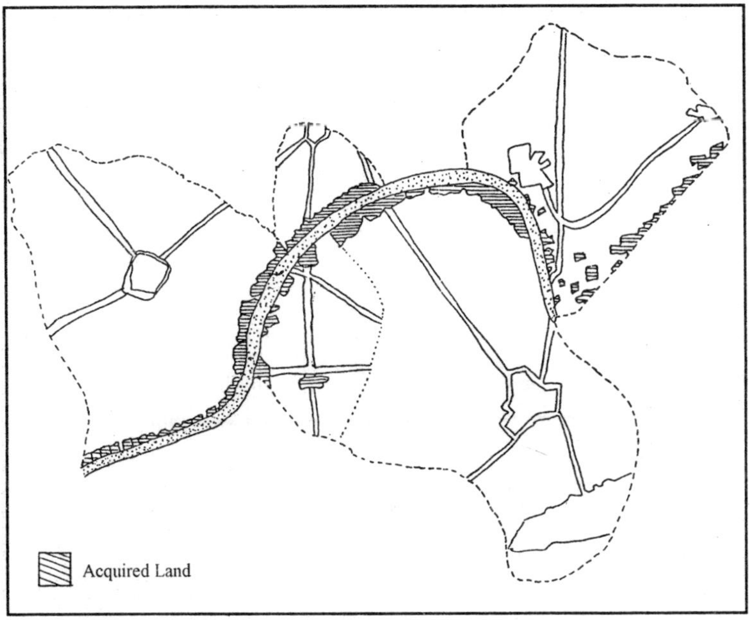

Fig. 17.3: Plotwise Acquisition of Land in the Sample Study: Before 1981.

and also some parts in pockets were acquired in some fringe villages. The main reason was to widen the drain and to build a road along the drain. This was done mainly by the Government, but the land along the Najafgarh Road in certain pockets was acquired haphazardly by the private colonisers (Fig. 17.3).

After 1981, villages in K division lying between the Najafgarh Road and the railway line to Alwar and east of the drain were notified for acquisition by DDA for Dwarka Project. To the north of Najafgarh Road and south of Rohtak Road villages were swallowed by the process of urbanisation. The expansion of urban land uses proceeded unevenly and often aimlessly. It has mainly followed the linear pattern along the transportation networks (Fig. 17.4).

Land Quality and Urban Encroachment

The rate of conversion of rural land and the quality of land being converted has been studied in detail. For this purpose, a group of five villages has been chosen lying in the north of Najafgarh Road. This area has observed 1129.31 increase in population from 1961-91. The

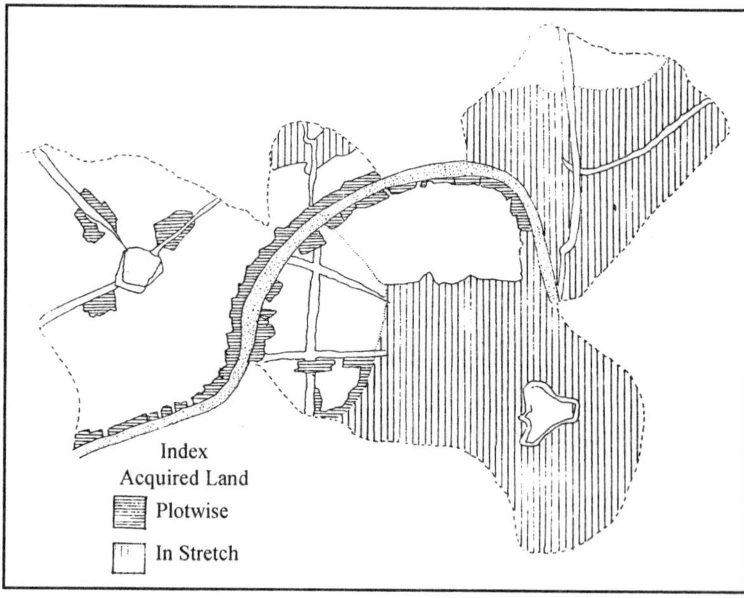

Fig. 17.4: Plotwise Acquisition of Land in Sample Study after 1981

rate of encroachment calculated for the period 1981-91 is the highest for Hastsal village being 10.36 hectare per year. The corresponding figures for Tilongpur Kotla, Razapur Khurd and Baprala village are 0.82, 9.56 and 3.45 hectare/year respectively. The rate of encroachment has been the highest for those villages lying near the urban limits. Hastsal village lies near the Najafgarh Road and Nilothi near the Rohtak Road. Both the villages are well linked by the road. Therefore, it is accessibility which is responsible for the acquisition and development of an area negating the land quality factor.

Initially the rate of encroachment was high along the Najafgarh Road and Najafgarh Drain, but, from the survey it was found that after 1981, acquisition process also spread in the interior of the village.

Land qualities are attributes of land units that have distinct influence on their suitability for different land uses. A detailed examination of the plotwise areas of five villages was made. The entire land quality was divided qualitatively into three categories of A, B and C type (Fig. 17.5). The A type or good quality land is fit for cultivation or irrigated land or land near the road, drain or *laldora*. The B type or medium quality land is unirrigated while the C type or poor quality land is the waterlogged area, rugged land and the land under Gram Sabha.

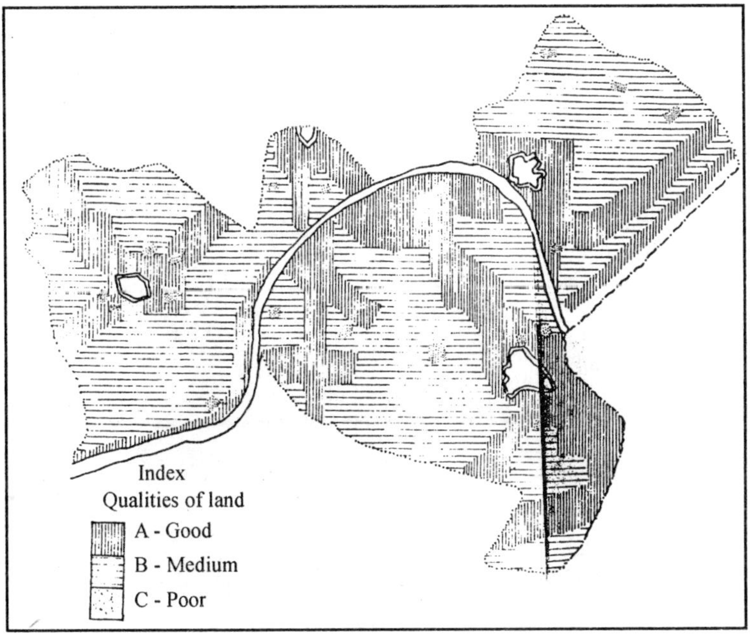

Fig. 17.5: Land under Different Qualities of Land.

It is evident from Table 17.2 that the maximum per cent of land falls under the A and B quality which is considered as fertile and capable of high yields and it is this land which has fallen prey under the process of encroachment.

Table 17.2: Percentage of Total Area under Different Qualities of Land.

Land Quality	*Total Area (In Ha)*	*Percentage of Total Area*
A	926.32	49.44
B	793.88	42.37
C	153.47	8.19
	1873.67	100.00

Source: Village Patwari Records, 1991.

Urban Encroachment and Land Quality : An Evaluation

A detailed examination of the relationship between land quality and urban growth was carried out by Best and Swinnerton (1970) wherein

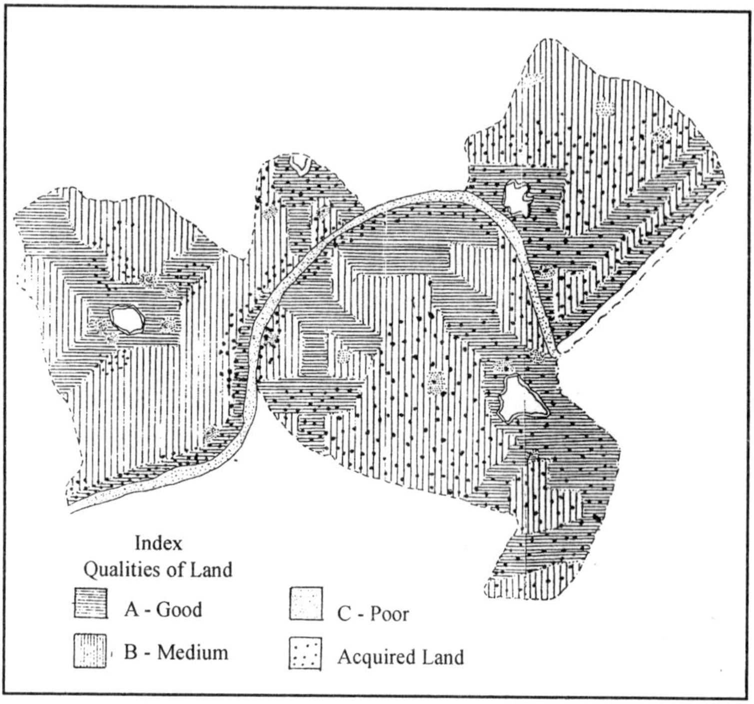

Fig. 17.6: Superimposed Map of Land under Various Qualities and Acquired Land.

they analysed sample of towns with considerable expansion. In India, this kind of work has been done by Siddiqui and Hussain (1977) for Ghaziabad city. In the study area, the land qualities falling under the encroached area are shown in Table 17.3.

Table 17.3: Encroached Area under Different Land Qualities.

Villages	*Land Quality*		
	A	*B*	*C*
Baprala	380.12	143.20	03.23
Tilongpur Kotla	10.10	20.00	0.00
Nilothi	130.20	75.00	47.00
Razapur Khurd	66.60	30.00	0.00
Hastsal	144.00	318.00	13.50

Source: Village Patwari Records, 1991.

From the table 17.3, it would be clear that of the total land in the study area, 731.02 hectare or 78.91 per cent of the A quality land, 73.64 per cent of the B quality land and 41.52 per cent of C quality land have been acquired. A quality land is found along the drain which has been acquired for construction of road by the government (Fig. 17.6). The highest amount of land acquired falls in this category. On observing the superimposed map (Fig. 17.6) it is quite interesting to note that hundred per cent A quality land has been acquired in Nilothi and Baprala followed by Hastsal (60%). The C quality category is the least acquired land, i.e. only forty two per cent. This indicates that the highest quality land is being acquired maximum, whereas the lowest quality is acquired the least. It also means that the acquisition process is not associated with the quality of land. It is any type of land which fetches the maximum profit is acquired by the private colonisers. The following section on changing land value pattern provides a better picture on this relationship.

Changes in Land Value Pattern

The lure of urban lifestyle, better transportation, housing, education and medical facilities, white collared jobs attract the farmers towards the city and force them to sell their farms and enjoy the urban lifestyle; whereas the urban rich is moving towards the periphery for a quiet and peaceful life. This leads to an increase in the land values along the margins.

Land values in and around the study area in Hastsal, Razapur, Khurd, Matola are increasing every year. The compensation given to the farmers on these acquired lands has also increased. In 1955, the compensation per acre given by the Government Revenue Department was Rs. 20,000 which increased to Rs. 4.60 lacs per acre by 1981.

During the field survey, it was found that in actual practice the ratable value of land along the Najafgarh Road was more than Rs. 5,000 in 1981 which has escalated between Rs. 8,000 to Rs.10,000 in Hastsal. The main factor deciding the land value is the accessibility from the road. As we move to the interior from the main road, the value of land goes down proportionately irrespective of the fertility of land.

The process of urban encroachment has certain impacts – the rural Delhi has shrunk and has made rural character of villages a thing of the past. The disappearance of greenery has disturbed the ecology. The people who sold their land in lure for money have lost their money in gambling, weddings, etc.

Conclusion

The problem of urban encroachment is an unnoticed one. It is in operation from the last three decades taking away thousands of hectares of good quality land for urban uses. The rate of conversion analysed in the study area was maximum for the villages adjoining the urban limits in a linear pattern along the Najafgarh Road and drain Acquisition was also observed in the interior of the villages in small pockets which later on coalesced to form a continuous patch. It was found that the acquisition process is not associated with the quality of land. The hypothesis stating that the better quality land is more resistant to development has been rejected. In fact, it is the land which fetches the maximum profit is acquired by the private colonisers.

Land being a finite resource, all necessary steps should be taken to preserve it specially the good productive agricultural land. It has been strongly felt the need for decentralised pattern of urbanisation. To check the increasing population, other ring towns or satellite towns in National Capital Region need to be developed so that they can act as counter magnets to check migrations to Delhi. Proper amendments in the existing rules and procedures or Acts passed by the government should be made. The less fertile land could be used for construction purposes but the more fertile land should be protected at all costs. Steps should be taken to keep a check on the colonisers to prevent the good quality land from being encroached upon.

REFERENCES

Aziz, A. (1968), "Impact of Urbanisation on Rural Land", *The Geographer*, Vol. XV, (1), p. 30.

Best, R.H. and Champion, A.G. (1970), "Regional Conversions of Agricultural Land to Urban Use in England and Whales; 1945-1967", *Transactions of the Institute of British Geographers*, Vol. 49, p. 1532.

Bryant, C.R. (1981), "Agriculture in an Urbanising Environment; A Case Study for the Paris Region, 1968-1976", *Canadian Geographer*, Vol. 25 (1), pp. 27-45.

Dunford, R.W. (*et. al.*) (1985); "A Case Study of Rural Land Prices at the Urban Fringe including Subjective Buyer Expectations", *Land Economics*, Vol. 61 (1) pp. 10-16.

Hart, J.F. (1968), "Loss and Abandonment of Cleared Farmland in the Eastern U.S." *Annals of Association of American Geographer*, Vol. 58, pp. 417-39.

Harvey, R.O. and Clark, W.A.B. (1965), "Nature and Economics of Urban Sprawl", *Land Economics*, Vol. 41, (1), pp. 1-9.

Hussain, I.M. and Siddiqui, M.F. (1978), Urban Encroachment on Rural Lands: A Case Study of Modi Nagar", *The Geographer*, Vol. xxvi, (1), p.14.

Hussain, I.M. and Siddiqui, M.F. (1979), Urban Encroachment on Rural Lands: A Case Study of Ghaziabad City, *The Geographer*, Vol. xxiv, (2) p. 112.

Libery, B.W. and Evans, N.J. (1989), "Estimating Land Loss in the Urban Fringe: A Comparison of the Agricultural Census and Aerial Photograph", *Geography*, Vol. 74.

Lockertz, W. (1989), "Secondary Effects on Mid Western Agriculture of Metropolitan Development and Decrease in Farmland", *Land Economics*, Vol. 65, (3) pp. 205-16.

Rikkinen, K. (1972), "Boundary Changes in Rural Urban Fringe Areas", Acta *Geographica*, Vol. 26, p. 25.

Russwurm, L.H. (1973), "*The Urban Fringe in Canada : Problems, Research Needs, Policy Implications*". A report submitted to the Research Branch, Ministry of State, Urban Affairs, Ottawa, p. 5.

Shonkwiler, J.S. and Reynolds, J.E. (1980), Vol. 62, (1), *Land Economics*, p. 58.

Sinclair, R.J. (1967), "*Von thunen and Urban Sprawl*", *Annals of the Association of American Geographer*, Vol. 87, pp. 72-87.

Singh, R.L. (1967), "The Rural Urban Fringe Studies", *The Geographer*, Aligarh, Vol. 14, p. 921.

Thakur, B. (1991), *Land Utilisation and Urbanisation : The Rural-Urban Fringe*, Deep and Deep Publications, New Delhi, p. 40.

18

Sprawling Cities and Loss of Agricultural Land
A Case Study of Indore

VIJAYA PHANSE, VENU TRIVEDI and RAJESH AWASTHI

Urban sprawl refers to the areal expansion of urban concentrations beyond which they have been (Northam, 1979, 467). Ultimately, it involves conversion of peripheral agricultural land to urban uses. The most serious problem associated with an urban sprawl is the inefficiency of landuse. There exists a constant conflicting demand for land between rival claimants and city expansion takes place in the form of new residential, commercial and industrial use of farmland beyond the existing limits of the city. It tends to occupy "space" which is most accessible and acceptable for development at a minimum cost.

After 1961, the country has witnessed uneven, unplanned, uncontrolled and haphazard growth of urban centres. Though the level of urbanisation in the state of Madhya Pradesh is low (20.29 % in 1981 and 23.21 % in 1991) as compared to Maharashtra, Tamil Nadu, Karnataka, Gujarat as well as the national average (25.72%), yet cities like Indore and Bhopal are fast growing in population and area and have attained the status of million cities in 1991. Similarly, the number of towns in Madhya Pradesh has increased from 250 in 1971 to 325 in 1981 and 433 in 1991.

Objective and Data Base

In the present study an endevour has been made to assess how much fringe land has been included in the sprawling city of Indore between 1961-76 and what changes have occurred in landuse of the incorporated villages thereafter. The problems associated with urban sprawl have also been highlighted.

The study is based on primary and secondary data. Since the landuse data regarding six villages before and after inclusion was not available, it was collected by the researchers through an extensive field investigation. The field survey was conducted in August, 1990 to incorporate the recent changes in landuse. The village *Patwari*, cultivators and senior persons of the villages were interrogated to collect the data. For the remaining seven villages, the data was collected from the Statistical Office, Collectorate, Indore. Data thus obtained was analysed.

Methodology

The study pertains to the changes in the landuse pattern of 13 villages included within the corporation limits of Indore metropolis. The landuse for all the villages before and after inclusion was mapped and compared to examine the changes. Band-graph is also prepared to show the changing trend in landuse.

Study Area

Situated along north-flowing Khan and Saraswati rivers of the Malwa plateau and N.H. No. 3, Indore is the largest metropolitan city of Madhya Pradesh with a population of over 11 million, according to the 1991 census. Indore (Lat 22° 43′ N and Long 75° 50′ E) stands 448.64 above MSL and covers an area of 113.52 sq. km. National Highway No. 3 connecting Agra with Bombay passes through the middle of the city. State Highway Nos. 27 and 22 joining Indore with Ujjain and Neemuch, respectively, pass through the city. Although Indore is not a railway junction, yet it lies on Ajmer-Khandwa metre-gauge line of the Western Railway. It is also connected with Delhi, Bombay, Howrah, Madras and Kochin by broad-gauge railway line. There are regular flights from Indore for Bombay, Delhi, Ahmedabad via. Gwalior and Bhopal. The degree of connectivity has made Indore the commercial capital of the State (Fig. 18.1).

Favourably situated on the Deccan trap with rolling topography, it has a gentle north-easternly slope. Towards south of the city height increases towards Vindhyan escarpment, reaching a maximum of 881.18 m. To the south-east of the city lie Deoguradia and Ralamandal hills truly representing the trap topography. Indore

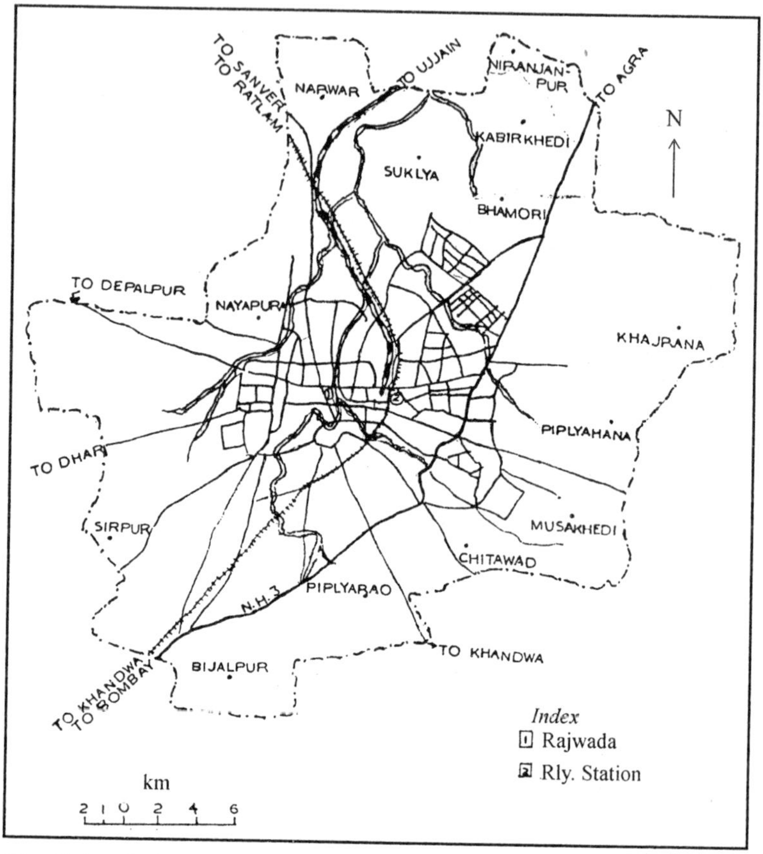

Fig. 18.1: Indore City Base Map

enjoys the typical monsoon type of climate with hot summers and cool winters. The rich black soil better known as "regur" is best suited for cotton cultivation.

Landuse of the 13 Villages Incorporated within the Municipal Corporation Limit of Indore

Before Inclusion

A study of the general landuse (Appendix 1) reveals that the area occupied by forests before inclusion was insignificant covering 150 hectares, out of a total of 7,000 hectares land of 13 villages. The maximum area under forests may be noted in Bijalpur (61 hectares), where they occupied 10.39 per cent area. Khajarana stands next with 53 hectares of forestland. Chitawad and Piplyarao had 20 and 16 hectares of forest land respectively.

Agriculture was the main occupation of the people with over 53.55 per cent of land under plough. Nayapura, Niranjanpur, Khajrana and Musakhedi situated away from major transport arteries (Fig. 18.1) had more than 70 per cent land under plough. Narwar had the lowest (42 hectares) cultivated land. Out of 3,749 hectares of net sown area, 1,419 hectares were double cropped. Suklya, followed by Bijalpur and Kabirkhedi, had larger shares of double-cropped area (Appendix 1). Suklya shows the maximum intensity of landuse where 320 hectares out of a total of 323 hectares of net area sown were double-cropped.

Area not available for cultivation was an insignificant 17.84 per cent. Narwar followed by Piplyahana and Suklya claimed large shares of 78.64, 35.57 and 35.56 per cent area respectively. This is significant of smaller demands for other urban uses of land specially residential. Piplyarao, Kabirkhedi, Khairana and Sirpur had less than 6 per cent land under this category. Wasteland occupied a substantial share of 26.45 per cent area. Sirpur claims the lion's share (54.30%), followed by Chitawad (37.85%), Piplyahana (29.02%) and Kabirkhedi (25.53%). This shows low demand of land for other productive uses.

In all the villages, wheat was the dominant crop followed by jowar. Sugarcane and groundnut were other major crops with respect to area. Table 18.1 provides the relative importance of agricultural, residential and other uses of land. The last category includes forests, industries, brick kilns and transport (Table 18.1). Before inclusion of villages the residential area occupied 3,370.34 hectares (41.3%). The highest residential area (84.05% of the village area) may be traced in Narwar. Narwar is situated along State Highway No. 27 joining Indore with Ujjain. Bijalpur with 64.53 per cent and Piplyahana with 63.94 per cent of residential areas stand next to Narwar. Bijalpur is favoured by its location on A.B. Road to the south of the city and Piplyahana enjoys proximity to the city centre. It may be assumed, therefore, that accessibility has the major contribution in city sprawl.

Table 18.1: Landuse Before Inclusion (1961).

Sl. No.	*Name of Village*	*Total Area Hec.*	*Agricultural Area %*	*Residential Area %*	*Others* Area %*	*Total Area %*
1.	Piplyarao	110	54.54	30.00	15.46	100.00
2.	Chitawad	140	33.57	32.00	34.43	100.00
3.	Bhamori	64	62.5	31.25	6.25	100.00
		Landuse Before Inclusion (1976)				
1.	Kabirkhedi	305	67.57	31.05	1.38	
2.	Khajrana	1520	77.36	18.42	0.74	100.00
3.	Musakhedi	407	71.25	27.86	0.89	100.00
4.	Sirpur	1524	41.86	57.07	1.07	100.00
5.	Suklya	793	40.73	58.41	0.86	100.00
6.	Narwar	276	15.21	84.05	0.74	100.00
7.	Nayapura	64	82.82	15.62	2.56	100.00
8.	Niranjanpur	555	78.9	20.19	0.91	100.00
9.	Piplyahana	655	35.41	63.93	0.66	100.00
10.	Bijalpur	587	34.58	64.53	0.89	100.00

* Other area includes forest, industries, transport, etc.

Before inclusion of 13 villages, 178 handicraft and other small-scale industries were functioning in the villages. Poultries (53), brick kilns (35), leather goods (28), jaggery making (18), piggeries (14) were some of the important small-scale industrial units.

After Inclusion

Agriculture in the fringe is inevitably influenced by urban sprawl, with the result that there is loss of productive farmland. Anticipatory purchase of fringe land with price rise in view leads to fragmentation of holdings, inefficiency of landuse, demoralisation of farmers and growth of hobby farmers or pseudo-farmers. The real loss of farm output, as a result of urban sprawl, comes from commercial farmers whose operations are threatened by urban expansion.

In 1981, the population of Indore urban agglomeration was 829,327, which increased to 1,104,065 in 1991. Though the growth rate between 1981-91 is slower (33.13%) than the growth rate recorded during 1971-81 (47.85%), yet there is a marked increase in absolute population. Population growth is always associated with the demand for land for non-agricultural uses. Consequently, three surrounding villages in 1961 and 10 villages in 1976 were included within the corporation limits of the city. After the inclusion, a number of changes occurred in the landuse of the villages, the major being putting the agricultural land to non-agricultural uses. This should be taken not merely as a shift of land from one category to another but has led to a number of serious problems associated with it. The fertile, valuable farmland is used for industries, roads, residential buildings, brick kilns, etc. This process is repeated as the city sprawls at the cost of agricultural land.

It is obvious (from Appendix 2) that there is no change in the geographical area of the 13 villages but there are drastic changes in the landuse. The forest area has been reduced from 150 hectares before inclusion to 45 hectares after inclusion. In Khajrana village, so much deforestation has taken place that practically no forests are left.

There is net loss of 872 hectares of agricultural land after the inclusion of the villages. It has been reduced from 3,749 to 2,877 hectares after inclusion. Bhamori and Suklya are the major victims of the urban sprawl. In Bhamori the net sown area is reduced from 40 to 24 hectares while in Suklya it is reduced from 323 to 100 hectares. The village which is least affected is Bijalpur where an insignificant loss of over 2 per cent is recorded. This may be attributed to its distance from the city centre.

There is close association between urban sprawl and high pressure of population on remaining arable land. It has a two-fold effect — one, double-cropped area is increased in response to population pressure on farmland and the other farmers are discouraged and the double-cropped area is reduced. The former result has been experienced in Chitawad, Bhamori, Muskhedi, Sirpur, Narwar and Nayapura, all situated away from highways. Piplyarao, Kabirkhedi, Khajrana, Suklya, Niranjanpur, Piplyahana and Bijalpur have witnessed reduction of double-cropped area. These villages enjoy better access to highways and hence are better suited for residential colonies.

Before inclusion 1,495 hectares of land were double-cropped, while after inclusion it was reduced to 881 hectares — a net loss of 614 hectares. In Suklya the double-cropped area has been reduced from 320 hectares before 1976 to 60 hectares in 1990. Piplyahana and Khajrana also show a reduction of 95 and 64 hectares of land respectively.

Wasteland has also been reduced remarkably in all the villages showing an accute demand for land for non-agricultural uses. In Khajrana it has been reduced from 238 to 25 hectares followed by Sirpur where a net reduction of 328 hectares is noticed. Suklya, Piplyahana and Bijalpur have also witnessed the same trend. Wasteland before inclusion accounted for 1,852 hectares but after inclusion it was reduced to 1,203 hectares.

Table 18.2: Landuse After Inclusion (1961).

Sl. No.	*Name of Village*	*Total Area Hec.*	*Agricultural Area %*	*Residential Area %*	*Other* Area %*	*Total Area %*
1.	Piplyarao	110	36.36	51.00	12.64	100.00
2.	Chitawad	140	28.57	55.00	16.43	100.00
3.	Bhamori	64	37.5	58.37	4.12	100.00
			Landuse After Inclusion (1976)			
1.	Kabirkhedi	305	84.27	52.80	2.93	—
2.	Khajarana	1520	59.21	38.01	2.78	100.00
3.	Musakhedi	407	61.43	34.39	4.18	100.00
4.	Sirpur	1524	34.77	62.00	3.23	100.00
5.	Suklya	793	12.61	86.39	1.00	100.00
6.	Narwar	276	12.68	85.32	2.00	100.00
7.	Nayapura	64	51.56	47.80	0.64	100.00
8.	Niranjanpur	555	72.07	26.93	1.00	100.00
9.	Piplyahana	655	20.53	68.47	1.00	100.00
10.	Bijalpur	587	32.36	66.43	1.21	100.00

* Other area includes forest, industries, transport, etc.

It is apparent from Table 18.2 and Fig. 18.2 that practically in all the villages the agricultural land has been reduced considerably. Nayapura closely followed by Suklya, Bhamori and Kabirkhedi have lost 31.26, 28.12, 25.0 and 23.30 per cent of agricultural land respectively. Simultaneously, there is a substantial increase in the residential area almost in all the villages. The major change may be noticed in Nayapura, Suklya and Bhamori with an increase of 32.18, 27.98 and 27.12 per cent in the residential area.

Wheat has retained its dominance but Soyabean is emerging as the second dominant crop specially in Kabirkhedi, Khajrana,

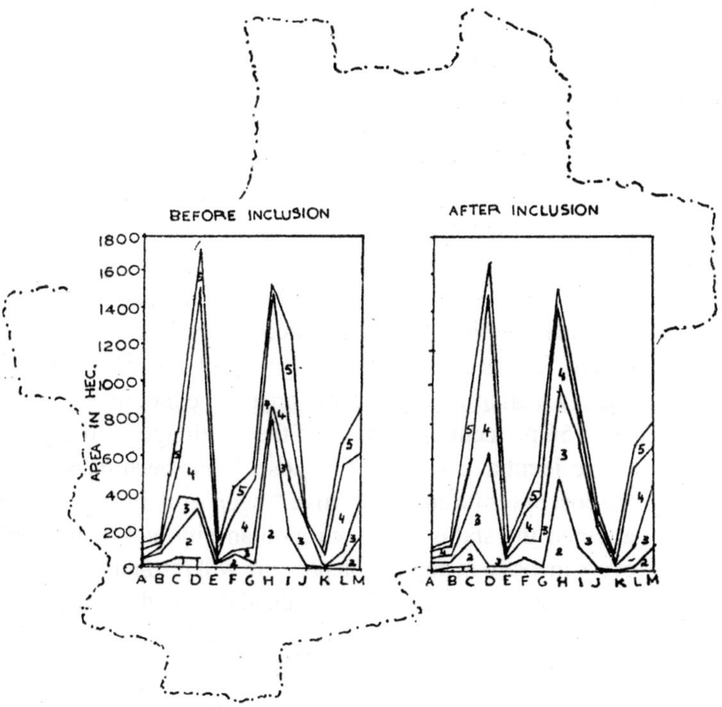

A — Piplyarao
B — Chitawad
C — Bijalpur
D — Khajrana
E — Bhamori
F — Kabirkhedi
G — Musakhedi
H — Sirpur
I — Suklya
J — Narwar
K — Nayapura
L — Niranjanpur
M — Piplyahana

1 Forest
2 Wasteland
3 Land not available for cultivation
4 Net sown area
5 Double-cropped land

Fig. 18.2: Indore City Changes in Landuse Pattern.

Musakhedi, Narwar, Suklya, Nayapura, Piplyahana, Niranjanpur and Bijalpur. Zaid crop area has been increased specially in Suklya and Bijalpur.

One hundred and eighty-nine small-scale industrial units are functioning in the study area after inclusion. Poultries (62), leather goods (33), basket-making (20), piggeries (20) are among the important units.

Conclusion

From the above study, the following conclusions may be drawn:

(i) After the major inclusion of 1976, the total agricultural land lost was 872 hectares, which happens to be rich black cotton soil. In 1981, per capita agricultural land within the corporation limits recorded was .0034 hectares, which is further reduced to .0026 hectares in 1991. This is quite alarming and needs attention of the geographers and planners alike.

(ii) An equally serious matter of concern is the loss of forestland. Today, forests cover a very insignificant area of 45 hectares out of a total of 11,352 hectares area of the Indore metropolis.

(iii) Over 56 per cent of the land of the 13 villages comes under the residential category of landuse. It has been increased at the cost of fertile and productive "regur" soil. In 1991, Indore attained the status of a metropolitan city and, henceforth, every increase in population pressure will result in conversion of more agricultural land to residential and other urban uses. The alternative lies in intensive use of arable as well as urban land to check further expansion of the city.

REFERENCES

Carter Harold (1981), *The Study of Urban Geography* (3rd Edn.), pp. 323-24.

Census Handbook of M.P., Village and Town Directory, General Economic Tables Pt. III, Social and Cultural Tables Pt. IV, General Population Tables Pt. II-A.

Provisional Population Table 18.4, 1991.

Ray, M. Northam (1979), *Urban Geography* (2nd edition), p. 467.

Singh, L.R., "Changing Landuse Pattern in the Urban Fringe of Allahabad: Umarpur Niwar: A Case Study", *Trans of the Indian Council of Geographers* No. 38, pp. 32-45.

Singh, L.R. and Shrivastava, M.K. (1970), "Changing Landuse Patterns in the Urban Fringe of Allahabad : Baghara Balan – A Case Study, *Urban Geography in Developing Countries*, (ed.) R.L. Singh, pp. 194-98.

APPENDIX 1

Landuse of 13 Villages Before Inclusion (1961)

Sl. No.	Name of Village	Total Area Hec.	Forest Area Hec.	%	Wasteland Hec.	%	Land Not Available Hec.	%	Net Area Sown Hec.	%	Double-cropped Area Hec.
1.	Piplyarao	110	16	14.5	30	27.2	4	3.7	60	54.54	25
2.	Chitawad	140	20	14.29	53	37.85	20	14.29	47	33.57	25
3.	Bhamori	64	—	—	20	31.25	4	6.25	40	62.75	20
	Landuse of 13 Villages Before Inclusion (1976)										
1.	Kabirkhedi	305	—	—	80	26.53	18	5.90	207	67.57	146
2.	Khajarana	1520	53	3.48	238	15.65	53	3.48	1176	77.36	214
3.	Musakhedi	407	—	—	27	6.63	90	22.11	290	71.25	70
4.	Sirpur	1524	—	—	828	54.30	58	3.84	638	41.86	29
5.	Suklya	793	—	—	188	23.71	282	35.56	323	40.73	320
6.	Narwar	276	—	—	17	6.15	217	78.64	42	15.21	7
7.	Nayapura	64	—	—	1	1.56	10	15.62	53	82.82	3
8.	Niranjanpur	555	—	—	40	7.32	77	13.87	438	78.90	120
9.	Piplyahana	655	—	—	190	29.02	233	35.57	232	35.41	220
10.	Bijalpur	587	61	10.39	140	23.85	183	31.18	203	34.58	150
13.	**Total**	**7000**	**150**		**1852**		**1249**		**3749**		**1495**

APPENDIX 2

Landuse of 13 Villages After Inclusion (1961)

Sl. No.	*Name of Village*	*Total Area Hec.*	*Forest Area Hec.*	*%*	*Wasteland Hec.*	*%*	*Land Not Available for Cultivation Hec.*	*%*	*Net Area Sown Hec.*	*%*	*Double-cropped Area Hec.*
1.	Piplyarao	110	6	5.23	40	36.14	25	22.50	40	36.14	20
2.	Chitawad	140	17	12.14	43	30.72	40	28.57	40	28.57	30
3.	Bhamori	64	—	—	35	54.68	24	7.81	24	37.51	27
	Landuse of 13 Villages After Inclusion (1976)										
1.	Kabirkhedi	305	—	—	70	22.95	100	32.78	135	44.27	50
2.	Khajarana	1520	—	—	25	1.64	595	39.15	900	59.21	150
3.	Musakhedi	407	—	—	20	4.91	137	33.66	250	61.43	80
4.	Sirpur	1524	—	—	500	32.80	494	32.43	530	34.77	55
5.	Suklya	793	—	—	140	17.65	553	69.74	100	12.61	60
6.	Narwar	276	—	—	10	3.62	231	83.70	35	12.68	20
7.	Nayapura	64	—	—	10	15.62	20	32.82	33	51.56	14
8.	Niranjanpur	555	—	—	35	6.31	120	21.62	400	72.07	110
9.	Piplyahana	655	—	—	150	22.91	305	46.56	200	30.53	125
10.	Bijalpur	587	22	3.70	125	21.29	250	42.55	190	32.36	140
13.	**Total**	**7000**	**45**		**1203**		**2875**		**2877**		**881**

19

Capability Classification and Reclamation of Wasteland in Etawah District, U.P.

B. THAKUR and MANU BAJPAI

Wasteland is receiving a lot of attention in the recent decades. The ever increasing population pressure and consequent demands for food, fuel and fodder have made it essential to look for additional land that could be utilised for satisfying some of those demands. The vast stretches and varieties of wasteland that exist throughout the length and breadth of the country are the areas, which attract attention as the potential lands that could be identified, developed and put to gainful uses. This dimension therefore demands efforts to study the ways and means to utilise wasteland by multidimensional approach in order to derive information to be used purposefully to overcome the menace of wastelands.

Nature of Problem

Land comprising soil, water and associated plants and animals involving the total ecosystem is the most important among major resources of the country. Its utilisation determines the pace of India's economic growth. Land resources fixed land resources. Ill effects of development like destruction of forest cover, mismanaged irrigation, waterlogging, salinity and unscientific agricultural practices are responsible for degradation of land resources. Adverse effects of environmental forces are also adding to the increasing menace of degradation.

Productivity of present agriculture land cannot be increased beyond certain limits, moreover, application of technology on a mass scale is limited in our country. In order to meet the demands of ever-increasing population and to achieve the cherished goal of sustainable development, reclamation and utilisation of wasteland is inevitable. Our country is facing severe problems of wastelands and if allowed to continue it may reach alarming heights. According to National Wasteland Development Board, 37.5 per cent of Indian geographical area is wasteland. Evaluation of problem on district level points out that 148 districts of India are critically affected by the problem of wasteland. In Etawah district alone which is the study area of the present work, more than 600 sq. km. of land are turned into different types of wastelands which are posing threats not only to environment and ecological security but also to sustainable socio-economic development. Various projects on saline soil and ravine land reclamation are running from last ten years but the problem remains unchanged.

Keeping in mind the correct assessment of the problem and solution in the broad objective, wasteland capability classification is given special attention in this work. The capability class determine the potential use of any piece of land and further helps in utilisation. Such studies may help in reducing pressure of population on land and also to meet agricultural demand for the growing population. Utilisation of land particularly in plain areas has almost reached at the saturation point. It is, therefore, desirable to utilise wasteland and evaluate it on the basis of its potential so that intelligent decision can be taken to utilise it for different purposes and needs.

Review of Literature

The literature available on wasteland capability classification is scarce. However, few scholars have attempted in this direction. Yadav (1984) presented capability classification of wasteland in Sultanpur district considering various environmental and anthropogenic factors which quantify the extent of influence of these factors. In another study, Yadav (1987) discussed the conceptual framework of wasteland capability classification by defining reclamation cost as the suitable base for categorising capability classes. Kumar (1989) adopted integrated approach to wasteland capability determination for landuse planning of Deoghar district. The author encoded environmental attributes, assigned capability value and gave them weighted and computed rating for each wasteland. The present work was an attempt to present an outline on the utilisation of wasteland.

Studies on reclamation of wastelands are vast and diverse. Singh (1978) studied the economics and geography of agricultural land reclamation in Koil tehsil of Aligarh district. Jha (1984) presented the effective utilisation of different types of wasteland in the Birbhum district. The Academy of Science of USSR (1958) has discussed the application of drainage and other scientific reclamation measures for the saline soils. Sharma, Tripathi and Sharma (1988) described various reclamation measures for *usar* land in the Sangrampur block of Sultanpur district. Prasad and Yadav (1969) have prepared a list of questions and answers for the alkali and saline soil reclamation and have also pointed out water quality, drainage measure, cultivation, sowing time, nutrient requirements, control of pest and diseases for reclamation. Kumar and Panday (1983) determined the inherent characteristics of wasteland, causes and hazards for reclamation and controlling the destructive tendency operating over them. This work also provides necessary information required for conservation programme under which wasteland presently degraded by erosion can be improved.

Singh (1962) presented the possibilities of reclamation of cultivable wastelands, which are being used as threshing grounds and sites for social and religious performance. Yadav and Padmaja (1985) attempted to reclaim usarland of Sultanpur district revival and development by correlating various morphometric, climatic, hydrogenic and pedogenic factors and presented reclamation structures on the basis of quantifying values. Yadav and Aggrawal (1959) provided measures to reclaim saline and alkali soils along with assessment of their potential. Shukla (1988) studied the eroded lands of the Yamuna and Chambal rivers of Etawah district and provided empirical structure for ravine reclamation.

In the field of wasteland reclamation most of the work is done by governmental organisations. Their works are based on the physical characteristics of soils, geology, morphometric and other factors. However, anthropogenic factors and their correlation with the problem are neglected in their studies. If we incorporate wasteland capability classification and its reclamation, then hardly any studies can be found. On the other hand, the combined aspects may provide an ideal model for the utilisation of wasteland, which is the core of this study.

Objectives of the Study

1. To show distribution of wasteland in Etawah district.

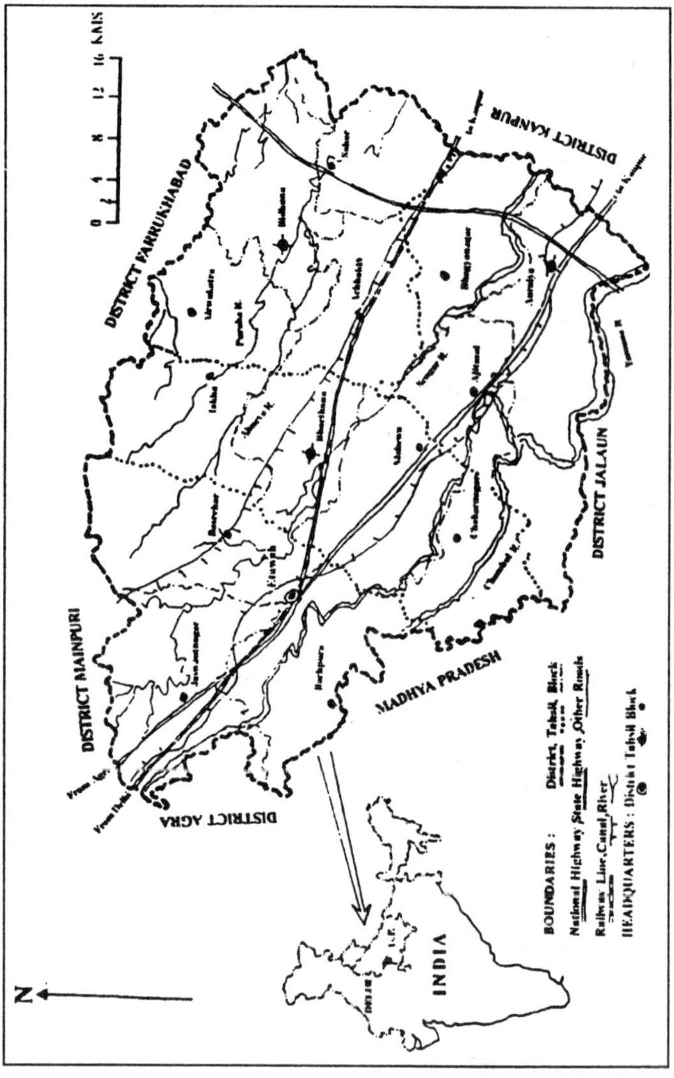

Fig. 19.1: Administrative Divisions (Etawah District).

2. To present wasteland capability classification by measuring the intensity of various physical and anthropogenic factors chosen for present study.
3. To determine association between various factors (physical and anthropogenic) and wasteland.
4. To suggest reclamation measures for different kinds of wastelands found in the Etawah district.

Study Area

The Etawah district lies in the south-western part of Uttar Pradesh in fertile soils of the Ganga-Yamuna doab. It extends between 26°21'N latitude and 78°45E and 70°45E longitude (Fig. 19.1). It is bounded on the north by Mainpuri and Farrukhabad districts, on the east by Kanpur district, on the South of Jalaun district, on the west for some distance by Agra district and the rest by the state of Madhya Pradesh. The district lies entirely in the Ganga plain but its physical features differ considerably and are determined by the Yamuna and Chambal rivers.

The present work chooses district as a unit of study because the present work is application-oriented and structured in such a way for which a district is more suitable for study area. Secondly, as a district is a recognised unit of socio-economic development in various government plans, it is more practical to correlate wasteland with this aspect. All the Government policies and projects on wasteland are applicable on administrative units which make easier to collect data and information.

The Etawah district aptly represents the problem of wastelands faced by U.P. Few districts of Uttar Pradesh are so severely affected with the problem of wastelands as the Etawah district. Six-hundred sq. km. of area is badly affected with various types of wastelands. Ravine land affected the southern region of the district. Saline soil is evenly distributed in northern part of the district. Waterlogged land, upland and degraded forestland are other types of wasteland found in the district. Only the middle portion of the district which is sandwiched between northern saline soil and southern ravine soil are agriculturally productive regions. This spatial distribution of wasteland makes Etawah peculiar and complex as compared to other districts of the state. Northern adjoining district of Etawah like Mainpuri. Etah and Fatehpur are mainly saline affected, whereas southern adjoining districts like Bhind are ravine affected districts.

The other main objective to choose the district is that the district is poor in socio-economic development, which can be correlated with poor utilisation of land and distribution of wasteland. The area has a potential for agricultural development and suitable measures for utilisation of wasteland will serve as input into the economic betterment of the area as a whole. The Etawah district is now divided into two districts. This study is conducted by taking into account the older administrative unit of Etawah district.

Data Sources

Data for present study were collected from both primary and secondary sources. General information related with people, agriculture and socio-economic aspects were collected from District Census Handbook. Village map of the district was obtained from Primary Census Abstract. Detailed geological map on 250,000 scale of the district was collected from Geological Survey of India, Lucknow. Soil maps of the district have been collected from National Bureau of Soil Survey and Land Use Planning, which also provide information on soil characteristics of the district. The district is a plain area therefore slope map could not be prepared. Slope map was derived from soil map of the district, which had detailed slope information. Land erosion map of the district has been obtained from National Bureau of Soil Survey and Landuse Planning. Information on natural regions was obtained from Census Atlas of Uttar Pradesh. Wastelands maps of the district on 1:50,000 scale has been obtained from Remote Sensing Application Centre. Ravinous soil map on 250,000 scale of the district was collected from Geological Survey of India. Saline soil map was collected from National Remote Sensing Agency, Hyderabad. Data on various determinants used for capability classification were collected from secondary sources. Rainfall data has been collected from Indian Meteorological Department. Data on groundwater level and Ph value of groundwater were collected from Central Groundwater Commission, Lucknow. Sodicity, electrical conductivity of soil and surface soil drainage information were collected from National Bureau of Soil Survey and Landuse Planning and Remote Sensing Application Centre. Some valuable information was also obtained from Usar Land Reclamation Project, sponsored by World Bank. Canals & tubewell data were obtained from Deputy Director, Agriculture Office, District Etawah. Suitable strategy of reclamation of wasteland has been recommended by surveying the village personally. Twenty-nine villages were selected on the basis of simple random sampling to find correlation between wasteland and various factors. One village from each type of wasteland was been selected to present reclamation remedies for each type of wasteland.

Methodology

Distribution, identification and correct assessment of wasteland is done on 1:50,000 scale by the maps obtained from Remote Sensing Application Centre. Correct demarcation of wasteland is the basic task of this research as accurate analysis of other objectives depends upon this work. Wasteland is mapped for every administrative unit, that is, from village to district level. Demarcation of wasteland is done on the basis of criteria given by National Wasteland Development Board. Maps of different types of wastelands on district, tehsil and block level have been prepared. Attempt has also been made in present study to correlate wasteland with various physical and socio-economic conditions of area. Wasteland area of each and every village is calculated with the help of graph method and curve meter since reliable data is lacking.

Capability classification of wasteland is done on the basis of various parameters mentioned below:

1. Sodicity of the soil
2. Surface soil drainage
3. Electrical conductivity of the soil
4. Land erosion
5. Slope
6. Water table depth (in dry season)
7. Electrical conductivity of groundwater
8. Mean annual rainfall
9. Number of tubewells (state and private tubewells)
10. Length of canals.

The study is conducted on two levels, namely, district and village. Capability classification on district level is the difficult task because of a selection of unit for a micro-level analysis. Few conceptual and practical problems arise in the selection of micro unit for capability classification such as:

(a) Villages cannot be selected for capability classification as a unit because there are 1563 villages and it is impractical for present purpose.
(b) All types of wastelands found in Etawah are so evenly distributed in patches of varied size that it is difficult to categorise these wastelands into grid. Size of patches ranges from less than one hectare to more than 1000 hectares.

Therefore, to overcome these problems whole of the Etawah district is divided into grids. Distribution of wasteland in various grids is properly demarcated. Intensity of various ingredients taken for the capability classification is quantified grid by grid and on the basis of that wasteland of grids is grouped into four categories which are:

(a) Possibilities of utilisation are better
(b) Possibilities of utilisation are moderate
(c) Possibilities of utilisation are poor
(d) Possibilities of utilisation are very poor.

Where the intensity of all the ingredients are unfavourable, it implies lack of appropriate conditions for utilisation of wasteland. In the areas where intensity of factors is favourable it means wasteland can be utilised through various inputs and efforts. This is the basic theme behind the categorisation of wasteland into various capability classes. It is important to mention that capability classification is done on the basis of severity of wasteland. Some factors which do not show significant variations and because area of study is not large enough, e.g. depth and texture of soil etc. are excluded from capability classification.

The micro level study is conducted by selecting villages from each type of wasteland to formulate reclamation methods for wasteland improvement. Almost 860 villages of the district are affected with the problem of wasteland. Out of which, 29 villages have been selected by using simple random sampling to present the correlation between various factors and wasteland. This correlation further helps to provide a fair idea as to which particular factor is responsible for genesis of wasteland. Seven villages each are chosen from ravines and saline soil as these constitutes the most dominant types of wasteland. Five villages each are chosen from other three types of wasteland. To investigate the genesis of wasteland 10 factors are taken into consideration, which are given below:

1. Slope
2. Mean annual rainfall
3. Water table depth [(in dry season)]
4. Sodicity in esp
5. Electrical conductivity in (ecg (dsm-1)
6. Soil reaction
7. Texture
8. Number of tubewells and wells
9. Length of streams and canal.

Correlation between factors and wasteland is done by using Carl Pearson co-efficient of correlation formula.

$$r = \frac{\sum x, y - \frac{\sum x, \sum y}{N}}{\sqrt{\left\{\sum x^2 - \frac{1}{N}\left(\sum x_1\right)^2\right\}\left\{\sum y_1^2 - \frac{1}{N}\left(\sum y_1\right)^2\right\}}}$$

Out of these 29 villages, field survey is done for five villages, namely, Mirgowan, Kanmau, Hardu, Ramnagar and Garha Kosda representing saline, upland, waterlogged, degraded forest and ravine types of wasteland, respectively. Reclamation measures for these five villages are suggested taking into account the techniques of reclamation suggested by various institutes, such as Saline Research Institute, Karnal and Local conditions and perception of farmers related with the utilisation of wasteland.

Distribution of Wasteland

The Etawah district is prone to the problem of wasteland. More than 1/10th of its area is subjected to the wasteland (Fig. 19.2). Though, the distribution of wasteland is not uniform in the district, yet all the 14 blocks of the district constitute one or the other type of wasteland. The Barhpura block has more than 22 per cent of its total area most of which is ravinous. Besides this, blocks like Bidhuna, Chakarnagar and Auriya have 15 per cent to 20 per cent of their area under wasteland. Ajitmal is the least affected block

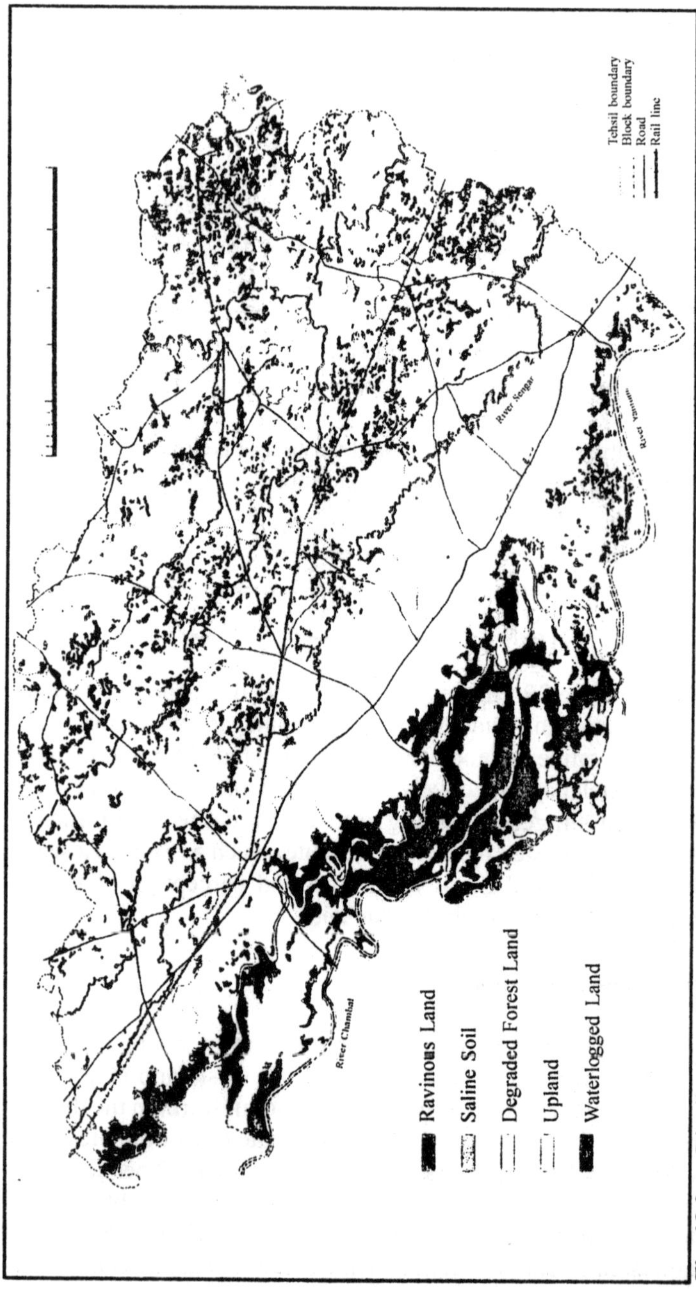

Fig. 19.2: Distribution of Wasteland (Etawah District, 1990)

Source: Wasteland maps of 1:50.000, Prepared by R.S.A.C., Lucknow.

Table 19.1: Distribution of Wasteland in Etawah District, 1990.

Sl. No.	*Block*	*Area (in hectares)*	*Type of wasteland*					
			Ravines	*Saline*	*Waterlogged land*	*Degraded forestland*	*Upland*	*Total*
1	Jaswantnagar	38800	2472	1786	10	—	—	4250
2	Bidhuna	31000	39.5	4648.2	810	0.15	—	5459.35
3	Basrehar	37500	39.5	3046.1	378.5	—	—	3464
4	Bharthana	25900	—	2534.2	29	—	12	2575.2
5	Takha	27500	—	2218	492	—	—	2710
6	Mahewa	32400	2514	246	—	32	—	2792
7	Chakarnagar	37200	5911	—	—	133	—	6044
8	Airwakatra	22400	—	1760.3	240	—	—	2000.3
9	Sahar	28400	—	2263	108	164	635	2598.3
10	Auriya	41400	7877.6	229	—	199	—	8225.6
11	Ajitmal	20700	753.5	30	—	30	—	813.5
12	Bhagyanagar	27600	—	3406.6	60	129	15	3676.5
13	Achalda	28200	—	2731.5	5	25	96.5	2858
14	Bharhpura	32900	6918.2	128.54	—	67	—	7107.54
15	Etawah Dist. Total	432600	26551.8	2132.5	2132.5	778.3	758.3	54460.4

Source: Primary data calculated with the help of maps of wasteland acquired from Remote Sensing Application Center, Lucknow, 1990.

of the district having only 4 per cent of its area under wasteland (Fig. 19.3). Rest of the block constitute about 10 per cent of its area under wasteland. Out of total 54604.39 hectares of wasteland in Etawah district, more that 15 per cent i.e., 8255 hectares are located in Auriya block, Bidhuna and Chakarnagar have more than 23 per cent of the total wasteland of Etawah district which is 5459 and 6044 hectares, respectively. The Ajitmal block is relatively less prone to wasteland.

The wasteland problem of the district is depicted more clearly if number of villages affected by wasteland in each block is analysed. Out of total 1543 villages of Etawah district around 925 (60% of total villages) are affected by wasteland, which implies to the gravity of the problem of wasteland in the district. In Chakarnagar tehsil no village is observed to be free from wasteland, whereas in Bidhuna, Takha, Sahar and Barhpura blocks, most of the villages are prone to wasteland. The Ajitmal and Mahewa blocks are found relatively free from wasteland (Table 19.2). Most of the wasteland prone villages in the district are affected by Saline wasteland, but their overall percentage wise area is less than ravines wasteland. Saline soil affected villages are small in area as compares to those affected by ravinous soil. Only 216 of all villages are affected by ravinous soils while 638 villages are affected by saline wasteland.

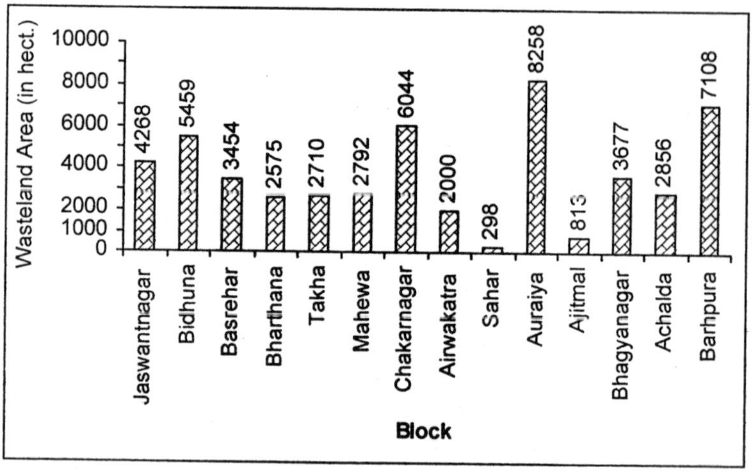

Fig. 19.3: Distribution of Wasteland (Etawah District, 1990).

Table 19.2: Villages Affected by Wasteland in Etawah District, 1990

Block	*Number of villages affected with wasteland*						
	Block area (in %)	*Saline*	*Ravines*	*Water-logged*	*Degraded forest land*	*Upland*	*Total*
Jaswantnagar	131	50	15	1	—	—	66
Bidhuna	105	84	—	24	1	—	109
Basrehar	142	74	4	13	—	—	83
Bharthana	82	57	—	3	—	1	57
Takha	78	54	—	14	—	—	55
Mahewa	118	8	22	—	3	—	33
Chakarnagar	38	—	31	—	7	—	38
Sahar	96	72	—	5	8	6	75
Auriya	192	12	73	—	4	—	92
Ajitmal	112	5	15	—	3	—	19
Bhagyanagar	134	80	5	6	4	4	83
Achalda	118	69	—	1	1	14	74
Bharhpura	94	10	51	—	4	—	63
Airwakatra	115	63	—	15	—	—	78
Etawah Dist. Total	1543	638	216	81	38	25	925

Source: Primary data generated with the maps of wasteland obtained from Remote Sensing Application Centre. Scale of maps is 1:50,000, 1990.

Types of Wasteland

In Etawah district five types of wastelands are found. These are: ravines land, saline land, waterlogged land, degraded forest land and upland with or without scrub. Detailed distribution of these lands is given below.

Ravinous Wasteland

Ravinous land, the major wasteland category covering 268551.8 hectare area of the district account for 51.75 per cent of the total wastelands and 6.13 per cent area of the district. All the nine rivers in the district have 26551.8 hectare of eroded land. In which, Yamuna, Chambal and Kunwari river basin covers 23,551.8 hectares of land. Major proportion of this ravinous land occurs in Auriya tehsil. At most of the places, these ravine patches are located near the reserved or protected forest and are infested with grasses, babul or other thorny bushes, and scrubs as cover.

The ravine wasteland on the basis of the magnitude of the problem of soil erosion can be demarcated into two categories:

1. Areas where the problem is more acute and which are dominated by a network of deep gullies and wild ravines.
2. Areas where the gullies and ravines are not very deep and dense and which occupy a comparatively narrow zone along the river banks.

The eroded land of the catchment areas of the river Yamuna, Chambal and Kunwari come in the first category where as those of the river Sengar, Ahaniya, Puraha and Arind in second category.

Block where the largest area is affected with ravinous wasteland is Auriya followed by Barhpura but in terms of percentage of the total block area, Chakarnagar is covered with 16 per cent of the total area by ravine wasteland. Bidhuna, Bharthana, Takha, Airwakatra, Sahar Achalda blocks one found free from ravine land. Eight blocks of Etawah with total 216 villages are affected with ravinous wasteland. Auriya block has the largest number of villages affected with wasteland followed by Barhpura, Chakarnagar and Mahewa.

Land Affected by Salinity/Alkalinity

Saline wasteland is the second largest category in the district which occurs over 25024.34 hectares accounting for 40.42 per cent of total wastelands and 5.78 per cent of the total district area. The northern part of the district, especially in north-east, is severally affected with salinity and alkalinity. All the blocks excepting Chakarnagar are affected with salinity with varying degree of magnitude. Proportions of saline soil vary from 14.99 per cent total area of Bidhuna to.75 per cent of the total area of Mahewa block.

In terms of Tehsil-wise distribution, the salt affected patches occur in all the four tehsils. These wastelands occur mostly along and around the Arind river where large patches can be seen between Sahar distributary, Kansa distributary and also south-west of Bela town in Bidhuna tehsil. Other salt affected wasteland patches occur along and around the Etawah branch of the Ganga canal starting from village Kanchousi upto the end of Bidhuna. The western portion of Bidhuna tehsil is also infested with the salt affected land between Bidhuncs-Umrain and Bidhuna Kudarkot Road. Large patches also occur on the north of Kudarkot town. A cluster of big and small salt affected lands occur along and around kuthipur and phaphund distributary.

The land affected by salinity/alkalinity in Bharthan tehsil is mostly located in the central part of the tehsil. It is also found towards the western portion of the tehsil extending from north to south. Some patches are found between Puraha nadi, Bansok distributary, Aheniya Nadi, Boharpur distributary, Balunda distributary and Sengar river. In the eastern portion of the tehsil, small salt affected patches are seen between Sengar river and Puraha Nadi.

In Etawah tahsil, small patches of salt affected land occur along the Sengar river on the right bank. Large salt affected patches occur in the northern portion the tehsil along Bansok distributary, Etawah branch of the Ganga canal and along the Aheniya Nadi.

In Auriya tehsil, salt affected patches are localised mostly towards the eastern part of tehsil. These patches occur between Sengar river, Mangalpur distributary and Phaphund town.

Block level pattern shows that saline soil varies from 14.9 per cent in Bidhuna to 0.75 per cent in Mahewa block. Chakarnagar is the only block of the district which is free from saline soil. Blocks which are severally affected with saline soil are Bidhuna, Bhagyanagar and Basrehar. Area of saline soil in these blocks are 4649.2, 3406.6, 3046.1 hectares respectively. Least affected blocks are Ajitmal, Mahewa Auriya and Barhpura. In Etawah district, 638 villages are affected with saline soil of which highest number is found in Bidhuna (84) followed by Basrehar (74), Bhagyanagar (80), Sahar blocks. Chakarnagar is the only block where no village is affected with saline soil. Least number of villages affected with saline soil is found in Ajitmal, Mahewa and Barhpura.

Waterlogged and Marshy Land

This is the third category of wasteland in the district and which occurs on 2132.5 hectare of land accounting for 4.72 per cent of total wasteland and 0.48 per cent of the total district area. The northern portion of the district particularly north-east is affected with this type of wasteland. Nine blocks out of 14 are affected with waterlogged land. Waterlogged land in these blocks varies from 5 hectares in Achalda to 810 hectares in Bidhuna, Mahewa, Chakarnagar, Auriya, Ajitmal, Barhpura are the blocks which are observed to be totally devoid of this type of wasteland. Generally, there seems to be a close relationship between drainage density and waterlogged land. In case of Etawah district it is found that blocks having more drainage density have no waterlogged land. Interestingly, Chakarnagar, Auriya, Barhpura where Chambal, Yamuna river flow from west to east are devoid of waterlogged land. The obvious reason for not having waterlogged land along these major rivers is the depth of river bed and slope of surface towards river. In the north of the district, no major river flows and the slope is level to level plain. As such, waterlogged land is found along canals and highway. Tehsil level analysis shows that waterlogged land mostly occurs in Bidhuna tehsil. Some waterlogged lands in this tehsil are found near and around the Lower Ganga Canal (West Allahabad Branch). Some patches are also found in the northern part of Bharthana and Etawah tehsils. Thus, marshlands occur on both the sides of Etawah-Fatehgarh Road and have water hyacinth growing on it. Bidhuna block has the highest number of villages affected with waterlogged land that is (24). Takha (14) and Airwakatra (15) also have some share of villages affected with waterlogged land. Blocks, namely, Ajitmal, Auriya, Chakarnagar, Mahewa have no villages affected with waterlogged land. Total 8.1 villages of Etawah district are found affected with waterlogged and marsh land.

Underutilised/Degraded Forestland

This pertains to the land notified under the Forest Act, in which denudation of vegetative cover has resulted in degraded land. The vegetative cover has less than 20 per cent. This category occupies 779.15 hectares of land which is 1.67 per cent of the total wasteland and 0.180 per cent of the total land area of the district. In terms of forest area it covers 3.37 per cent of the reserved/protected forest in the district. In Etawah tehsil the distribution of this category is restricted along left bank of the river Yamuna near Bathi, Nagla Salahi, Nagla Tor, Sarongpur and Bulokipur village and Purha, Dangl and Bohri village on the right bank. In tehsil Bharthana, these lands occur on the south of Yamuna near Miholi and Khitriti villages and between Chambal and Kunwari.

In Bidhuna, tehsil degraded forests occur near village Jamalpur and Singh Purwa Mana on the north of Arind river whereas in Tehsil Auriya, these lands occur near Debiapur, and also along Yamuna river near Akbarpur, Asta and Nagla Banadas village. Nine blocks of

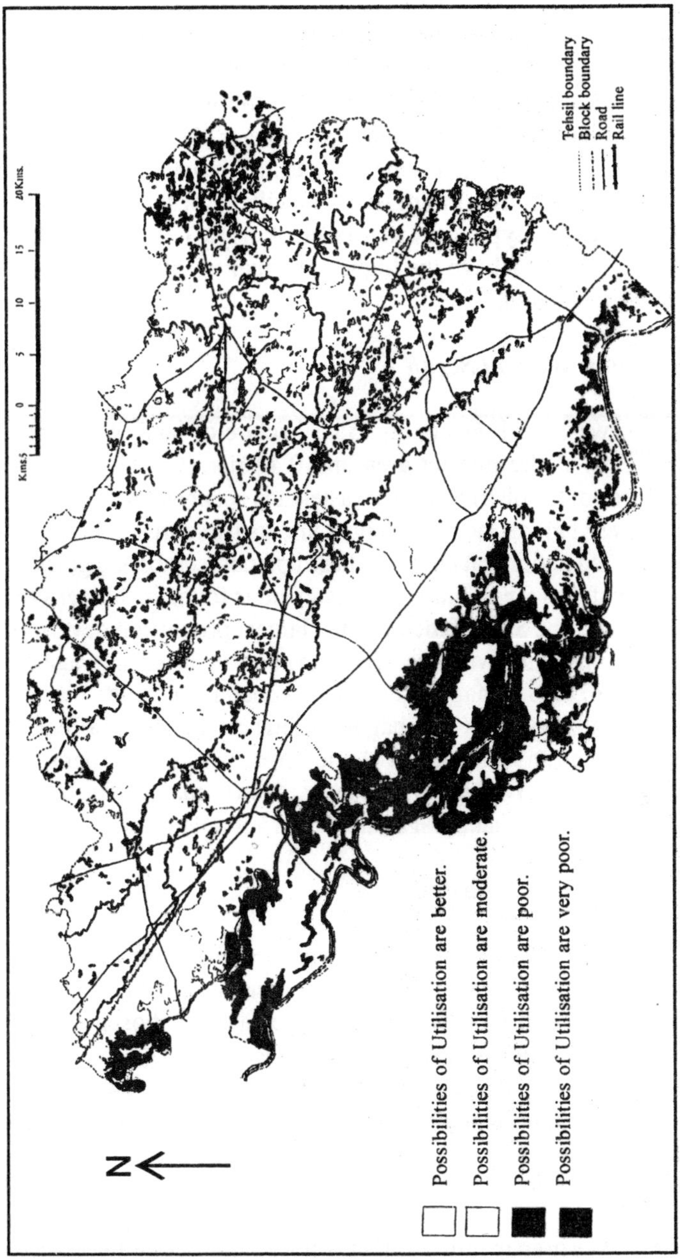

Fig. 19.4: Capability Classification of Wasteland (Etawa District.*Source*: Prepared and compiled by author.

Source: Prepared and compiled by author.

Etawah district are affected with degraded forestland, which varies from 199 hectares of Auriya to 0.15 hectares of Bidhuna. Degraded forestland is the highest in Auriya (199 hect.) followed by Sahar (164 hect.), Bhagyanagar (129 hect.) and Chakarnagar (133 hect.) blocks. Jaswantnagar, Basrehar, Bharthana, Takha and Airwakatra are totally devoid of wasteland. Total number of 39 village of the district are affected with this type of wasteland. Auriya (7), Sahar (8), Chakarnagar (7) block have the highest number of villages affected by degraded forestland.

Upland With or Without Scrub

This type of wasteland occurs on 758.3 hectare of land, which are 0.174 per cent of the district area and 0.44 per cent of the total wasteland area of the district. These wastelands occur as small patches and are scattered along the river Sengar and its tributaries. These are generally eroded peripheral lands. In these uplands, scrubs and grasses in scattered patches are found. This wasteland is found in 4 blocks of Etawah district. The Sahar block has the largest proportion covering an area of 635 hectares whereas Bharthana has the least coverage of 13 hectares of land. Bhagyanagar and Achada are other blocks reported to be affected with 25 villages, the highest number of affected villages are found in Achalada (14) and Sahar (4). Blocks, namely, Auriya, Ajitmal, Barhpura, Airwakatra, Jaswantnagar, Bidhuna, Takha, Mahewa and Chakarnagar are free from uplands.

Capability Classes of Wasteland

The wasteland of Etawah district is classified into four capability classes on the basis of factors mentioned earlier. Wasteland capability Class I includes those areas where aggregate value of chosen factors are above 90, this class is represented by green colour on capability map (Fig. 19.4). Only 7.2 per cent (i.e. 3967.5) of the total wasteland area is covered by this category, found mostly in few separate areas lying along Yamuna and in north central part of the district.

Capability Class II is represented by light orange colour on capability map which represents the areas with aggregate value between 80-90. This capability class includes saline and waterlogged lands which are found scattered in the district lying north of Sengar river. Seventy per cent of wasteland in north-eastern part of the district belongs to wasteland capability Class II.

Table 19.3: Area under Wasteland Capability Classes.

Sl. No.	*Categories*	*Aggregate Point*	*Area (in hectares)*	*Area (in %)*
	I	Above 90	3967.5	7.2%
	II	90-80	14851	27.7%
	III	80-70	13195	24.1%
	IV	Below 70	23325	42.7%
	Total		**54601**	**100%**

Source: Area calculated from the wasteland capability map prepared by author.

Wasteland capability Class III, shown by brown colour, covers about 24 per cent of the total wasteland in the district. Areas of this class are found scattered in north of the river Sengar, but we also find large stretch in north-eastern part of the district. Most of the saline and waterlogged lands represent this capability class.

Wasteland capability Class IV, represented by red colour, constitutes more than 40 per cent of the total wasteland of the district. Possibilities of utilisation are very poor in the wasteland of this class. Ninety per cent of ravine land are included in this class which are located in southern part of the district along river Yamuna and Chambal. Some saline areas of central and north central part of the district also represent this class of wasteland. Table 19.4 shows block level distribution of wasteland capability classes.

Wasteland Capability Class I

Wasteland of this class covers 3967.5 hectares of the total wasteland of the district. The possibilities of utilisation are better in the areas under this class. These wastelands are found in all blocks except Takha, Chakarnagar, Bhagyanagar and Achalda blocks (Fig. 19.5). Along the Yamuna river, two patches of 208 hectares are found in Ajitmal block. These are degraded forest land, where initial forest has been degraded.

Table 19.4: Block Level Distribution of Wasteland Capability Classes.

Block	*Wasteland Area (in hectares)*	*Wasteland Capability Classes*							
		Class I		*Class II*		*Class III*		*Class IV*	
		Area		*Area*		*Area*		*Area*	
		(in hect.)	*(%)*	*(in hect.)*	*(%)*	*(in hect.)*	*(%)*	*(in hect.)*	*(%)*
Jaswantnagar	4250	935	22	1147.5	27	2167.5	51	—	—
Bidhuna	5459.35	163.78	3	1910.77	35	3384.79	62	—	—
Basrehar	3464	242.48	7.1	1558.8	45	866	25	796.785	23
Bharthana	2575.2	128.77	5	1158.98	45	849.92	33	486.96	17
Takha	2710	—	—	596.2	22	1084	40	1029.8	38
Mahewa	2792	223.36	8	83.76	3	67.52	6	2317.36	83
Chakarnagar	6044	—	—	—	—	—	—	6044	100
Airwakatra	2000.3	600.04	30	500.07	25	900.15	45	—	—
Sahar	2598.3	519.66	20	2078.64	80	—	—	—	—
Auriya	8255.6	165.11	2	—	—	—	—	8090.40	98.1
Ajitmal	813.5	207.44	25.5	16.27	2.5	20.33	2.5	568.75	70
Bhagyanagar	3676.5	—	—	3235.3	88	367.65	10	72.53	2
Achalda	2858	—	—	1714.8	60	1114.62	39	28.58	1
Barhapura	7107.54	781.82	11	852.90	12	2274.41	32	3909.14	55
Total	**54601**	**3965.5**		**14851**		**13195**		**23325**	

Source: Area calculated from the wasteland capability map prepared by the author.

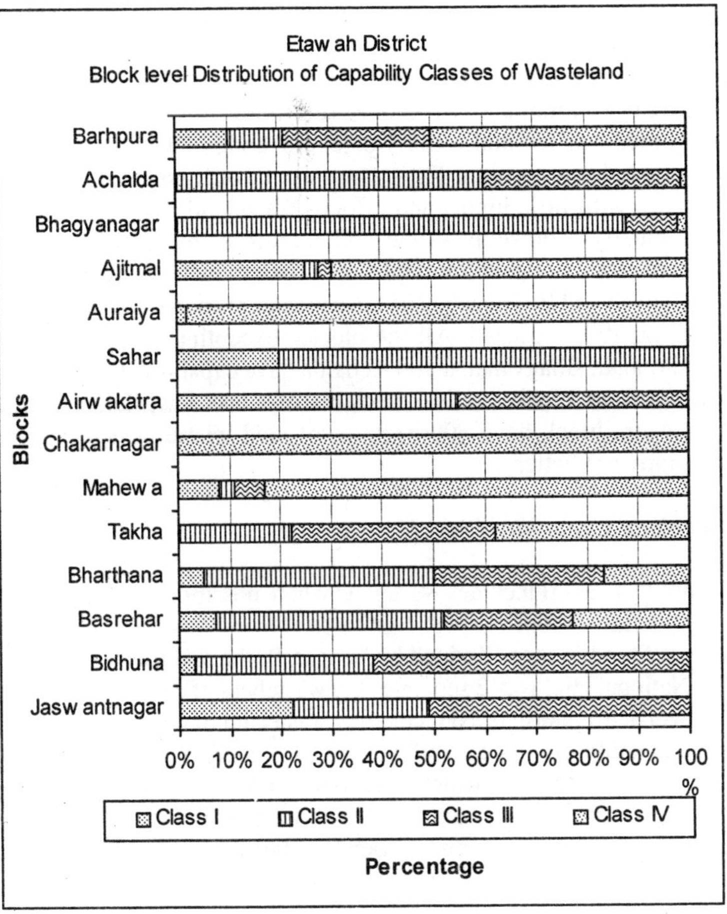

Fig. 19.5: Etawah District: Black Level Distribution of Capability Classes of Wasteland.

Jaswantnagar along the Yamuna River on an area of 935 hectares is included in this class. This area is mostly ravenous land, but the intensity of ravines is not very severe, thus, possibilities of reclamation are better. We also find numerous patches of Class I wasteland along Sengar river in Mahewa block covering an area of 120 hectares. These patches are covered mainly by saline soil. The region surrounding Kharaghpur town in Airwakatra block also contains Class I category of wasteland which roughly covers 500 hectares of land.

Wasteland under Class I is mostly free from land erosion, except few patches along river Yamuna. Slopeout, electrical conductivity of groundwater are also favourable in this class while sodicity and electrical conductivity of soil is relatively high. Other factors mentioned earlier make the wasteland under Class I relatively favourable for utilisation. These lands can be utilised for agricultural purposes with little inputs and efforts. Wasteland belonging to this class mostly represents saline soil therefore, by growing saline resistant crops along with suitable inputs can show favourable results.

Etawah and Bidhuna tehsils have greater proportion of wasteland under capability Class I. All the blocks, excepting four i.e. Takha, Chakarnagar, Bhagyanagar and Achalda have capability Class I land. Jaswantnagar block has 935 hectares of Class I wasteland while Airwakatra block have 600 hectares of land where possibilities of utilisation are better.

Wasteland Capability Class II

Capability Class II includes wasteland where possibilities of utilisation are moderate. It consists 14851 hectares of land. Wasteland under this class is found in areas drained by Sengar, Purha and their tributaries. The National Highway 2 separates the wasteland of capability Class II and IV. District north of NH_2 contains wasteland of capability Class II while in south of NH_2 capability class IV is predominant. Eighty per cent of the wasteland of Class II are found in central eastern part of the district, which is mostly saline and waterlogged land.

In these areas groundwater depth is favourable, land erosion in minimum and surface soil drainage is good. Agriculture is possible in these wastelands where only salinity resistant crops can be grown. Dense network of canals and tubewells adds to the possibility of utilisation, because canal water is good for leaching process in the saline soil.

Wasteland of western and south-western part of Bharthana town represents Class II land. Some ravine soils in eastern part of the district are also included in this category.

Although salinity resistant crops can be grown in capability Class II wasteland, but social forestry of economically profitable tree species would be more beneficial and practical because salinity is relatively higher than capability Class I and utilisation of wasteland would require larger input.

In Bidhuna tehsil more than 45 per cent of total wasteland belongs to the category of moderate possibilities of utilisation. In Auriya and Etawah tehsil, 25 per cent and 23.9 per cent of total wasteland fall under the capability Class II. All the blocks except Auriya and Chakarnagar have wasteland under class II. In Bhagyanagar, Sahar, Bidhuna, Achalda block more than 90 per cent of the total wasteland represent wasteland of capability Class II. More than 40 per cent of the total wasteland representing a capability Class II is in Jaswantnagar, Barhapura, Takha, Bharthana blocks.

Wasteland Capability Class III

Possibilities of utilisation are poor in capability Class III. This class covers an area of 13195 hectares of land, i.e. 24.1 per cent of the total wasteland of the district. Wastelands of this class are found in three regions lying in eastern, central and north-western parts of the district. The State Highway 21 separates these three regions. Mostly, saline and waterlogged land constitute wasteland of this class, but in north-western part near Yamuna and Chambal river ravine lands are included in capability Class III. Wasteland belonging to this category located east of State Highway 21 is unique because very high groundwater level which is responsible for high sodicity and salinity of the soil. Salinity more than 30 ec/dsm^{-1} is found in this particular region. Similarly, sodicity above 40 esp makes wasteland unsuitable for any type of agricultural activity. Groundwater level in this region is only 2 meters below the surface which increase the water salt content in the surface soil during dry season. This region looks like hard light grayish pan in dry season. Only few sodicity resistant tree species can be grown here. Moreover, lack of agricultural infrastructure and low economic prosperity of this region hinder the reclamation measures. User and Reclamation Project sponsored by World Bank have surveyed 3 villages in this region but results are not favourable. It was observed that very high salinity and sodicity makes any reclamation strategy futile, as land reverts back to its original state within 2 to 3 months. Saline soil expands very rapidly, therefore, it is necessary to check its growth to adjoining areas.

Wasteland of capability Class III, lying in central part of the district, represents more or less similar conditions, as mentioned for the region east of SH. 21, but severity of impact if relatively less. Land erosion, slope, surface drainage and groundwater depth are unfavourable in this region. Besides this, the region also lacks the irrigation sources because of the infavourability of these factors.

Block level picture shows that all the blocks have presence of wasteland of capability Class III, except Chakarnagar, Sahar and Auriya. In Bidhuna, Basrehar and Barhapura more than 70 per cent of wasteland falls in capability Class III.

Wasteland Capability Class IV

Possibilities of utilisation are very poor in the wasteland capability Class IV. This class covers an area of 23325 hect. (i.e. 42.7%) of the total wasteland of the district. Thus, this class covers majority of wasteland in the district. Major portions of this class are found along the river Yamuna and Chambal in southern part of the district. More than 90 per cent of wasteland of capability Class IV consists of ravines soil. Rest of the wasteland includes saline soil of central and north central part of the district. Ravines soil has a rugged topography, which makes its utilisation almost impossible. Severe land erosion and slope are main drawbacks of wasteland of this class. Wasteland of this class is isolated region of the district. The depth of ravines is more than 15 meters in certain areas which manifest the severity of land erosion. Groundwater is below 20 meters. It is also below 30 meters in a few areas. Land erosion is more than 70 tonnes per hectare per year. In the wasteland of capability Class IV slope is more than 15 per cent in most area.

Development of infrastructure for agriculture and economy is very poor, which further diminishes the possibility of utilisation. Though salinity, soil texture is relatively better but severity of other phenomena, such as land erosion, slope makes their favourability insignificant and renders their utility. Deforestation in these areas also leads to severe land erosion. It is imperative to provide vegetation for this region so that land erosion can be controlled.

Wasteland of capability Class IV is not found in Bidhuna, Airwakatra and Sahar. In Chakarnagar block wasteland of capability Class IV covers 6044 hectares while in Auriya block it covers 8090.78 hectares. These figures show that block in southern part of the district contains major share of wasteland under capability Class IV.

Overall, picture shows that most of the wastelands located in central part of the district have better possibility of utilisation because intensity of factors in this part are the least. Infrastructure development, economic conditions also increase the possibilities of utilisation of wasteland. Severity of wasteland increases as we move from central to northern and southern part. In the south, severity of wasteland increase because of intense land erosion, moderate, low slope, groundwater depth. The importance of this capability classification lies in the fact that reclamation and utilisation can be applied according to the possibilities of utilisation of each class. Thus, the classification gives a fair idea for the planner as to how to implement the reclamation and utilisation plans.

Factors Promoting Wasteland

Wastelands are types of land which are formed due to the complex interaction of various processes at varying intensity. The combined impact of man and nature interaction and the varying levels of technology inputs are considered to be the main factors in the formation of wastelands.

Factors of Ravines

A correlation of ravinous land with factors like morphometric, climatic, pedogenic, hydrological and anthropogenic reveals that a positive correlation exists with morphometric and climatic factor. Out of these factors, ravine land is highly correlated with slope, the value of correlation being 0.96428. The slope is found to be high in most of ravine land. The belt along river Yamuna and Chambal in a linear stretch from west to east is badly affected with ravine land. In these areas slope of 15 per cent is found, which is highest in whole of the district.

A low positive correlation exists with respect to factors like number of tubewells and wells, rainfall and electrical conductivity. Rest of the factors, such as soil reaction in Ph, texture of soil, hydrological depth of groundwater in dry season and other anthropogenic factors, shows a negative correlation with ravenous land. Of these, soil reaction in Ph has a high negative correlation with ravinous land, followed by hydrological depth of groundwater in dry season.

Factor of Saline Land

Saline soil shows a very high positive correlation with sodicity in ESP (0.94) and electrical conducting (0.96). This is followed by a high positive correlation with number of tubewells and wells (0.84). The saline soil of the region has a moderate positive correlation with soil reaction in Ph (0.66).

The correlation between saline soil and factors like sodicity and electrical conductivity and soil reaction is simple to explain. These factors are responsible for the formation of saline soil. Therefore, it is but natural that high correlation exists between these factors and saline soil. Rest of the factors, especially the slope of the region and depth of surface groundwater in dry season represent a high negative correlation with saline soil. Where the slope is low, saline soil concentration is high. The northern part of the district is level to level.

Factors of Waterlogged

A correlation of the waterlogged land with the above-mentioned factors reveals a very high positive correlation with the length of streams and canals. A general picture of Etawah district emerges which shows that concentration of waterlogged land is mainly in the villages located in the northern part. One of the reasons is that northern part is criss-crossed by various branches of Lower Ganga Canal while in southern part the canals are totally absent. Waterlogged as observed in the sample villages is mainly confined to natural low laying areas where there is every possibility of water accumulation, through seepage and more inflow of water either through sub-surface drainage or due to over-irrigation. The slope is the determining factor for waterlogging conditions of the district as all the waterlogging patches are confined to low lying areas. Therefore, correlation analysis between slope and waterlogged land of sample village shows high negative correlation (Table 19.3). The areas having low slope have more waterlogging conditions than the areas having high slope. This analysis logically explains the presence of waterlogged land in northern part that is mostly level to level plain and absent in southern part where slope is more than 15 per cent.

Waterlogged land shows a negative correlation with depth of groundwater. The villages like Sadanpur where depth of groundwater is low, the presence of waterlogged land is high. Waterlogging decreases with high groundwater depth.

Low positive correlation exists between waterlogged land with soil reaction in Ph. In the villages where saline soil is missing, the presence of waterlogged land is also missing which is difficult to explain logically and scientifically. Low positive correlation of waterlogged land is followed by number of tubewells and wells (0.23) and rainfall (0.17). The rainfall increases from middle part towards east and presence of waterlogging is also found high in the east and with rest of the factors waterlogged land has negative correlation. A moderate negative correlation can be seen with respect to factors like electrical conductivity, soil texture and sodicity in ESP.

Factors of Upland

A correlation of upland with the above-mentioned factors reveals interesting result. A high positive correlation (0.83) with slope in the region is obvious. The correlation is negative with other factors, except electrical conductivity and length of streams and canals, where the correlation values were 0.32 and 0.06 respectively.

Uplands are topographically high located areas, therefore, it is natural that slope is a dominant and influencing factor regarding upland. The length of streams and canals shows low positive correlation but these only explain peculiar conditions of the sample village. The correlation with streams and canal length or electrical conductivity with upland cannot be logically explained; therefore, it would not be proper to apply this correlation all over the district.

Factors having negative correlation with upland are number of tubewells and wells, rainfall, soil reaction and sodicity. High negative correlation exists with soil reaction (-0.69) in the Ph. and sodicity (-0.63). This is followed by moderately negative correlation of soil texture and groundwater depth.

Factors of Degraded Forestland

Correlation of degraded forestland with these factors shows that, it is highest with the groundwater depth in dry season, the correlation value being 0.68. A low positive correlation is found with the slope and soil texture.

The degraded forestland of the region has a negative correlation with rest of the factors. A high negative correlation can be seen with respect to electrical conductivity where the correlation value goes to –0.57. This is followed by the correlation with factors, such as length of streams and canals, rainfall (mm) and soil reaction in Ph, where correlation ranges from –0.17 to –0.25. A low negative correlation is visible with respect to sodicity in ESP and number of tubewells and wells.

Reclamation of Wasteland

An attempt to reclaim the wastelands in Etawah district can be described as essential from the point of view of the development of national wealth through increasing agricultural production by increasing the cultivated area. The remedies for the reclamation of different wasteland types in totality cannot be uniform because for different wasteland types, different reclamation methods are required.

Remedies for Reclamation of Saline Soil

To present the reclamation remedies for saline soil, Mirgawan village located in eastern part of Bidhuna tehsil in Achalda block has been surveyed. This village is severally affected with the problem of saline soil which is evident from the fact that 131 hectares of village land (i.e. 33.25%) are affected with saline soil where agricultural activity is not feasible.

The problem of saline soil is more chronic in eastern part where almost all the lands are left barren and look like a hand pan. Agricultural activity is therefore located in the western part of the village. Thirteen patches of saline soil are found in this village. These patches are small in western part but as we move in eastern direction, saline soil problem becomes more acute.

The important measures of saline soil reclamation are as follows. Proper levelling of saline soil areas, providing bunds along the field should be done to obtain a uniform spread of water. The field should be divided into convenient size plots with small earthen embankments in accordance with slope. Small suitable drains should be made for the removal of excess irrigation water. Leaching with irrigation water should be done so that salt content of soil decreases.

The water used for leaching should be of good quality. Either going for sprinkler irrigation or alternate pounding and drying the field could improve the efficiency of leaching operation. The application of mulches considerably checks upward movement of water and improves the efficiency of leaching. The western part of the village where saline soil is found in small patches can be reclaimed with leaching and drainage.

It is interesting to mention that some farmers are already utilising some saline areas with their own indigenous knowledge. They used to grow crop during the period of rainy season. Rainwater helps to wash the salt and reduces it by making the soil better for cultivation. Throughout the monsoon period, there is considerable fall in the salt content of the soil. In dry season when salt starts rising due to capillary action, but, by that time plants become mature to handle the salinity of the soil. This method is useful, but, farmers are unable to get substantial return as crop yields are poor and only a few crop species are possible to grow.

Deep drainage by tubewells also helps in reclaiming the saline soil as it helps in bringing down the groundwater table to a safe limit. This list is crops can be used to utilise saline soil for agricultural purpose.

1. *High Salt tolerant crops*: dhainha, paddy, sugarcane, oats, and barseen in kharif and Lucerne, senji (*trigownella*) melhi (*ferugreck*) and barley during rabi.
2. *Medium Salt tolerant crops*: castor, cotton, jowar, bajra, maize in the kharif season and mustard and wheat during rabi.
3. *Low Salt tolerant crops*: moong, urad, arhar and sunhemp in kharif and gram, peas, rinseed and sesamum during rabi.

Besides this, vegetable crops, such as the garden beet, asparagus, spinach, cabbage and tomato are high tolerant vegetable crops. The salt sensitive vegetables are raddish, calory and green beans. These crops can be grown for cultivation according to the Ph value of the soil.

Green manure, crop residues and weeds can also be used for reclaiming saline soil. They serve as important source of readily available nutrients on decomposition. Along with that they also help in solubilising calcium and neutralising the soil. The consequent improvement in soil permeability and increased biological activity helps in slowly regenerating the soil. Of all the plants used as green manure, dhiancha or janter has been most successful in saline alkali soils.

Some saline affected areas in eastern part where Ph value is high and land cannot be reclaimed for agriculture, high to medium salt tolerant trees like sheesham, Mahua (*Moduca Indica*), Dak, Neem (*Azadirah-Taindica*), Eucalyptus, Arjun (*Terminalia Arjuna*), Bakkain (*Azadiraita Extica*), Bobul (*Acacia Nolitical*), Bilayti Babool (*prosopis Tuliflora*). Siris (*Albizia Labbeck*) etc. should be grown. Research work carried out at the Central Institute of Medicinal and Aromatic plants, Lucknow has shown that oil bearing plants like Palmarosal (*Symbopogon stopt martini*), Lemongrass (*Cymbopogon flexuous steud wats*) and Citronella Java (*symbopogon winterianus jowtt*) can be raised in soils affected with salinity/alkalinity up to ec level of 11.5.

Salt tolerant grasses like Karnal grass (*Diplachnefusca*) Rhodes gram (*chloris gayana*), bermveda gram, paragrass (*Brachiaria mutica*), Blue panic (*Panicum antidotale*). Doob (*Cynodon dactylon*) etc. can be grown successfully either along with trees or individually, on saline and alkaline soil.

Remedies for Reclamation of Ravine Land

Garha Kosda village is situated on the southern bank of rive Yamuna. The belt along the Yamuna in which Garha Kosda village is located is severally affected with land erosion. Ravine land covers half of the village area and extends from north to south in the western part. This village is situated few kilometres away from the river Chambal and 15 km. away from the confluence of river Chambal and Yamuna. In other words, this village is located in the triangular belt between Yamuna and Chambal where severe ravine formation has been found.

Several factors are responsible for the formation of ravines in the triangular belt between the Yamuna and Chambal. Clearance of forest in this belt seems to be mostly responsible for ravine formation which reduces the infiltration and increases the runoff in this area. Further uncontrolled grazing, low annual rainfall, downpour frequency of greater intensity, pronounced vertical difference between the general ground level and the river bed level and extreme dryness of the soil due to easy drainage and high porosity are responsible in one way or the other in the formation of ravine land.

Ravine land of Gasha Kosda village can be demarcated into two belts: northern belt along the river Yamuna and southern part of the ravines. The basic difference between these two belts is severity of land erosion. The former belt is severely eroded while later is less eroded, with shallow gullies. Thus, ravine in southern part can be suited for utilisation if some authentic steps are taken. The reclamation strategies to reclaim and to utilise these lands are given below.

In the southern portion of the village where depth of ravine land is shallow, mechanical (tractor) reclamation measures can be used for levelling the ravine land and bullock power can also be used for terracing and levelling of the land. According to the slope and terraces, the field should be divided into different plots and strong bunds should be constructed around the field to check the erosion and water flow. Suitable drains should be made for removing excess water. Land preparation and ploughing should be done according to the nature of slope. Diversion channels can be constructed to reduce the erosion by gullies. Testing of the soil shall be done to know the deficiencies of the soil. *Shakti khad*, potash and compost may be applied in both kharif and rabi reasons to overcome the deficiency.

In the first kharif season maize fodder crop should be sown and in the rabi season barseem fodder crop should be sown with light but frequent irrigation, as this soil lacks moisture retaining capacity. In the summer season, after the rabi crop, dhaincha as a green manure crop should be sown. The farmers are advised to follow this crop rotation for at least 3 years and not to leave the land fallow for a long time.

Areas of village where gullies/ravines land have acquired severe form can be utilised for plantation. Erosion resistant trees can be grown here which in the long run not only checks erosion but also helps to meet the fuel demand and wood supply of the village. In the gullied and/or ravine land, the flat tops generally have clayey soil. These could be planted with *prospis juliflora* and *acacia nilotica*. The ravine bottom should be planted with *syzygivm cumini, halplophlogma asdchophyllum, dalbergia sissoo* and *cassia siamea*. The gentle slope should be planted with *cassia auriculta, acacia noilotica, pongamia pinnata, pilhese lobium dulce* and *holotelea integrifolia*.

Steeper slopes can be planted with tufts of *eulaopsis binata* and *socchrum mumja* on southern and western aspects. *Dendro colamus strictus* can be planted on slopes on the eastern and northern aspects.

In the agriculture field which are prepared, farming should be done on scientific basis. Crop rotation should be given emphasis as it helps to check soil loss and runoff. A mixed cropping and inter-cropping practices have also been found very effective agronomic measures for soil erosion control. Gora paddy-jowar combination was found to reduce the soil less by 4.1 tonnes/hectare and runoff by 17.2 per cent of the rainfall and was considered the most suitable mixed crop for reducing erosion. Inter-cropping of cowpea with cotton reduced soil loss from 6.4 to 4.1 tonnes/hect. and runoff from 30 to 19 per cent when compared with cotton alone. So, the farmers should adopt these combinations of crops.

In order to retain and safe disposal of rain water, contour bunding with varying cross-section depending upon length and degree of slope, soil texture and vegetable cover at different vertical intervals have been found very effective. For stabilising these bunds *aicanthium annulatum, amphilophis glabra, cenchrus ciliaris* and *Andro pogen ischaemum* with good fodder value have been proved very effective.

For gully plugging where runoff is expected from the top, earthern gully plugs 1.1 sq. km. cross-section with graned ram 22.5 cm. below the top and spaced at 45-60 in horizontal intervals will prove to be effective. On the other hand, where runoff is expected from the top, earther gully plug of 2.2 m. cross-section with a pipe outlet will be effective. The size of pipe would depend on the expected discharge. In extreme northern part where sometime catchment area is larger than 16 hectares, a composite check dam of earth and brick masonry is considered essential. Brush wood has not been found very useful because of its maintenance cost.

Remedies for Reclamation of Waterlogged Land

To present the reclamation remedies for waterlogged land, Hardu village, situated in the north-eastern corner of the district in Bidhuna block, has been surveyed. Hardu village lies between two canals, that is, Kansua distributary and Lower Ganga Canal. The former makes the north-eastern boundary of Hardu village while the latter exists in south-west, half kilometre away from the village. The severely village is affected with the problem of waterlogging which is evident from the fact that waterlogged covers 182 hectares (18.2) of village land and is found in eight patches. Though the village economy is comparatively better and is well connected with metal road unlike other villages, but, waterlogging has severely affected the agricultural activity in the village.

To present the reclamation remedies for waterlogged land, Hardu village, situated in the north-eastern corner of the district in Bidhuna block, has been surveyed. Hardu village lies between two canals, that is, Kansua distributary and Lower Ganga Canal. The former makes the north-eastern boundary of Hardu village while the latter exists in south-west, half kilometre away from the village. The severely village is affected with the problem of waterlogging which is evident from the fact that waterlogged covers 182 hectares (18.2) of village land and is found in eight patches. Though the village economy is comparatively better and is well connected with metal road unlike other villages, but, waterlogging has severely affected the agricultural activity in the village.

It is important to remove the excess water with sufficient drainage and adopt proper framework to utilise. The levelling of land should be done, providing strong bunds along the field and the farmers will be trained for proper implementation of the correct amount of fertiliser, irrigation and other technological inputs. As 15 cm. below from the level of the field, in accordance with the slope, small drains can be made nearer to the field for the outflow of the excessive water. After testing the soil, and according to the requirements of the soil to produce paddy crop and keshari crop, nitrogen, phosphate and potash fertilisers can be applied. Deep-water paddy can be grown because of shallow water depth to use the land for agriculture. Paddy should be grown as a first crop during the kharif season selecting high yielding tolerant variety and timely plantation of relatively older seedling. In the second rabi season keshari should be preferred mainly because in the field, paddy takes more time and more moisture of the soil, while keshari crop does not need much water and is found to be ready within the limited time. The crop rotation will be followed for at least three years and the farmers are advised not to leave the land fallow for a long time.

Waterlogged areas lying in north-eastern part of Hardu village can be reclaimed with afforestation of suitable species. The preparatory step is to remove the excess waster by constructing proper drainage and then species may be selected from *dolbergia sissco, eucalyptes saligna, e. robusta, acocia nilotica, a. auriuiliformis, cassia siamea, pongamia pinnota, terminalia arjuna, azad jrachta Indica, hoplophrogma adenophyllum, syzygium wmirui, salix tetrasperma, populas deltodes, albizzia lebbcck etc.*

The canal slope should be planted with grasses and shrubs to prevent erosion. Species, such as *eulaliopis binata, cymbopogon jwarankusha, adhatodo vasica* and *cassia auriculata* will be a good choice. A row of salix *tetrosperma* planted through sets along the canal helps in reducing canal seepage as the fibrous root system clog the soil pores. The field boundaries where conditions of waterlogging do not exist should be planted with eucalyptus species. Due to the fast growth planting adequate rows of trees could control the rise in water table. Such farms can utilise these trees for meeting their energy needs and simultaneously maintain their productivity for agricultural crops.

The planting in these areas should be done in the months of February to May when the problem of waterlogging is not so acute. Seedlings of good size should be used for planting. Irrigation may also be provided during summer months to accelerate the growth of plants so that by ensuring rains the plants are sufficiently established to withstand waterlogging. Planting may be done on earthern mounds of varying heights depending upon the level of water stagnation. This technique has successfully been tried in Tawa command area of Madhya Pradesh.

Reclamation Remedies for Degraded Forest Land

To present the reclamation remedies for degraded forest land, Ramnagar village of Ajitmal block of Auriya tehsil has been surveyed. The village is situated on the bank of river Yamuna and contains an area of 294 hectares. Ravine and degraded forest lands both are found in the village, while former accounts only 7 per cent of the area the later covers 22.25 per cent and is mostly located in the eastern part.

Two areas of degraded forest land have been found. One is along the village built-up area and second is the larger one, which is located in the southern part. The degradation of forest land in Ramnagar village modifies the structure and composition of natural forest community and transforms the forests into bush, scrub and grazing lands. Grazing and demand of the wood for construction and for fuel seems to be the obvious reason for the degradation of such lands. The density of trees within the forest has declined to almost 60 per cent and clean patches are emerging within the forest. As degraded forest land and ravine land co-exist, forest degradation, in turn, increases the rate of soil erosion in this village. The following measures are taken to protect the forest land and regenerate it.

Area should be protected from grazing by digging cattle proof trenches and other methods depending upon the budgetary provisions and availability of fencing materials. Grazing is the major threat to this forest due to the cattle population of Ramnagar village and near by village, such as Taiya Pur and Mandia village, this forest is the only source of fodder. Fencing of plantations and rehabilitation is a very costly operation, yet degraded forest of Ramnagar village can only be revived unless grazing is meticulously excluded by creating an effective fence. Fence may be of trench and ridge (cattle proof trench) or barbed wire. Thorny bushes of *carissa spinarum* were also fascinated along ridge to afford extra protection against cattle grazing.

Along with above-mentioned protective measures, rehabilitation of degraded forests is very essential as areas of forest are lying unproductive. The task of rehabilitation is not that difficult provided the area is relieved of the biotic pressure. The cost involved is also low as compared to reclamation of other types of wasteland, such as saline, waterlogged, ravine land, etc. The original tree species which are found in these forests of Kanmav village are sal (*shorea robust*), shisham (*Dol bergia sisso*), jamum (*Eugenia jambolana*), prosopis tree, eucalyptus, semal, kair etc. These trees should be given priority in plantation and afforestation programmes.

Ravine land which covers 7 per cent of the village area should also be used for afforestation and erosion resistant trees, such as *prosopis juliflora, acacia nilotica, hoplophragma adenophyllum, dal bergia sissoo* and *cassia siamea, pongamia pinnota and holoptelea integrifolia*, should be used for afforestation programmes, which in the long run not only protect the valuable soil but also meet the fuel and fodder demand of the village.

Fast growing tree species, such as *dolbergia sissoo, teetona gran dis, albizzia procera, azadirachta India, pongamia pinnato, eucalyptus species, gmelina arborea* etc. can also be used for the afforestation programme to get quick results.

Reclamation Remedies of Upland

Kanmav village situated on the bank of the Arind river in Bhagyanagar block has been taken into consideration to present reclamation remedies of uplands.

Upland areas are found in eastern portion of the village which is 10-15 meters high from the surrounding land and are dominated by kankars. Problems of water retention and soil erosion due to sheet wash are two such problems, which make it uneconomic to utilise these lands for agriculture. Therefore, the first step to reclaim upland area should involve, checking the soil erosion due to sheet wash, help accumulation of silt from the slopes and growing fuel, fodder and economic species of trees. Uneven sloppy surface is the main hindrance to utilise uplands. *Kankar* nodules dominate the upland, which further poses difficulties for the utilisation of upland area for agriculture purposes. Providing irrigation to these lands is another difficult task because of high slope and low groundwater retention.

The upland areas located in south-western part of Kanmav village are areas with lesser slope. They contain good soil depth, which can be taken up for growing agricultural crops after bunding and constructing suitable water harvesting. These areas should be ploughed and divided into sufficient compartments according to the slope, *kankars* should be thrown out of field and soil testing should be done to know the physio-chemical status of soil. Suitable agriculture crops can be grown here. Strip cropping should be given preference to check sheet wash and erosion. Contour cultivation can also be given emphasis if required. In contour cultivation, each furrow, which lies approximately on the level, intercepts flowing water, hold it and allow it to soak into the soil. Thus, it is possible to reduce run off and soil loss from 54 to 41 per cent of rainfall and from 29 to 19 tonnes of soil respectively through contour cultivation.

Patches of upland which are located on the bank of Arind river are high, uneven and predominant with Kankar can be utilise for afforestation. These patches are difficult to reclaim for cultivation as heavy expenditure will be needed and nature of soil cannot ensure for good economic returns. Afforestation is only suitable way to reclaim these upland patches. Tree species, such as *acacia noilotica, cassja siamea, pilhecelobiumdulce, eulalopsis binata etc.* should be planted here which in long-term bind the soil and reduce the rate of erosion and provide fuel and fodder to villagers.

The analysis portrasys the acuteness of wasteland problem in the Etawah district. Wasteland covers 12 per cent of the district area mostly by saline and ravines land. Others type of wasteland cover small proportion in comparison of saline and ravenous land. Wastelands located in central part of the district have better possibility of utilisation because intensity of factors in this part are the least. Infrastructure development, economic conditions also increase the possibilities of utilisation of wasteland. Severity of wasteland increases as we move from central to northern and southern part. In the south, severity of wasteland increase because of intense land erosion, moderate low slope, groundwater depth. The importance of this capability classification lies in the fact that reclamation and utilisation can be applied according to the possibilities of utilisation of each class. Thus, the classification gives a fair idea for the planner as to how to implement the reclamation and utilisation plans. Correlation analysis show that environmental factors dominantly influence the wasteland while anthropogenic factors dominate in only one or two types of wasteland. It should very clearly be understood that like other development programmes the wasteland development programme is the programme of the people, by the people and for the people, and they expect a greater contribution to the development process and objectives. Need of hour is not to take isolated efforts but to create a system which should involve every one including common people to government for action role to play. Then only we can succeed to mitigate this problem and judiciously utilise our land resources.

REFERENCES

Academy of Science of U.S.S.R. (1958), *The Application of Drainage in the Reclamation of Saline Soils*, Translated from Russian, Moscow.

Aggarwal, R.K. (1980), "Physio-chemical states of soils under khejri prospopis cineraria linn" in Mahn., H.S. and Saksena, S.K. (ed.), *Khejri in the Indian Desert,* Monograph No. 11, Jodhpur, CAZRI: 31-36.

Aggarwal, R.R. (1957), "Delhi soils can be reclaimed", *Indian Farming*, 7(9): 24-48.

Bennet, H.H. (1939), *Soil Conservation*, McGrow Hill Company, New York.

Bhumbla, D.R. and Khare, A. (1984), *Estimate for Wasteland in India*, Society for Promotion of Wasteland Development, New Delhi.

Clark, A.N. *Longman Dictionary of Geography*, Longman, London.

Davidson, D.A. (1992), *The Evaluation of Land Resources*, Longman, London.

Uttar Pradesh District Gazetters, Etawah (1981), Government of Uttar Pradesh, Lucknow: XI.

Gautam, N.C. (1983), "Satellite Remote Sensing techniques for natural resource survey" in Singh L.R. and *et. al.* (ed.). *Environmental Management*, Allahabad Geog. Society: 177-81.

Haussen Builler, R.L. (1963), *Principles of Soil Science*, Orient Longman, Bombay.

Govinda Rajni, S.V. (1966), *Soil Survey Manual*, All India Soil and Landuse Survey Organisation, New Delhi.

Jha, B.N. (1965), "Land capability classification of the Kosi region", *Geog. Outlook*, (4): 76-77.

Jha, V.C. (1984), "Wasteland types and their effective utilisation in Birbhum district", *The Deccan Geographer*, (41) : 231-42.

Kumar, A. Pandey, R.N. and Singh, S.P. (1983), "Wasteland a case for environmental management" in Singh L.R. (ed.), *Environmental Management*, Allahabad Geographical Society, Allahabad: 229-42.

Kumar, A. and Pandey, R.N. (1989), *Wasteland Management in India*, Ashish Publishing House, New Delhi.

Kumar, A. (1989), "Integrated approach to wasteland capability determination in landuse planning : A case study of Deoghar district of Bihar", in Singh (ed.), *Problem of Wasteland and Forest Ecology in India*, Ashish Publishing House, New Delhi : 89-115.

Mohammad, N. (1988), "Wastelands in India : some issues and problems" in Sharma S.C. (ed.) *Utilisation of Wasteland for Sustainable Development in India*, Concept Publishing Company, New Delhi: 57-65.

Mukherjee, A.B. (1971), "Geographical patterns of changes in agricultural wastelands in Uttar Pradesh", *Geographical Review of India*, 2(33) : 120-32.

National Remote Sensing Agency (1991), *Wasteland Atlas of India*, Hyderabad.

N.R.S.A. (1986), Manual of Procedure for Wastelands Mapping Using Remote Sensing Techniques, Department of Space, Govt. of India.

Prasad, R. and Yadav, J.S.P. (1969), *Alkali Soil Reclamation, Question and Answers*, Central Soil Research Institute, Karnal, India.

Remote Sensing Application Center (1990), District Wasteland Report : Etawah district, Lucknow.

Russell, E.J. (1950), *Soil Conditions and Plant Growth*, Longman Green and Co. London.

Shafi, M. (1968), "Problems of Wasteland in India", *The Geographer* (15): 1-15.

Shafi, M. (1969), "Can India support five times her population?" *Science Today* (2): 21-27.

Sharma, H.S. (1968), "Land-capability classification of the lower Chambal Valley" *Proc. Symp.* Landuse in developing countries, Aligarh Muslim. University: 173-79.

Sharma, H.S. (1980), *Ravines Erosion in India*, Concept Publishing Corporation, New Delhi.

Sharma, S.C. Tripathi, D.N. and Sharma, M.L. (1988), "Usar land reclamation measures in India with special reference to Sangrampur block, District Sultanpur, U.P." in S.C. Sharma (ed.), *Utilisation of Wasteland for Sustainable Development in India*, Concept Publishing Company, New Delhi: 285-302.

Shukla, P.N. (1988), "Yamuna and Chambal rivers eroded land and its reclamation in Etawah district, U.P." in S.C. Sharma (ed.), : *Utilisation of Wasteland for Sustainable Development in India*, Concept Publishing Company, New Delhi: 359-68.

Singh, A.L. (1978), *Economics and Geography of Agriculture Land Reclamation*, B.R. Publishing Corporation, New Delhi.

Singh, B. (1962), "Distribution of cultivable waste and possibilities of its reclamation in Chakia, Banaras", *The National Geographer of India*, 1(7): 59-72.

Singh, R.L. (1971), *Indian Regional Geography*, National Geographical Society of India, Varanasi.

Soil Survey Manual and U.S.A. Development of Agriculture (1954), A manual on Conservation of Soil and Water, Deptt. of Agriculture, U.S.A.

Spate, O.H.K. (1954), *India and Pakistan*, Mathuen, London.

Stamp L.D. (1954), *The Land of Britain, its use and misuse*, Longman, London.

Storie, T.E. (1933), "An index for rating the agricultural value of soils" *California Experimental Station* Bulletin, No. 556: 221-29.

U.S.A. Deptt. of Agriculture (1951), *Soil survey manual*, U.S.A.

Yadav, H.R. (1987), "Agroforestry management systems in wastelands" in *Workshop Proceeding* of Agroforestry for Rural Needs: II, Khosala P.K. and Khurana, D.K., I.S.T.S., Solan, India: 108-22.

Yadav, H.R. (1987), "Suitability classification of the wasteland", in H.R. Yadav (ed.) *Wastelands: Diagnosis and Treatment*, Concept Publishing Company, New Delhi: 93-98.

Yadav, H.R. (1984), *Genesis and Utilisation of Wasteland*, Concept Publishing Company, New Delhi.

Yadav, J.S.P. and Aggrawal, R.R. (1959), "Dynamics of soil changes in the reclamation of saline, alkali soils of the Indo. Gangetic Alluvium", *Journal of Indian Society of Soil Sciences* (7) : 77-86.

Young, A. and Goldemith, P.F. (1977), "Soil Survey and land evaluation in developing countries: A case study of Malawai", *Geographical Journal*, (43): 407-38.

20

Loss of Agricultural Land and Changing Utilization of Resource Base

A Case Study of Three Selected Villages of Srikakulam District, A.P.

K.L. NARASIMHA MURTY

Land, labour, capital and organisation are considered as important basic resource factors of production which yield rent, wages, interest and profits whether it is agricultural or industrial economy. Keeping this in view, the factors of production should be put to use for profit maximisation, the imputed costs should be at the minimum level. This can be achieved only if the factors of production are kept for better utilisation. But geographers consider land, water and air are three basic resources besides the human resources (human factor). Human resources are essentially considered for building up the economy at the micro level. They play a vital role in deriving, to a maximum possible extent, utilities on different scales depending upon the "geographic potentialities" and "production opportunities". With the maximum utilisation of geographic potentialities and production opportunities coupled with basic resources and human intervention, maximum profits may be derived even at the village level. The pressure of population compels man to overutilise resources in order to feed the teeming hungry population. Man has been utilising the resources for different uses depending upon the geographic viability and possibility. According to Heady, "Land as a factor of production has no unique characteristics which should cause it to be set aside by itself in an economic analysis. The principles defining the optimum use of land are those drawn from the general production economic principles. From the standpoint of contemporary production economics or economic efficiency, land differs from other factors, only in the sense that its opportunities are restricted. If the land or land resource is put to different uses, the other factors of production can be utilised at the minimum level as an alternative, so as to derive maximum profits, viz. shifting the use of resource to other purposes without disturbing it. Now, a humble attempt is made in this paper to examine Heady's principles of economy in the Indian village context to prove whether it is true or there is any deviation that takes place in a backward economic sites of different farming situations like agricultural farming, poultry farming and dairy farming.

Spatial Characteristics

Ragolu, Kusalapuram and Pedapadu villages of Srikakulam district have been identified as the representative villages of three cultures, viz. agricultural farming, poultry farming and dairy farming. Ragolu is located at a distance of about 8 km. away from the Srikakulam urban centre (18°14′N and 83°52′E longitude). It has a population of about 8,000 and an area of 5 km^2. It is an important village and is noted for the location of the Agricultural Research Station (Rice), and for building material (bricks and tiles) and two rice mills which have recently started functioning. Ragolu is known for paddy cultivation (kharif), groundnut and mango gardens. The soils are alluvial and black.

Kusalapuram is located at a distance of 8 km. from Srikakulam west-north-west direction on the right bank of the major river Nagavali (18°45′N and 83°52′E longitude). It has a population of 7,500 and an area of 7 km^2. It is an important village. On its eastern fringe is located the Government Polytechnic College and on the east-south-east the Narayanapuram channel helps irrigate the landscape leading to new cropping pattern. The red ferruginous soil spreads over all the village and is its predominant characteristic. The hillocks are located within the village and the hillocks are emelled and now used for building construction. This village is quite rich in poultry farming though there is no distinct poultry farm in this district. The village meets the demand for eggs of the urban and rural population of Srikakulam.

Peddapadu is located on the Madras-Calcutta Highway (N.H.5) at a distance of 5 km. east-south-east of the Srikakulam urban centre. It has a population of nearly 5,600 and an area of 3 km^2. There are nearly 600 houses in this village. The natives of the village mostly belong to the Yadava caste (Golla) herdsman community and their traditional occupation is dairy farming. This is the village's important feature. However, it is not properly organised, and yet, it is not only self-sufficient but is also considered as the backbone of the urbanites of Srikakulam, so far as the dairy production, that is, milk and milk products, are concerned. The A.P. Dairy Unit (Milk Chilling Centre) is located in this village. (This centre collects milk from villages other than Pedapadu village and supplies to the urbanites).

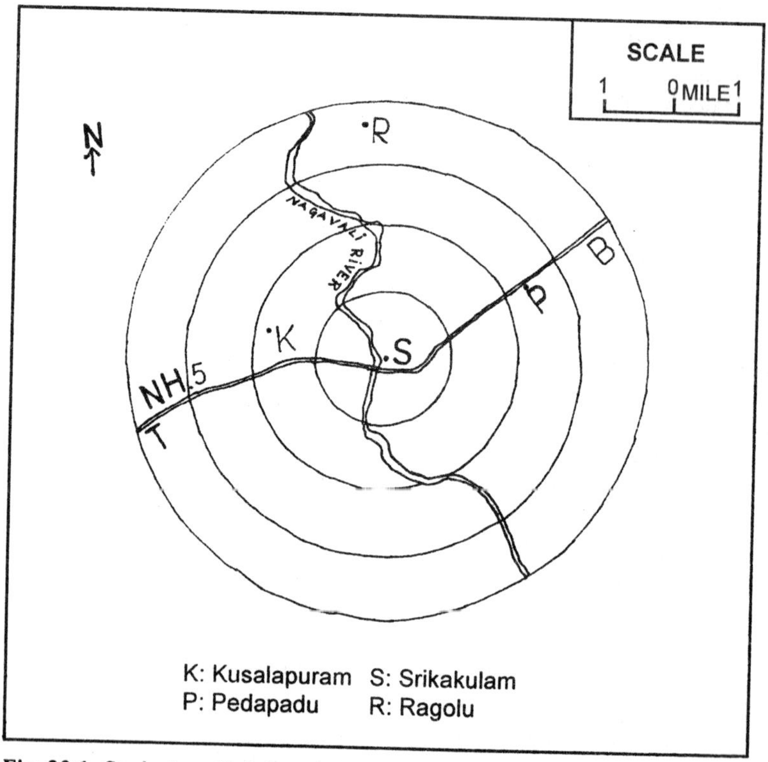

Fig. 20.1: Study Area K, P, R and S.`1

Agricultural Farming in Ragolu

A large portion of the area is cultivated in the kharif season. Due to scarcity of water the land cannot be cultivated during the Rabi season for rice. A significant transformation has taken place to the tune of 5 to 6 per cent of the total cultivated area. The farmer has to spend a lot money on many items, viz. ploughing, preparatory, sowing, weeding, harvesting, threshing and winnowing, seed expenses, fertilisers, insecticides, irrigation charges, carting of produce and others for cultivating his land. The total expenditure incurred on all these items by the farmer for one hectare is Rs. 2,024 (See Table 20.1). But he makes very good profits also. His net profit on one hectare of land is Rs. 9,216.

It can be seen from Table 20.2 that if a farmer substitutes cultivation by brick-laying his expenditure depending upon potentialities and the spatial specialisation, he will be in a position to get a net profit of Rs.1,00,000. The cost benefit ratio will be as high as 1:10.

Table 20.1: Agricultural Farming in Ragolu (Cost-Benefit Analysis).

Sl. No.	*Item of Expenditure*	*No. of Labourers*	*Expenditure*
1.	Ploughing	20	160
2.	Preparatory	33	264
3.	Sowing	10	80
4.	Weeding	15	150
5.	Harvesting	20	160
6.	Threshing/Winnowing	20	160
7.	Seed Expenses	—	80
8.	Fertilisers	—	600
9.	Insecticides	—	10
10.	Irrigation Charges	—	60
11.	Carting of Produce	—	100
12.	Others	—	—
13.	Total Expenditure		2,024
	Total Income at the rate of Rs.150	75 Bags	11,250
			(-) 2,024
		Net Profit	9,226

Case-1: Shri Chitti Bhadrayya (age 45 years) is an active farmer of Ragolu village. He possesses 1 hectare of land. Now, he is utilising his land for the making of bricks (9" × 4" × 3" dimensions). He can do brick-making in two ways. In the first, his family members and hired labour are engaged in the making of bricks. In the second, he leases out his land for the same purpose. Generally, the first type is in practice. The cost benefit analysis can be seen from Table 20.2. The farmer is able to get even Rs.1,00,000 within a couple of months, if his land is being transformed or transferred from agricultural use to brick-making. As a result, the agricultural land (potential one) will be decreased alarmingly (Plate 1 and 2).

Table 20.2: Brick-making (Expenditure and Income).

	Rs.
Expenditure	
1. Material size (9" × 4" × 3")	
2. Labour per 1000 bricks 3 males × 10 (Rs. 30)	3,000
3. Fire action per 1,000 bricks (Rs. 60)	6,000
Total Expenditure	9,000
Income	
1000 Bricks at the rate of Rs. 200 (1,00,000)	20,000
Rs. 20,000 – Rs. 9,000	11,000
Agricultural income per hectare	9,200
Income on bricks is (with in one week)	11,000

The second alternative for the above is that if the farmer makes or divides a parcel of one hectare of land into 10 bits, he will be able to get nearly Rs. 30,000 per bit and, ultimately, he gets a net profit of Rs. 3,00,000 in no time. The opportunity costs comparatively will be reduced and the marginal profit will be at the optimum level (only in the case of disturbing the land resource).

Agricultural Farming in Kusalapuram

Agricultural farming in Kusalapuram presents a different picture. Rice cultivation is by dry-farming technique. The slope sets a limit to the farming technique to a large extent (See Table 20.3). The expenditure on the farming items by the farmer is Rs.1,400. The farmer is able to get a net profit of Rs.1,000. The cost benefit ratio will be around 1:1 on one hectare of land.

Table 20.3: Agricultural Farming in Kusalapuram (Cost-benefit analysis in agriculture unirrigated rice).

Sl. No.	*Item of Expenditure*	*No. of Labourers*	*Per Person*	*Expenditure*
			Rs.	Rs.
1.	Ploughing	15	10	150
2.	Weeding	20	10*	200
3.	Harvesting	20	10	200
4.	Threshing	15	10	150
5.	Seed Expenses (40 kg.)	—	—	80
6.	Fertilizers (2 bags)	—	—	400
7.	Insecticides	—	—	20
8.	Carting of produce	—	—	200
		Total		1,400
	Production of 20 bags each at Rs.120 = 2,400-1,400 =Rs.1,000.			

* 1986 wages.

The alternative for the farmer is shifting the resource base to different use. The expenditure and income generation ratio is presented in Table 20.4. If the active farmer becomes an entrepreneur by investing his capital on poultry farming (shifting the culture as well as space without disturbing the soil) activity, the total cost is Rs.78,000 and his net profit comes to Rs. 3,36,000. The cost-benefit ratio will be around 1:4.

Case 1: Sri Modallavalasa Sukumar (age, 26 years) started Sri Vijayalaxmi Poultry Farming on December 6, 1980. He is a graduate. He started a poultry farm with 1,000 birds. He purchased the birds from Coastal Hatcheries, Visakhapatnam, at a cost of Rs. 8 per bird. The District Industrial Centre (D.I.C.), Srikakulam, has evolved a self-employment scheme for young and promising people to take up poultry farming in different villages of Srikakulam district. As a result, this activity came to the forefront. The Indian Overseas Bank has funded this activity. Sri Sukumar constructed a shed (120 m × 20 m) at a cost of Rs.30,000 for the birds for this activity. The feed cost him Rs. 40,000. The cost-benefit analysis can be seen in Table 20.4. The cost-benefit ratio will be 1:4 (Plate 4).

Plate 1: Brick-making in Ragolu village (loss of Agricultural land).

Plate 2: Agricultural land divided into plots for Settlement (loss of Agricultural land or its use of Agricultural land).

Plate 3: After Kharif Rice Crop, land put to fodder, Herdsman within cattle in the hectare of land (loss of Rabi Crop).

Plate 4: Poultry farming in Kusalapuram (loss of Agricultural land).

Table 20.4: Poultry-Farming (Expenditure and Income).

Sl. No.		*Rs.*
	Expenditure	
1.	1000 birds at Rs. 8 each	8,000
2.	Construction of shed	30,000
3.	Cost of feed Rs. 40 per bird	40,000
		78,000
	Less 20% (deaths)	1,600
		76,400
	Income	
1.	95% for 3 months: 950 × 3	2,850
2.	85% for 4 months: 850 × 4	3,300
3.	70% for 6 months: 700 × 6	4,200
4.	5 months – nil —	—
	Total Eggs	10,350
	Per egg 40 paise	4,14,000
	Less : Expenditure	76,400
	Net profit	3,37,600

Agricultural Situation in Pedapadu

The farmer has to follow the dry farming method to cultivate his parcel of land. During Rabi, he cultivates gram, oilseeds and fodder depending upon his taste as well as compelling needs. Basing upon the topographical features and surface water availability the farmer has to cultivate his small parcel of land for kharif crops (rainfed crops). To some extent, the farmer also depends on tanks for irrigating his small chunk of land (uneconomic).

The cost-benefit factor of agricultural farming for unirrigated paddy is presented in Table 20.5. The total expenditure on agricultural inputs is Rs.1,500, the net profit derived from one hectare of land is Rs.13,500. The cost-benefit ratio is 1:8.

The alternative for agricultural farming is Pedapadu is dairy farming. So he invests his revenue in farming. The cost factor involved in this farming is Rs. 8,052 which includes the cost of the animals and the daily cost factors like fodder and hay. The benefit factor will spread over to 300 days in a year. The net profit for an entrepreneur is Rs.16,800. The cost-benefit ratio will be 1:2.

Table 20.5: Agricultural Farming in Pedapadu (Cost-benefit analysis in rice).

Sl. No.	*Item of Expenditure*	*No. of Labourers*	*Per Person*	*Expenditure*
			Rs.	Rs.
1.	Ploughing	20	8*	160
2.	Preparatory	30	8	240
3.	Sowing	10	8	80
4.	Weeding	12	8	96
5.	Harvesting	15	8	120
6.	Threshing/ winnowing	20	8	160
7.	Seed expenses	—	—	80
8.	Fertilisers (2 bages) (Rs. 200 per bag)			400
9.	Insecticides	—	—	20
10.	Irrigation charges	—	—	60
11.	Carting of produce	—	—	80
	Total expenditure (round figure			1,490
	Total Income 100 bags × Rs. 120			12,000
	Black gram (5 bags) 5 × 400			2,000
	14,000 – 1,500		12,500	
	Fodder		1,000	
	Total		13,500	

Note: One hectare of land will be cultivated by family members (labour).
* 1983-84 wages.

Case 1: Sri K. Atchayya (40 years) belongs to the Yadava caste and is an active entrepreneur. He has two hectares of land. Besides, he has been leased out one hectare for agriculture for earning his livelihood. His family consists of two male and three female members. The leased out one hectare of land is utilised for growing one crop (kharif paddy) with the help of tank irrigation. On the other hectare of land dairy farming, the only non-agricultural occupation, is done. This gives him quick returns. He has four milch animals worth Rs. 2,000 each. Cow and she-buffalo are the two important milch animals in this village. The demand for milk (preferably buffalo) is from the urbanites. So she-buffalo outnumber cows. Besides buffaloes give more milk. The cost benefit analysis can be seen in Table 20.6.

Absolutely no man sells milk to the A.P. Dairy because of some technical constraints. A herdsman is looking after his buffaloes (It can be seen from Plate 3).

Table 20.6: Cost-Benefit Analysis in Dairy Farming.

Cost of the Item	*Rs.*	*No.*	*Total in Rs.*	
Cost Factor:				
Cost per animal	2,000	4	8,000	
Fodder (per bag)	6	4	24	Rs. 52/- per day
Hay	7	4	28	
Total cost	2,013		8,052	
Benefit Factor:				
1. Milk (36 litres per day) at Rs. 3 per litre (36×3 = Rs.108)				
For 300 days:	300×108	32,400		
	300×52	15,600 (-)		
	Net profit it	16,800		

Occupational Structure and Landuse three Villages

The village-wise occupational structure is presented in Table 20.7. In Ragolu, Kusalapuram and Pedapadu, the sample size is 25, 30 and 30 respectively for understanding the occupational structure at the basic level and the people's participation in new occupations. People belonging to Kalinga, Kinthali Kalinga and Golla (Herdsman) are predominant castewise as well as population wise and their basic occupations are agriculture farming and dairy farming. The transformation generally is accepted by people belonging to 15-45 age group. Now, people belonging to 20 and 20-40 age groups like to have a change from their basic occupation because of the promising returns from the new innovations. Sixty five people out of 85 samples welcomed the change from their basic occupations and nearly 76 per cent people have an inclination towards new and innovative occupations.

Landuse: Three hundred ninety-seven hectares have been utilised for different productive purposes without disturbing the soil and in Ragolu 119 hectares have been disturbed for productive purposes.

Table 20.7: Occupational Structure and Landuse in Three Villages (Ragolu, Kusalapuram and Pedapadu).

Sl. No.	*Name of Village*	*Caste*	*Age group*	*Total No. of people*	*Age group*	*Total No. of people*	*No. of acres utilised in New Innovations*
1.	Ragolu	Kalinga (Agrl.)	15-45	25	20-40	20	119
2.	Kusalapuram	Kintali Kalinga (Educated Unemployed Agrl.)	20-50	30	20 & above	20	10
3.	Pedapadu	Golla (Dairy/Agrl.) milk vendor	20-45	30	20 & above	25	268
				85		65	397

Heady's Concept *vs.* Murthy's Concept: A Geonomic Model

"Land can be shifted back and forth only between alternatives at one geographic location; capital, labour and management can be transferred between production opportunities in different locations as well as at one particular point."

Examining Heady's above concept we find that, if the landuse or land resource is put to different uses, the other factors of production can be utilised at the minimum level as an alternative so as to derive maximum profits (i.e. shifting the use of resource to other purposes without disturbing it).

But in different farming situations, Heady's concept does not hold good especially on the Indian landscape, people and cultures.

Murthy's Concept of Spatial Planning

Spatial planning concept with reference to different scales of economics within a geographic area, and at one point of time, can be examined. Now, we have examined the different farming situations with their cost-benefit ratios. The production opportunities and geographic potentialities will have a bearing on the occupational structure also at the village level. Geographers consider land as an irreplaceable resource and economists consider land as an immobile resource. If we examine these two characteristics geonomically, that the land resource should be kept for use without disturbing its structure (indestructibility) and changing the scales of economics, the profit, can be maximised. But, in the existing inevitable situations, man has been disturbing the land structure and utilising for achieving the maximum profits depending upon the geographic potentialities and production opportunities. To put the concept in a nutshell, "People are amenable in utilising or harnessing the resource base at the best to geographic potentialities and production opportunities but they are faster in disturbing the nature of resources, viz. land for their living in a backward economic situation due to the paucity of revenues for their traditional occupations which is presently a backlog and ultimately presents a transformation of resource base and occupational structure in a long way of urbanisation process".

Conclusion

1. Agricultural land (land used for agriculture) should not be transferred and put in for other than agriculture uses. Though

the changing resource base gives him (entrepreneur) more profit in a short period, the long-term disadvantages will be the gradual decrease of the soil (material) subsurface.

2. Uncultivated land or barren land should be used for poultry farming and dairy farming. For this, a legislation should be enacted, i.e. debarring the use of agricultural land for non-agricultural activities.
3. Organised activity will lead to a better or optimum use of resources for the people over a period of time. "Space" and "time" are the two wheels of the chariot that will take to an integrated and balanced development of a region.
4. The urbanisation process to some extent should be curtailed by encouraging urbanites to migrate to rural areas for a new type of living by way of purchasing the agricultural land from the farmer for settlement purposes.
5. The Government of Andhra Pradesh should undertake the survey of resources, identify the needs of the people, find out the methods for processing the raw material and finally market the commodities (final product) to the communities starting from the grassroot level, viz. rural areas, semi-urban areas, urban areas and metropolitan areas.

REFERENCE

Heady, E.O. (1964), *Economics of Agricultural Production and Resource use*, pp. 763-793, I.H.H., New Delhi.

21

Changing Pattern of Crop Failure in Rainfed Agricultural Lands

A Case Study of Hissar Tehsil of Haryana

DAYA RAM and S.S. YADAV

During the last two decades, intensive efforts have been made to increase agricultural production in different parts of the country. A variety of inputs and incentives have been provided both at national and grassroot level to promote growth of agriculture. It would certainly be a matter of great interest to agricultural economists and planners to understand the after-effects of these factors on the agricultural progress. In dryland farming, this can be adjudged by measuring the changing pattern of crop failure in the given area and given period of time.

Agriculture is not a static factor but a highly dynamic one, subject to frequent changes, which may be on either direction, on plus or minus side. Hence, this study throws light on the changing pattern of crop failure for a particular period in a given area. Yet, by and large, it may provide a good benchmark for future planning. The degree of change can be quantified and results measured from the benchmark. The construction of a benchmark, i.e. the baseline, calls for a great deal of caution, but once it is constructed, the changes which may occur could be described at any subsequent time.

Hissar tehsil (28°54′ to 29°25′N parallels and 75°23′ to 75°59′E meridians) in Hissar district of Haryana State comprises 122 villages with a total geographical area of 2,73,116 hectares. In this tehsil, 41 per cent of the total cultivated land was rainfed (1987-88). It lies on the north-eastern fringe of *Marusthali* (Thar desert) in Rajasthan State and has subtropical, continental, hot, semi-dry climate with monsoonal regime of rainfall. The greater part of the tehsil is an alluvial plain lying at 200 to 225 m. above mean sea level. The soils comprise loamy sands and sandy loams. At places, there are some broad ridges of sandy soil. The sub-soil water is deep-seated and mostly brackish.

Data Set and Their Processing

The study is based on the comparable data of crops sown and crops failed area (both summer and winter seasons) of the periods during 1968-69 to 1972-73 and 1983-84 to 1987-88 in 122 villages of Hissar tehsil taken from the records of the office of the Kanungo (Accountant) of the tehsil. Five-year totals for failure index were then calculated for all the crops. The index values for the first five years (1968-69 to 1972-73) were taken to represent the crop failure level at the commencement of the period under study. Likewise, the index values for the last five years (1983-84 to 1987-88) represent the crop failure level at the close of the period under study. The difference in the index values of the crop failure for the first and last five years provides the changing pattern of crop failure in rainfed cultivated areas of Hissar tehsil. The data were processed as under:

Crop Failure Index (CFI):

$$\frac{\text{Actual failed area in a sequence of years}}{\text{Total sown a rea in that sequence of years}} \times 100$$

The difference in the index values of all the crops was worked out. The differenced values thus derived were grouped and classified into 15 categories as presented in Table 21.1.

The question may arise that why totals of five-year data at the beginning (1968-69 to 1972-73) and the totals of five-year data at the end (1983-84 to 1987-88) of the period under study were considered. The reason is that generally there is probability of at least one bad, one good and one normal crop year out of every five consecutive years in this part of the country. During the five-year period at the beginning, 1968-69 was a bad and 1969-70 and 1970-71 were good crop years. The rest of the years (1971-72 and 1972-73) were normal. Similarly, during the five-year period at the end, 1987-88 was a bad and 1984-85 and 1985-86 were good crop years. The remaining years

(1983-84 and 1986-87) were normal. This has enabled the study to neutralise the effect of variability in rainfall by which a comprehensive picture of the changing pattern of crop failure for the whole period of study (1968-69 to 1987-88) in this tehsil has been brought out.

It is a prerequisite that higher the difference in the crop failure values, the higher would be the degree of change in the crop failure patterns in both positive and negative directions and *vice versa.*

Results and Discussion

The changing pattern of crop failure in both kharif and rabi seasons during the period 1968-69 to 1987-88 in Hissar tehsil has been shown in Tables 21.1 to 21.3 and Figures 21.1 to 21.11.

Summer (Kharif) Season Crops

The changing pattern of crop failure in the summer season is presented in Table 21.1 and Fig. 21.1 to 21.5. The results derived are discussed as under:

Bajra or Pearl Millet (Pennisetum americanum)

A moderate increase (30.01 to 45%) in the failure index of this crop was found in 4.94 per cent sown area (five villages), moderately low (15.01 to 30%) in 10.63 per cent area (12 villages) and low (0.01 to 15%) in 27.61 per cent area (31 villages). Thus, out of the total sown area of 122 villages under this crop, there was an increase in the crop failure index of 43.18 per cent area (48 villages). The remaining 56.82 per cent area scattered in 74 villages showed a decreased in the failure index of this crop. Low decrease (0.01 to 15%) in 34.26 per cent area (43 villages), moderately low (15.01 to 30%) in 15.55 per cent area (in 22 villages), moderate (30.01 to 45%) in 6.08 per cent area (seven villages), moderately high (45.01 to 60%) in 0.44 per cent area (one village) and high (60.01 to 75%) in 0.49 per cent area (one village) was observed (Table 21.1 and Fig. 21.1).

Guar or Clusterbean (Cyamopsis tetragonoloba)

A very high increase (75.01 to 90%) in the failure index of this crop was found in 1.37 per cent sown area (one village), moderate (30.01

Table 21.1: Changing Pattern of Kharif Crops' Failure in Dry Farming Areas of Hissar Tehsil (1968-1987).

Category No.	*Difference in Crop Failure Index*	*Bajra* Sown area		*No. of villages*	*Guar* Sown area		*No. of villages*	*Kharif pulses* Sown area		*No. of villages*	*Jowar* Sown area		*No. of villages*	*Degree of Change (increase or decrease in crop failure)*
		ha.	*%*		*ha.*	*%*		*ha.*	*%*		*ha.*	*%*		
1	2	3	4	5	6	7	8	9	10	11	12	13	14	15
I	+90.01 to +100	—	—	—	—	—	—	41	0.84	3	64	25.91	1	Extremely High (increase)
II	+75.01 to +90	—	—	—	1,686	1.37	1	201	4.13	4	66	26.72	1	Very High (increase)
III	+60.01 to +75	—	—	—	—	—	—	36	0.74	3	—	—	—	High (increase)
IV	+45.01 to +60	—	—	—	—	—	—	406	8.34	4	—	—	—	Moderately High (increase)
V	+30.01 to +45	16,971	4.94	5	13,531	10.97	6	101	2.08	4	—	—	—	Moderate (increase)
VI	+15.01 to +30	36,556	10.63	12	21,500	17.44	17	1,106	22.72	8	—	—	—	Moderately Low (increase)
VII	+0.01 to +15	94,929 (1,48,456)	27.61 (43.18)	31 (48)	36,153 (72,870)	29.33 (59.11)	35 (59)	429 (2,320)	8.81 (48.06)	3 (29)	— (130)	— (52.63)	— (2)	Low (increase)
VIII	±0.00	-	-	-	-	-	-	163	3.35	13	22	8.91	1	Neutral

(*Contd.*)

Table 21.1 (*Contd.*)

1	2	3	4	5	6	7	8	9	10	11	12	13	14	15
IX	-0.01 to −15	1,17,788	34.26	43	35,365	28.69	40	1,720	35.33	4	39	15.79	1	Low (decrease)
X	−15.01 to −30	53,447	15.55	22	8,913	7.23	13	226	4.64	5	—	—	—	Moderately Low (decrease)
XI	−30.01 to −45	20,885	6.08	7	4,389	3.56	8	400	8.22	3	56	22.67	2	Moderate (decrease)
XII	−45.01 to −60	1,516	0.44	1	,326	0.26	1	22	0.45	1	—	—	—	Moderately High (decrease)
XIII	−60.01 to −75	1,698	0.49	1	1,415	1.15	1	5	0.10	1	—	—	—	High (decrease)
XIV	−75.01 to −90	—	—	—	—	—	—	12	0.25	1	—	—	—	Very High (decrease)
XV	−90.01 to −100	— (1,95,335)	— (56.82)	— (74)	— (50,407)	— (40.89)	— (63)	— (2,385)	— (48.59)	— (15)	— (95)	— (38.46)	— (3)	Extremely High (decrease)
Crop not grown	—	—	—	—	—	—	—	—	—	—	—	—	116	Crop Not Grown

Note: Figures in parentheses in front of Category VII are totals for areas showing a positive change and those in front of Category XV are the totals for areas having negative change in the Crop Failure Indexed.

— Indicates Nil.

Source: Tehsil Office, Hissar (Haryana).

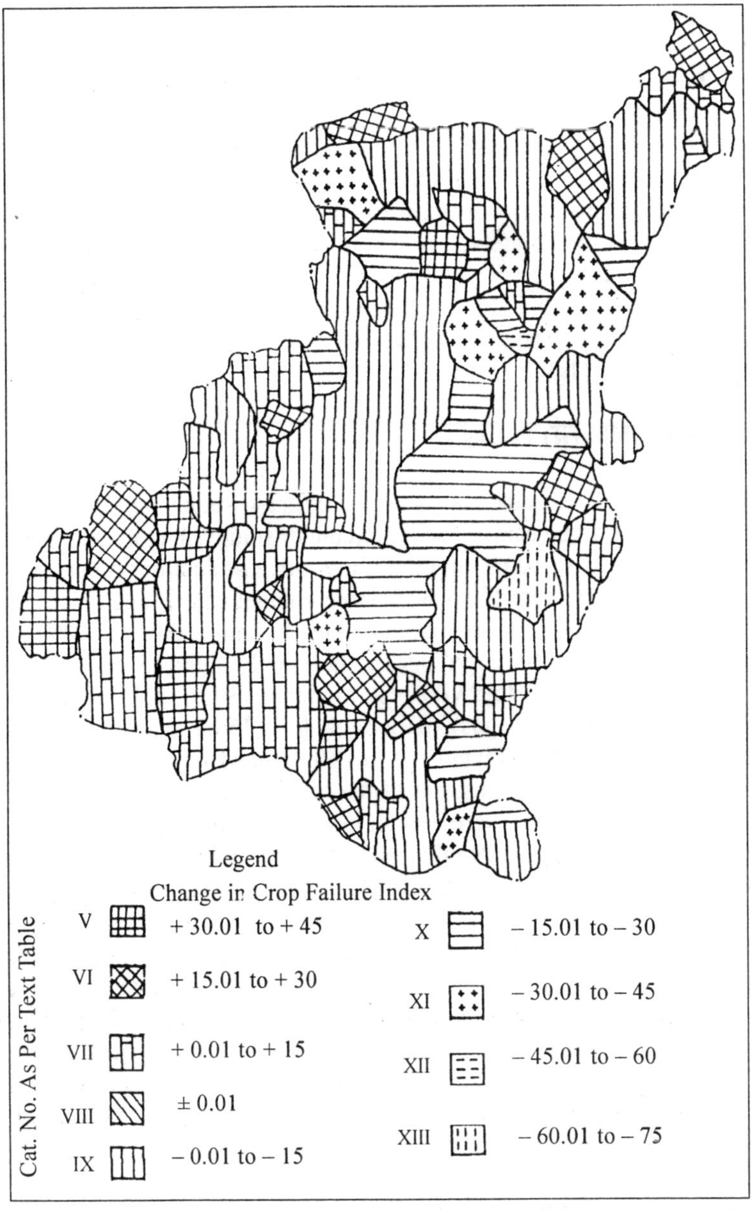

Fig. 21.1: Changing Pattern of Bajra Crop Failure — 1968-72 and 1983-87 (Hissar Tehsil).

to 45%) in 10.97 per cent area (six villages), moderately low (15.01 to 30%) in 17.44 per cent area (17 villages) and low (0.01 to 15%) in 29.33 per cent area (35 villages). Thus, out of the total sown area of 122 villages under this crop, there was an increase in the crop failure index of 59.11 per cent area (59 villages). The remaining 40.89 per cent area scattered in 63 villages showed a decrease in the failure index of this crop. Low decrease (0.01 to 15%) in 28.69 per cent area (40 villages), moderately low (15.01 to 30%) in 7.23 per cent area (13 villages), moderate (30.01 to 45%) in 3.56 per cent area (eight villages), moderately high (45.01 to 60%) in 0.26 per cent area (one village) and high (60.01 to 75%) in 1.15 per cent area (one village) was observed (Table 21.1 and Fig. 21.2).

Summer (Kharif) Pulses

In 65 villages, summer pulses were not grown either in the first or the last set of five years of the period under study. An extremely high increase (90.01 to 100%) in the failure index of summer pulses was found in 0.84 per cent sown area (three villages), very high (75.01 to 90%) in 4.13 per cent area (four villages), high (60.01 to 75%) in 0.74 per cent area (three villages), moderately high (45.01 to 60%) in 8.34 per cent area (four villages), moderate (30.01 to 45%) in 2.08 per cent area (four villages), moderately low (15.01 to 30%) in 22.72 per cent area (eight villages) and low (0.01 to 15%) in 8.81 per cent area (three villages). Thus, out of the total sown area of 57 villages under summer pulses, there was an increase in the crop failure index of 48.06 per cent area (29 villages). The crop failure pattern remains unchanged in 3.35 per cent area (13 villages). The rest of the area (48.59%) scattered in 15 villages showed a decrease in the failure index of summer pulses. Low decrease (0.01 to 15%) in 35.33 per cent area (four villages), moderately low (15.01 to 30%) in 4.64 per cent area (five villages), moderate (30.01 to 45%) in 8.22 per cent area (three villages), moderately high (45.01 to 60%) in 0.45 per cent area (one village), high (60.01 to 75%) in 0.10 per cent area (one village) and very high (75.01 to 90%) in 0.25 per cent area (one village) was observed (Table 21.1 and Fig. 21.3).

Jowar or Sorghum (Andropogon vulgare)

In 116 villages, sorghum was not grown either in the first or the last

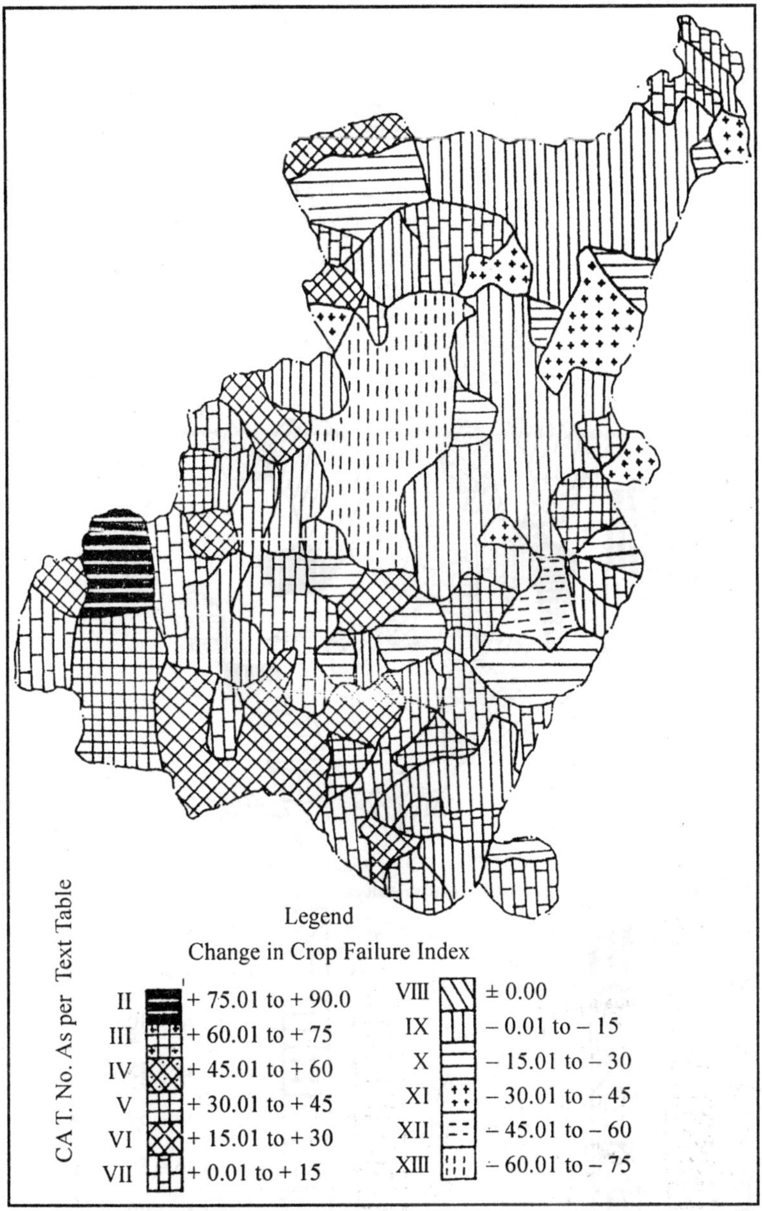

Fig. 21.2: Changing Pattern of Guar Crop Failure — 1968-72 and 1983-87 (Hissar Tehsil).

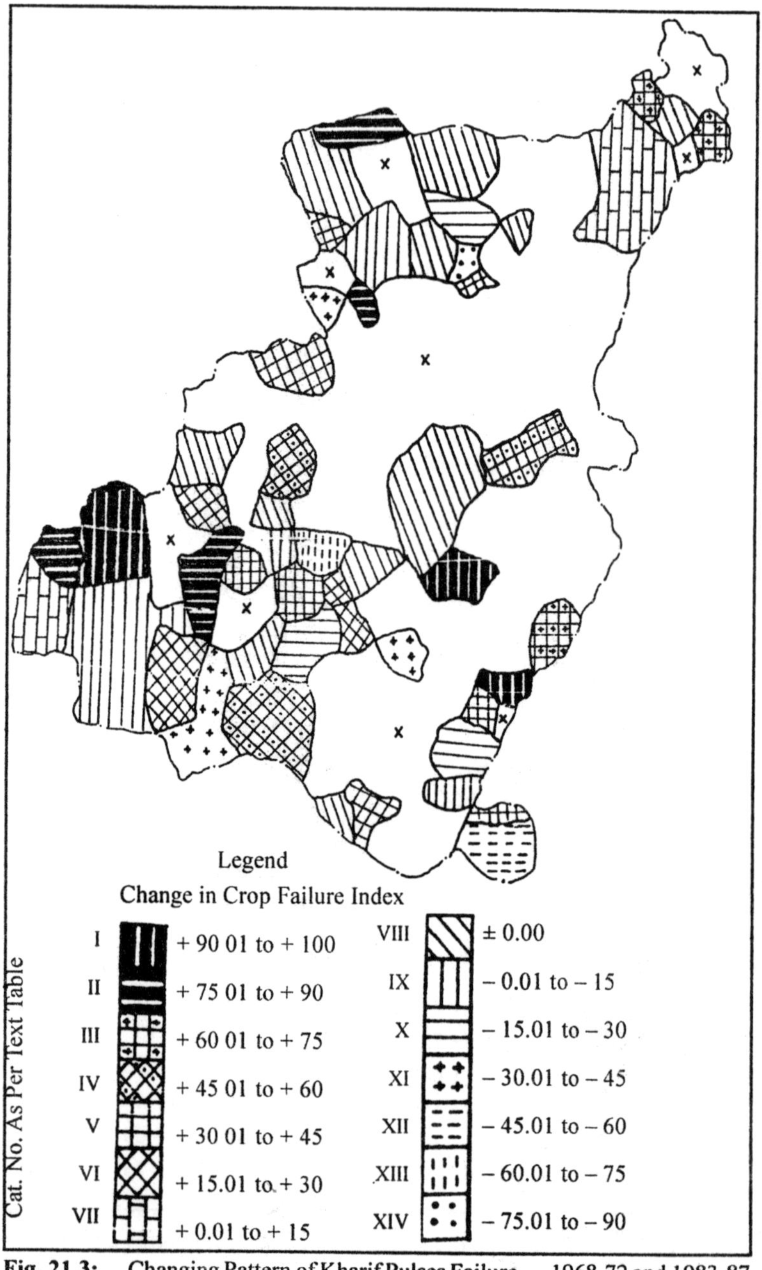

Fig. 21.3: Changing Pattern of Kharif Pulses Failure — 1968-72 and 1983-87 **(Hissar Tehsil).**

set of five years of the period under study. An extremely high increase (90.01 to 100%) in the failure index of this crop was found in 25.91 per cent sown area (one village) and very high (75.01 to 90%) in 26.72 per cent area (one village). Thus, out of the total sown area of six villages under this crop, there was an increase in the crop failure index of 52.63 per cent area (two villages). The crop failure pattern remains unchanged in 8.91 per cent area (one village). The rest of the area (38.46%) scattered in three villages showed a decrease in the failure index of this crop. Low decrease (0.01 to 15%) in 15.79 per cent area (one village) and moderate (30.01 to 45%) in 22.67 per cent area (two villages) was observed (Table 21.1 and Fig. 21.4).

Total of Summer (Kharif) Season Crops

A moderate increase (30.01 to 45%) in the failure index of summer crops altogether was found in 3.77 per cent sown area (three villages), moderately low (15.01 to 30%) in 14.03 per cent area (15 villages) and low (0.01 to 15%) in 30.28 per cent area (38 villages). Thus, out of the total sown area of 122 villages under summer crops, there was an increase in the crop failure index of 48.08 per cent area (56 villages). The remaining 51.92 per cent area scattered in 66 villages showed a decrease in the failure index of summer crops as a whole. A low decrease (0.01 to 15%) in 31.37 per cent area (38 villages), moderately low (15.01 to 30%) in 17.94 per cent area (23 villages), moderate (30.01 to 45%) in 2.16 per cent area (four villages) and moderately high (45.01 to 60%) in 0.45 per cent area (one village) was observed (Table 21.3 and Fig. 21.5).

Winter (Rabi) Season Crops

The changing pattern of crop failure in the winter season is shown in Table 21.2 and Fig. 21.6 to 21.10. The results obtained are discussed below:

*Gram or Chick Pea (*Cicer arietinum*)*

A high increase (60.01 to 75%) in the failure index of this crop was found in 0.84 per cent sown area (two villages), moderately high (45.01 to 60%) in 2.59 per cent area (three villages), moderate (30.01 to 45%) in 8.20 per cent area (eight villages), moderately low (15.01

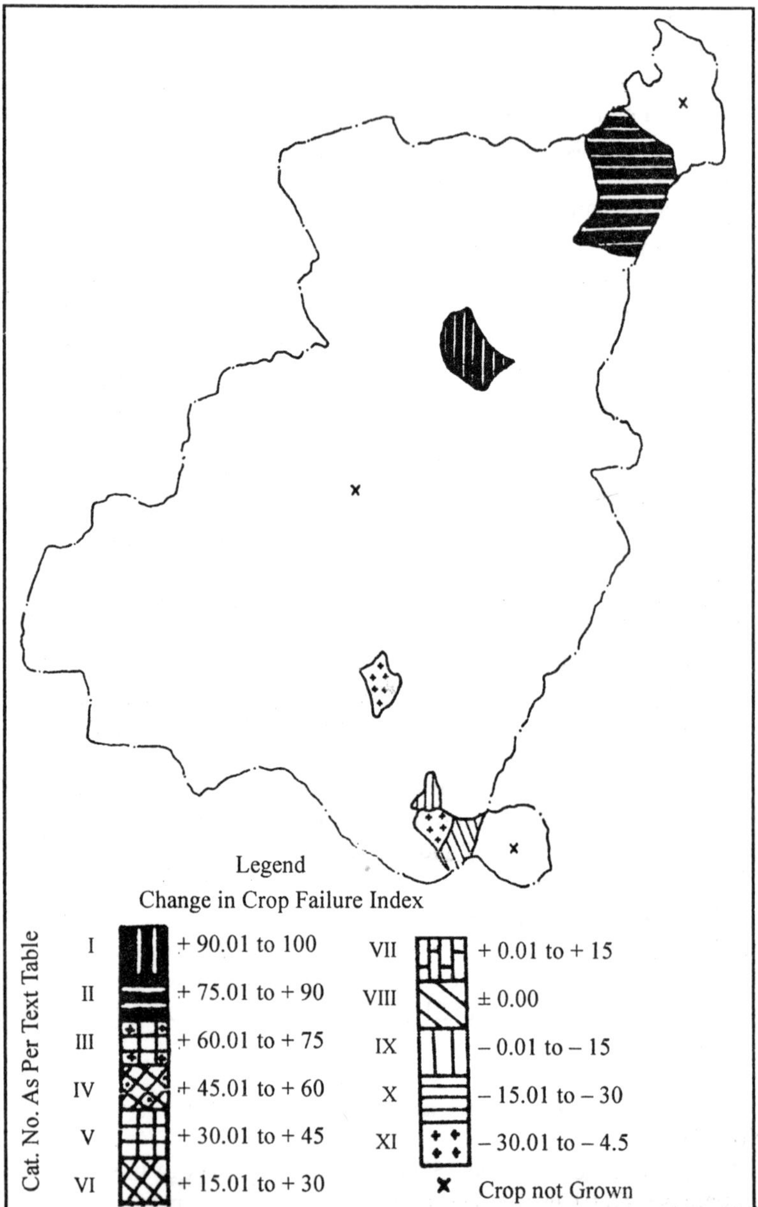

Fig. 21.4: Changing Pattern of Jowar Crop Failure — 1968-72 and 1983-87 (Hissar Tehsil).

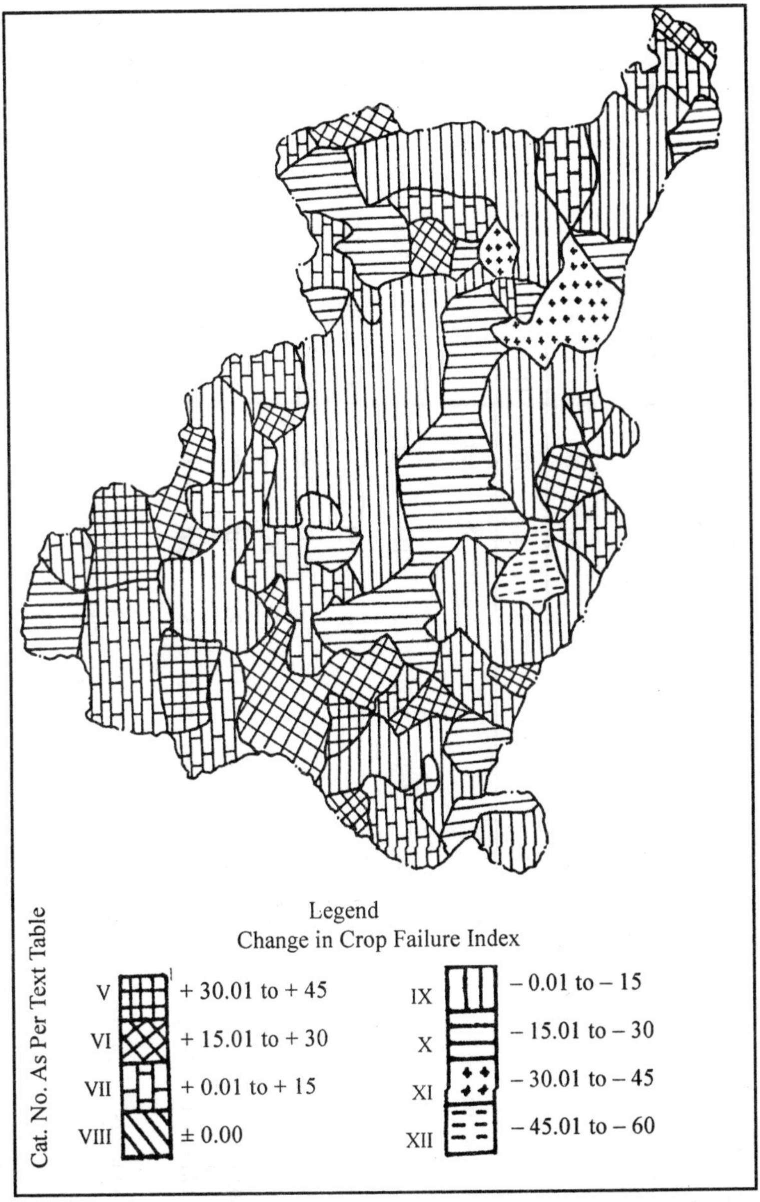

Fig. 21.5: Changing Pattern of Kharif Crops Failure — 1968-72 and 1983-87 (Hissar Tehsil).

Table 21.2: Changing Pattern of Rabi Crops' Failure in Dry Farming areas of Hissar tehsil (1968-69 to 1987-88)

Category No.	*Difference in Crop Failure Index*	*Gram* Sown area		*No. of villages*	*Mustard* Sown area		*No. of villages*	*Taramira* Sown area		*No. of villages*	*Barley* Sown area		*No. of villages*	*Degree of Change (increase or decrease in crop failure)*
		ha.	*%*		*ha.*	*%*		*ha.*	*%*		*ha.*	*%*		
1	2	3	4	5	6	7	8	9	10	11	12	13	14	15
I	+90.01 to +100	—	—	—	—	—	—	118	1.00	3	16	2.59	3	Extremely High (increase)
II	+75.01 to +90	—	—	—	703	2.98	1	111	0.94	2	46	7.44	3	Very High (increase)
III	+60.01 to +75	2,042	0.84	2	—	—	—	594	5.02	3	28	4.53	1	High (increase)
IV	+45.01 to +60	6,286	2.59	3	—	—	—	175	1.48	2	—	—	—	Moderately High (increase)
V	+30.01 to +45	19,893	8.20	8	316	1.34	4	318	2.69	3	78	12.62	4	Moderate (increase)
VI	+15.01 to +30	36,698	15.12	16	625	2.65	5	301	2.54	4	13	2.10	2	Moderately Low (increase)
VII	+0.01 to +15	24,782 (89,701)	10.21 (36.96)	15 (44)	3,992 (5,636)	16.92 (23.89)	18 (28)	1,878 (3,495)	15.86 (29.53)	14 (31)	64 (245)	10.36 (39.64)	1 (14)	Low (increase)
VIII	±0.00	—	—	—	41	0.17	3	10	0.08	1	4	0.65	1	Neutral
IX	-0.01 to –15	36,929	15.21	28	8,115	34.39	36	1,676	14.16	13	—	—	—	Low (decrease)

(Contd.)

Table 21.2 (*contd.*)

1	2	3	4	5	6	7	8	9	10	11	12	13	14	15
X	-15.01 to -30	75,352	31.04	30	3,039	12.88	18	698	5.90	7	39	6.31	3	Moderately Low (decrease)
XI	-30.01 to –45	21,930	9.04	10	3,531	14.96	15	653	5.52	11	50	8.09	2	Moderate (decrease)
XII	-45.01 to –60	12,695	5.23	7	1,232	5.22	6	1,884	15.91	15	70	11.33	2	Moderately High (decrease)
XIII	-60.01 to –75	5,382	2.22	2	1,161	4.92	5	2,117	17.88	13	—	—	—	High (decrease)
XIV	-75.01 to –90	736	0.30	1	448	1.90	2	492	4.15	7	27	4.37	1	Very High (decrease)
XV	-90.01 to -100	— (1,53,024)	— (63.04)	— (78)	393 (17,919)	1.67 (75.94)	2 (84)	813 (8,333)	6.87 (70.39)	7 (73)	183 (369)	29.61 (59.71)	3 (11)	Extremely High (decrease)
Crop not grown	—	—	—	—	—	—	—	—	—	17	—	—	96	Crop Not Grown

Note: Figures in parentheses in front of Category VII are totals for areas showing a positive change and those in front of Category XV are the totals for areas having negative change in the Crop Failure Indexed.

— Indicates Nil.

Source: Tehsil Office, Hissar (Haryana).

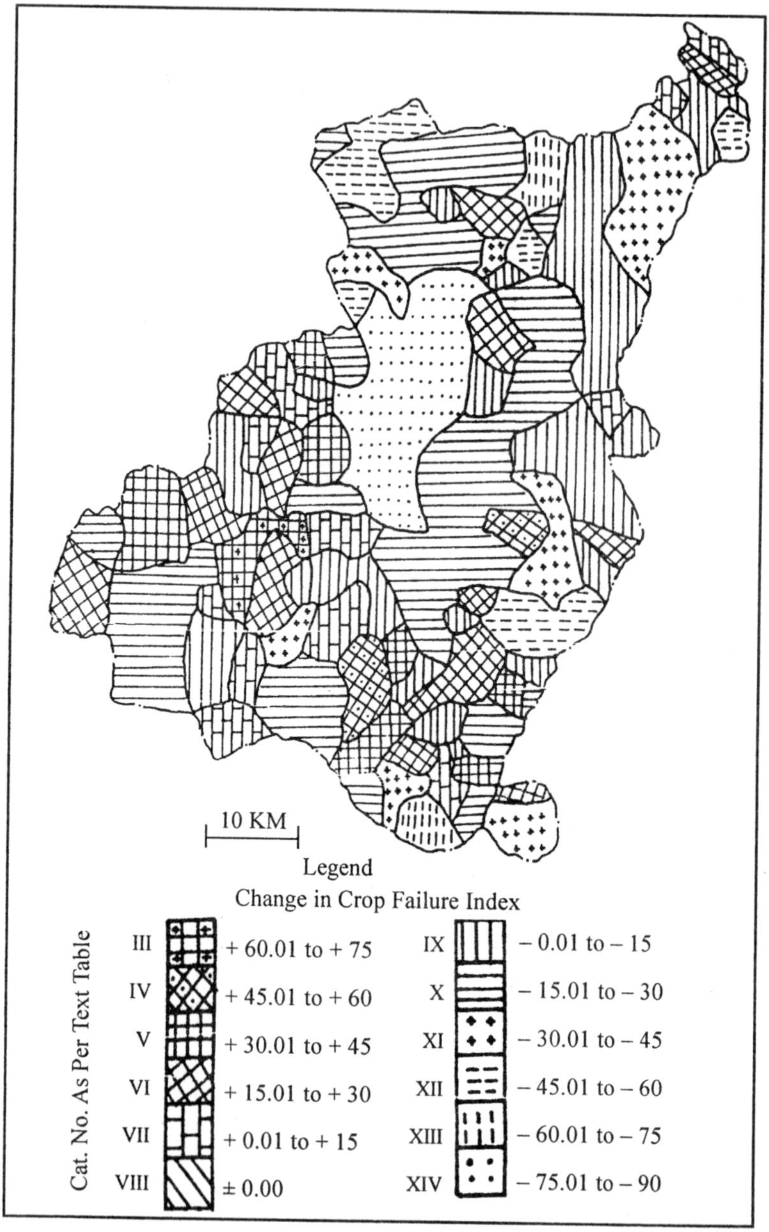

Fig. 21.6: Changing Pattern of Gram Crop Failure — 1968-69 to 1972-73 and 1983-84 to 1987-88 (Hissar Tehsil).

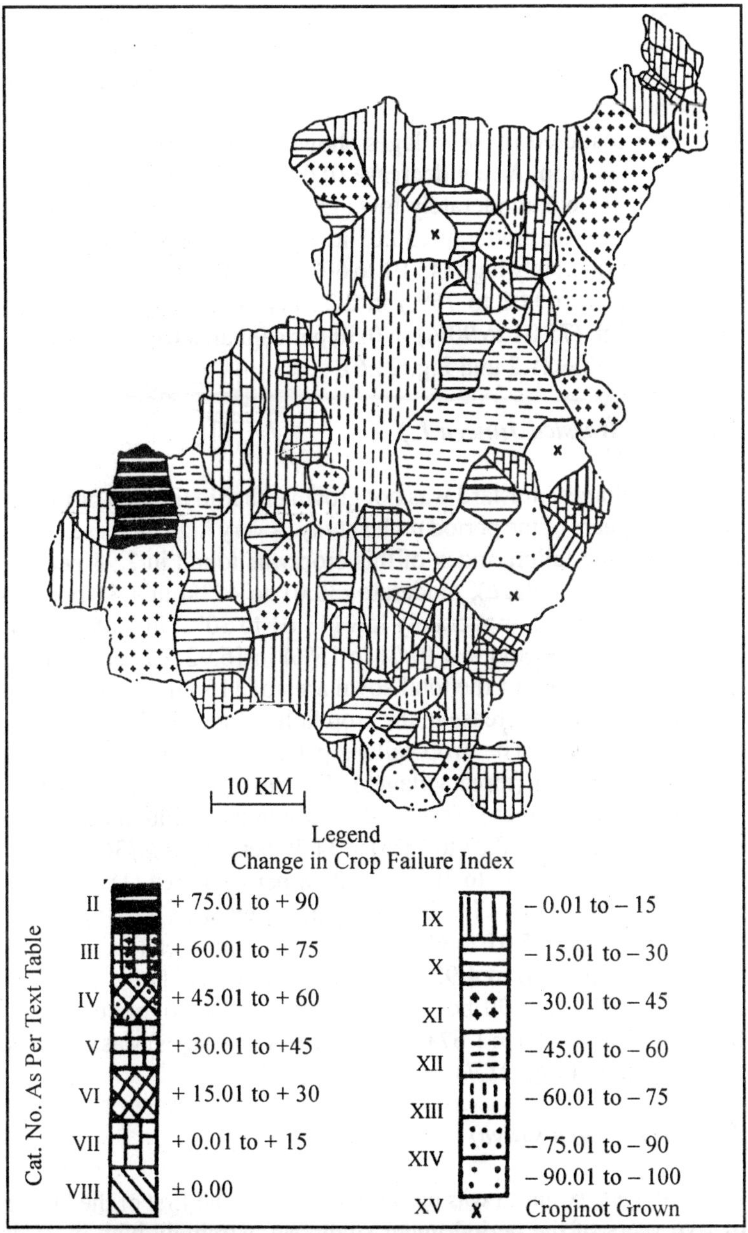

Fig. 21.7: Changing Pattern of Mustard Crop Failure — 1968-69 to 1972-72 and 1983-84 to 1987-88 (Hisar Tehsil).

to 30%) in 15.12 per cent area (16 villages) and low (0.01 to 15%) in 10.21 per cent area (15 villages). Thus, out of the total sown area of 122 villages under this crop, there was an increase in the crop failure index of 36.96 per cent area (44 villages). The remaining 63.04 per cent area scattered in 78 villages showed a decrease in the failure index of this crop. A low decrease (0.01 to 15%) in 15.21 per cent area (28 villages), moderately low (15.01 to 30%) in 31.04 per cent area (30 villages), moderate (30.01 to 45%) in 9.04 per cent area (10 villages), moderately high (45.01 to 60%) in 5.23 per cent area (seven villages), high (60.01 to 75%) in 2.22 per cent area (two villages) and very high (75.01 to 90%) in 0.30 per cent area (one village) was observed (Table 21.2 and Fig. 21.6).

Mustard (Brassica Juncea L.)

In seven villages, mustard was not grown either in the first or the last set of five years of the period under study. A very high increase (75.01 to 90%) in the failure index of this crop was found in 2.98 per cent sown area (one village), moderate (30.01 to 45%) in 1.34 per cent area (four villages), moderately low (15.01 to 30%) in 2.65 per cent area (five villages) and low (0.01 to 15%) in 16.92 per cent area (18 villages). Thus, out of the total sown area of 115 villages under this crop, there was an increase in the crop failure index of 23.89 per cent area (28 villages). The crop failure pattern remains unchanged in 0.17 per cent area (three villages). The rest of the area (75.94%) scattered in 84 villages showed a decrease in the failure index of this crop. A low decrease (0.01 to 15%) in 34.39 per cent area (36 villages), moderately low (15.01 to 30%) in 12.88 per cent area (18 villages), moderate (30.01 to 45% in 14.96 per cent area (15 villages), moderately high (45.01 to 60%) in 5.22 per cent area (six villages), high (60.01 to 75%) in 4.92 per cent area (five villages), very high (75.01 to 90%) in 1.90 per cent area (two villages) and extremely high (90.01 to 100%) in 1.67 per cent area (two villages) was observed (Table 21.2 and Fig. 21.7).

Taramira (Eruca sativa)

In 17 villages, taramira was not grown either in the first or the last set of five years of the period under study. An extremely high increase (90.01 to 100%) in the failure index of this crop was found in 1 per cent sown area (three villages), very high (75.01 to 90%) in 0.94 per cent area (two villages), high (60.01 to 75%) in 5.02 area (three villages), moderately high (45.01 to 60%) in 1.48 per cent area (two villages), high (60.01 to 75%) in 5.02 per cent area (three villages), moderately high (45.01 to 60%) in 1.48 per cent area (two villages), moderate (30.01 to 45%) in 2.69 per cent area (three villages), moderately low (15.01 to 30%) in 2.54 per cent area (four villages) and low (0.01 to 15%) in 15.86 per cent area (14 villages). Thus, out of the total sown area of 105 villages under this crop, there was an increase in the crop failure index of 29.53 per cent area (31 villages). The crop failure pattern remains unchanged in 0.08 per cent area (one village). The rest of the area (70.39% scattered in 73 villages showed a decrease in the failure index of this crop. Low decrease (0.01 to 15%) in 14.16 per cent area (13 villages), moderately low (15.01 to 30%) in 5.90 per cent area (seven villages), moderate (30.01 to 45%) in 5.52 per cent area (11 villages), moderately high (45.01 to 60%) in 15.91 per cent area (15 villages), high (60.01 to 75%) in 17.88 per cent area (13 villages), very high (75.01 to 90%) in 4.15 per cent area (seven villages) and extremely high (90.01 to 100%) in 6.87 per cent area (seven villages) was observed (Table 21.2 and Fig. 21.8).

Barley (Hordeum vulgare)

In 96 villages, barely was not grown either in the first or the last set of five years of the period under study. An extremely high increase (90.01 to 100%) in the failure index of this crop was found in 2.59 per cent sown area (three villages), very high (75.01 to 90%) in 7.44 per cent area (three villages), high (60.01 to 75%) in 4.53 per cent area (one village), moderate (30.01 to 45%) in 12.62 per cent area (four villages), moderately low (15.01 to 30%) in 2.10 per cent area (two villages) and low (0.01 to 15%) in 10.36 per cent area (one village). Thus, out of the total sown area of 26 villages under this crop, there was an increase in the crop failure index of 39.64 per cent area (14 villages). The crop failure pattern remains unchanged in 0.65 per cent area (one village). The rest of the area (59.71%) scattered in 11 villages showed a decrease in the failure index of this crop. A moderately low decrease (15.01 to 30%) in 6.31 per cent area (three villages), moderate (30.01 to 45%) in 8.09 per cent area (two villages), moderately high (45.01 to 60%) in 11.33 per cent area (two villages), very high (75.01 to 90%) in 4.37 per cent area (one village) and

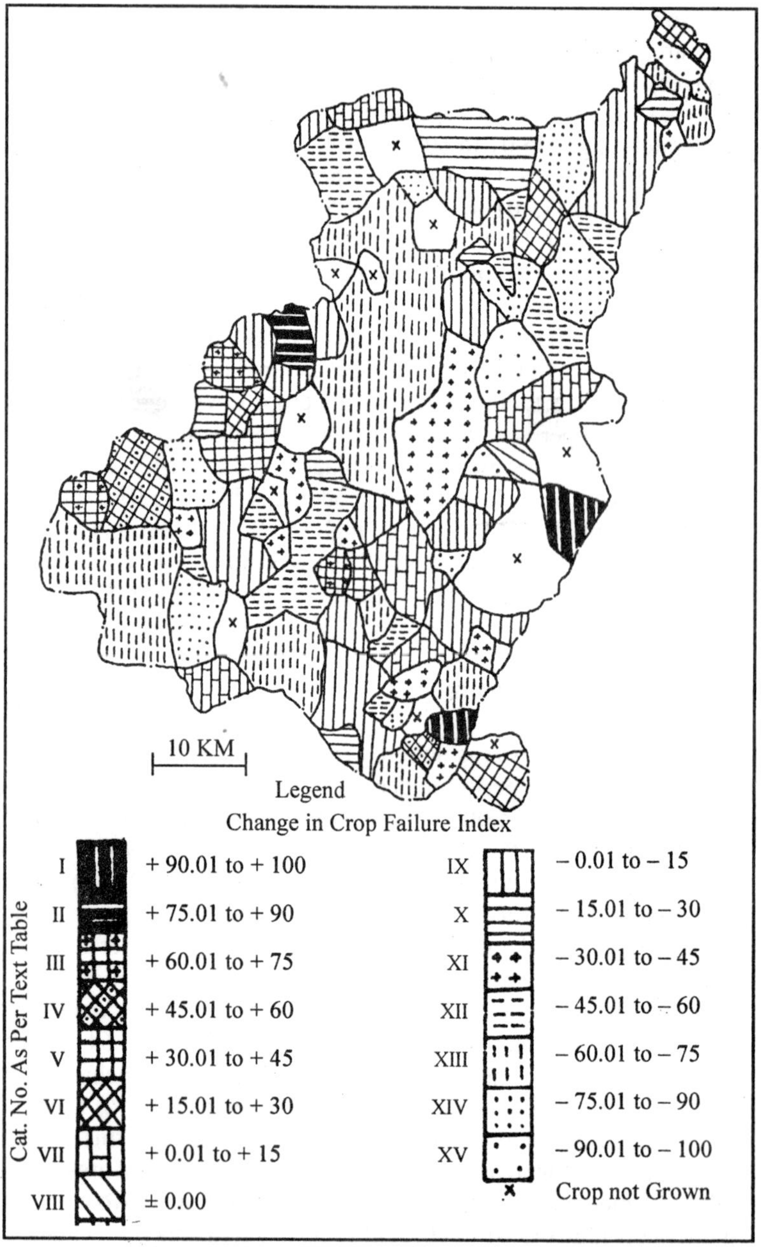

Fig. 21.8: Changing Pattern of Taramira Crop Failure — 1968-69 to 1972-73 and 1983-84 to 1987-88. (Hissar Tehsil).

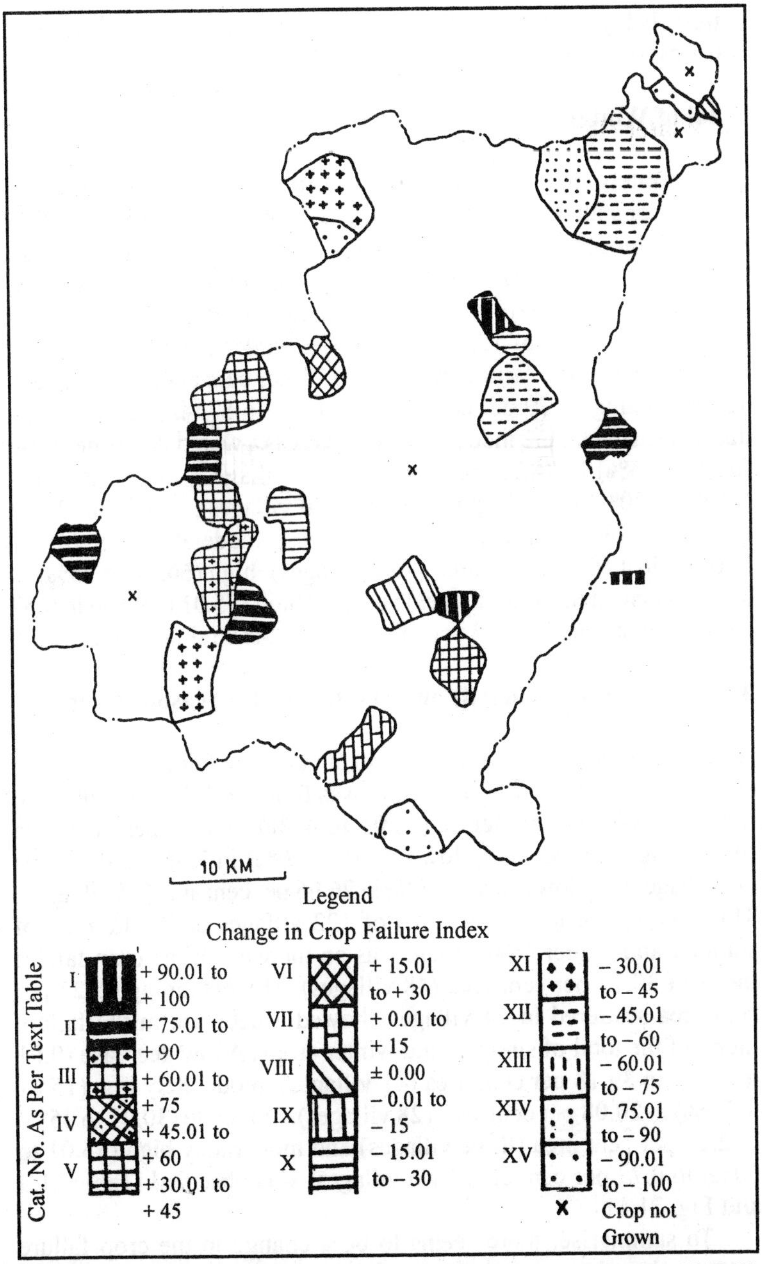

Fig. 21.9: Changing Pattern of Barley Crop Failure — 1968-69 to 1972-73 and 1983-84 to 1987-88 (Hissar Tehsil).

extremely high (90.01 to 100%) in 29.61 per cent area (three villages) was observed (Table 21.2 and Fig. 21.9).

Total of Winter (Rabi) Season Crops

A moderately high increase (45.01 to 60%) in the failure index of winter crops altogether was found in 3.73 per cent sown area (four villages), moderate (30.01 to 45%) in 4.76 per cent area (six villages), moderately low (15.01 to 30%) in 12.22 per cent area (11 villages) and low (0.01 to 15%) in 10 per cent area (14 village). Thus, out of the totai sown area of 122 villages under winter crops, there was an increase in the crop failure index of 30.71 per cent area (35 villages). The remaining 69.29 per cent area scattered in 87 villages showed a decrease in the failure index of winter crops as a whole. A low decrease (0.01 to 15%) in 18.75 per cent area (31 villages), moderately low (15.01 to 30%) in 31.13 per cent area (30 villages), moderate (30.01 to 45%) in 13.13 per cent area (16 villages), moderately high (45.01 to 60%) in 1.97 per cent area (Five villages), high (60.01 to 75%) in 3.74 per cent area (four villages) and very high (75.01 to 90%) in 0.57 per cent area (one village) was observed (Table 21.3 and Fig. 21.10).

Total of Summer (Kharif) and Winter (Rabi) Seasons Crops

A moderately high increase (45.01 to 60%) in the failure index of the summer and winter crops altogether was found in 1.30 per cent sown area (one village), moderate (30.01 to 45%) in 1.96 per cent area (two villages), moderately low (15.01 to 30%) in 7.55 per cent area (10 villages) and low (0.01 to 15%) in 26.96 per cent area (35 villages). Thus, out of the total sown area of 122 villages under the total of summer and winter crops, there was an increase in the crop failure index of 37.77 per cent area (48 villages). The remaining 62.23 per cent area scattered in 74 villages showed a decrease in the failure index of the total of summer and winter crops. A low decrease (0.01 to 15%) in 39.18 per cent area (41 villages), moderately low (15.01 to 30%) in 20.08 per cent area (28 villages), moderate (30.01 to 45%) in 2.24 per cent area (three villages) and moderately high (45.01 to 60%) in 0.73 per cent area (two villages) was observed (Table 21.3 and Fig. 21.11).

To summarise, there seems to be a change in the crop failure pattern when the average values of the crop failure index of the last

Table 21.3: Changing Pattern of Total of Kharif and Rabi Crops' Failure in Dry Farming Areas of Hissar Tehsil (1968-69 to 1987-88)

Category No.	*Difference in Crop Failure Index*	*Kharif Crops Sown area*		*No. of villages*	*Rabi Crops Sown area*		*No. of villages*	*Kharif + Rabi Crops Sown area*		*No. of villages*	*Degree of Change (increase or decrease in crop failure)*
		ha.	*%*		*ha.*	*%*		*ha.*	*%*		
1	2	3	4	5	6	7	8	9	10	11	12
I	+90.01 to +100	—	—	—	—	—	—	—	—	—	Extremely High (increase)
II	+75.01 to +90	—	—	—	—	—	—	—	—	—	Very High (increase)
III	+60.01 to +75	—	—	—	—	—	—	—	—	—	High (increase)
IV	+45.01 to +60	—	—	—	10,393	3.73	4	9,780	1.30	1	Moderately High (increase)
V	+30.01 to +45	17,804	3.77	3	13,274	4.76	6	14,699	1.96	2	Moderate (increase)
VI	+15.01 to +30	66,265	14.03	15	34,050	12.22	11	56,718	7.55	10	Moderately Low (increase)
VII	+0.01 to +15	1,42,980 (2,27,049)	30.28 (48.08)	38 (56)	27,876 (85,593)	10.00 (30.71)	14 (35)	2,02,442 (2,83,639)	26.96 (37.77)	35 (48)	Low (increase)
VIII	±0.00	—	—	—	—	—	—	—	—	—	Neutral
IX	-0.01 to – 15	1,48,125	31.37	38	52,267	18.75	31	2,94,181	39.18	41	Low (decrease)

(Contd.)

Table 21.3 (*contd.*)

1	2	3	4	5	6	7	8	9	10	11	12
X	-15.01 to –30	84,687	17.94	23	86,785	31.13	30	1,50,820	20.08	28	Moderately Low (decrease)
XI	-30.01 to –45	10,217	2.16	4	36,609	13.13	16	16,821	2.24	3	Moderate (decrease)
XII	-45.01 to –60	2,104	0.45	1	5,493	1.97	6	5,498	0.73	2	Moderately High (decrease)
XIII	-60.01 to –75	—	—	—	10,433	3.74	4	—	—	—	High (decrease)
XIV	-75.01 to –90	—	—	—	1,597	0.57	1	—	—	—	Very High (decrease)
XV	-90.01 to –100	—	—	—	—	—	—	—	—	—	Extremely High (decrease)
		(2,45,133)	(51.92)	(66)	(1,93,184)	(69.29)	(87)	(4,67,320)	(62.23)	(74)	

Note: Figures in parentheses in front of Category VII are totals for areas showing a positive change and those in front of Category XV are the totals for areas having negative change in the Crop Failure Indexed.

— Indicates Nil.

Source: Tehsil Office, Hissar (Haryana).

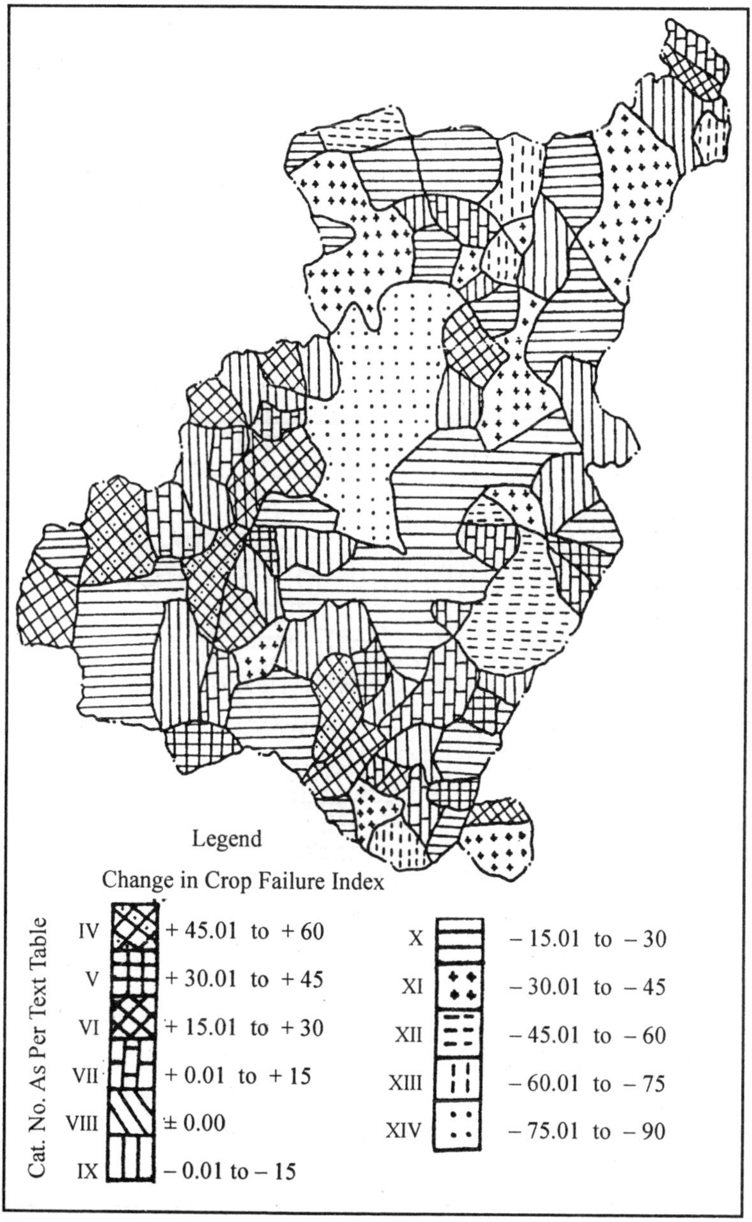

Fig. 21.10: Changing Pattern of Rabi Crops Failure — 1968-69 to 1972-73 and 1983-84 to 1987-88 (Hissar Tehsil).

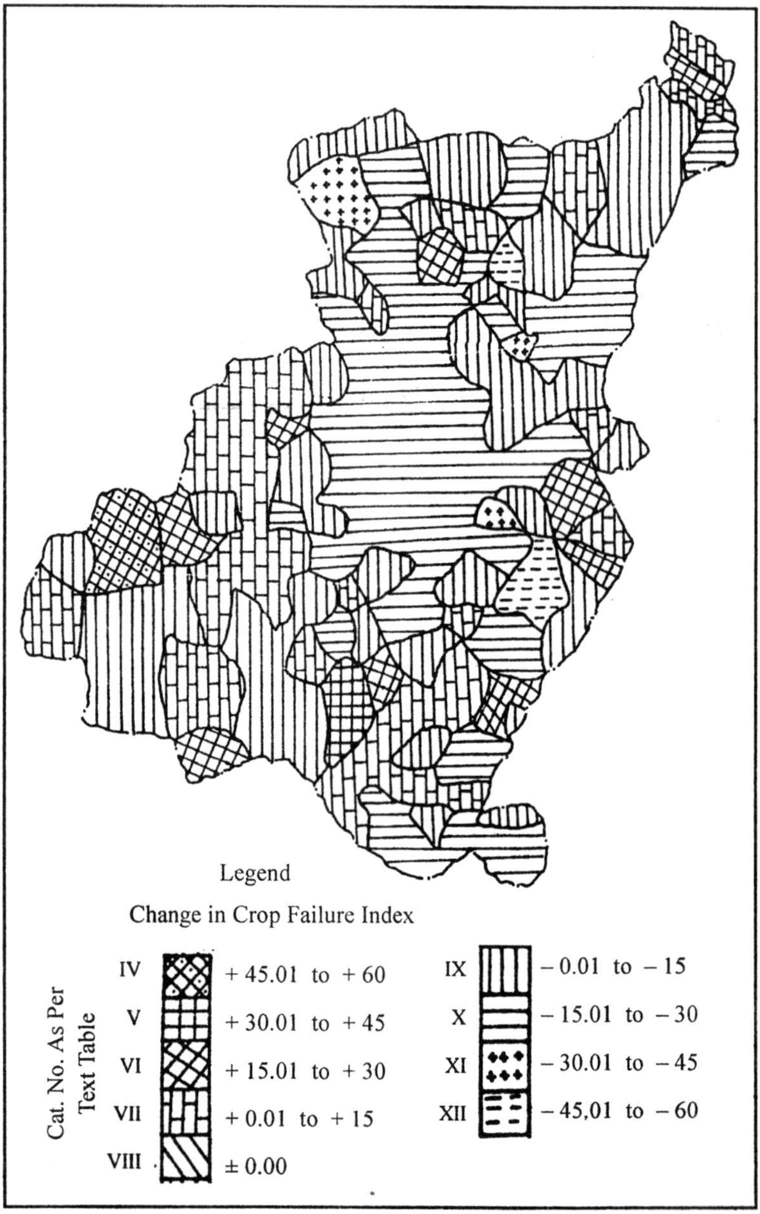

Fig. 21.11: Changing Pattern of Kharif and Rabi Crops Failure — 1968-69 to 1972-73 and 1983-84 to 1987-88 (Hissar Tehsil).

five years (1983-84 to 1987-88) was compared with that of the first five years (1968-69 to 1972-73). Though the index values of crop failure were downward, yet, these cannot be said to be significant keeping in view the innovations made in the field of agriculture and the initiatives provided by the government and other agencies in the past. This demands finding out the reasons behind it and the measures necessary to remove them. Only then it would be possible to increase foodgrains production of dryland agriculture.

Conclusion

The highest decrease in the crop failure index was recorded with pearl millet and mustard crops in the summer and winter seasons, respectively. The decrease in the crop failure index was higher in winter than in summer crops. Sorghum and barley crops are disappearing from the cropping pattern in their respective seasons, hence, their study deserves least importance. The south-western part of Hissar tehsil experienced an increase in the crop failure index while a decrease was observed in the north-eastern and central parts with a few exceptions. This clearly demonstrates that more attention should be focussed on checking the crop failure during the summer season and in the south-western part of the tehsil adjoining Rajasthan state. In the end, it may be concluded that an overall decrease in the crop failure index during the last two decades was not satisfactory. Therefore, there is an urgent need to investigate, improve, propagate and apply the latest technical know-how of rainfed agriculture in a new fashion to increase foodgrains production.

Acknowledgement

The Authors are thankful to the Professor and Head, Department of Soil Science, Haryana Agricultural University, Hissar, for providing all facilities to prepare this paper.

22

Land-man Relationship and Development Prospects for India

SAROJ K. PAL

The natural resources, based or land are basically cyclicale attributes, except the minerals. But since land itself is finite and growth is needed for development, so in a country like India where *land-man relationship* is intense a syndrome of decreasing land area being available for ecological resource attributes is emerging. The present paper has done the accounting of nine major identities of landuses of India since 1950-51 to 1985-86 based on a priori information to which a dynamics model of landuse charges has were applied. The model points out the need of a "synchronization" in the process of development taking peace in India so that the country carries on economic growth in harmonry with its ecology.

"Land-Man Relationship" has been a term commonly used especially in the field of geography. If this study puts its emphasis in land-centered direction, the focus of interest becomes the impact of land (or, the environment) upon man, and one moves into the relatively sterile fields of "physical determinism" that were ploughed by Huntngton[1], Ratzel[2] and Semple[3], the noted geographers of yesteryears. And if the converse view is considered, that is, in man-centered direction, then patterns of human activity on land is viewed as the result of the initiation and mobility of man operating within a framework of environmental forces. This is the concept of "possibilism" put forward by Blache and Brunhes in their books on human geography. This concept of possibilism occupied the attention of philosophical geographers[4] before he had quantification and models to keep him busy. Thus, it has been accepted in geography that the environment in terms of land provides possibilities, of which man selects one attribute of land regarding its use for his benefit. This selection of use of land depends on human perception and behaviour to which economic and political conditions also play an important role. Hence, in land-man relationship the use of the land must be perceived in the physical environment of the decision-maker and, secondly, it must be acceptable within his cultural milieu and it must be gainful to him.[5] Thus, an acre of land in the centre of a city has all the advantages of easy accessibility produced in consequence of the existence of an urban community and, thus, its market value rises very high compared to an acre of land at the edge of the city, though the latter is also socially important as a residential site. Hence, while social significance can place different values on similar units of land, land, in turn, also confers new and differential values on various members of the society. It is this capacity of land that serves as an intransitive object and yet it operates as a transitive medium of socio-economic variations that has forced a growing reevaluation of its role in the development process. Again among the possibilities inherent in the visible features and non-visible valuations of land regarding its mode of use, the choice which is the easiest of the possibilities is the most "probable".[6] Thus, the possibilities' view of man as an earth-influencing agent (akin to tectonic-, climatic and geomorphological processes) needs investigation and man's socio-ecologic-economic functions need to be understood.

Land-Man Relationships

As one of the simplest as well as an important index of spatial concentration it is expressed by density of population, i.e., the number of persons per unit area. Table 22.1 gives the statement of population of India, rural and urban, number of rural persons per urban man and population density as recorded at each decennial census from 1901 onwards. This statement also provides the availability of gross cropped area in hectare per capita since 1951.

Thus, the per capita availability of cropped land in India has declined to 0.216 hectare (=2,160 sq. metre) in 1991 but the size of the population living in rural India has become more than double in between 1951 to 1991. Hence, the natural resource capital in terms of land has been declining in India but the dependence on land for living has been on increase. Therefore, assessment of land-man relationships is very significant as an indicator both of the level of living and of the development prospects of our country and this necessitates a quantification of resource endowment as reflected by landuse. Hence, we start with intransitive and transitive character of land.

Table 22.1: Longitudinal Pattern of Population in India, 1901-1991[7].

Census Year	*Population (in millions)*			*Rural Man per Unit of Urban*	*Population Density per Sq. km.*	*Arable Land Per capita*
	Total	*Rural*	*Urban*			
1901	238.40	212.54	25.85	1 : 8.22	72	
1911	252.09	226.15	25.94	1 : 8.72	77	
1921	251.32	223.23	28.90	1 : 7.95	76	
1931	278.98	245.52	33.46	1 : 7.34	85	
1941	318.66	274.51	44.15	1 : 6.22	97	
1951	361.09	298.64	62.44	1 : 4.78	110	0.477 ha.
1961	439.24	360.30	78.94	1 : 4.56	134	0.402 ha.
1971	548.16	430.05	109.11	1 : 3.94	167	0.325 ha.
1981	683.33	523.87	159.46	1 : 3.29	208	0.266 ha.
1991	844.32	627.15	217.18	1 : 2.89	257	0.216 ha.

Land, Landuse and Land-Man Relationships

Land is used in a comprehensive and in an integrating sense[8]......It refers to "a specific geographic space on the earth's surface, it is an wide array of natural resource attributes in which it embraces all reasonably stable or predictably cyclic attributes from the atmosphere above the surface to some metres below the land surface in a profile form. In the process it includes the topography, its wilderness, climate, soil, flora, fauna and water. All these attributes of land in its private domain are ambient resources and in natural form they are excluded from accounting system since the latter deals with the economic values of natural resources. In this sense, no material use is made of these land attributes, and they are not used in the sense of being consumed and getting permanently destroyed. But they are valued at their factor cost and they may be modified as a result of human activities.[9] Again, in ecological sense, land by all its above attributes is one of the most important components of the life support system since human life depends very directly on the living soil and the ecosystems as without fertility of soil and microbial fauna that inhabit it. Food would not grow year by year and nutrients would not be recycled. Again land has been man's living habitat and living space. Stamp, the world's leading authority on land-man relationships was motivated by a concern primarily for the availability of land for agriculture.[10] In his book "*Land For Tomorrow*" in 1952, he remarked,

> .... "No problem confronting us today is more important than that of matching the world's use of its land with its needs of people". In a country like India, land as it possesses within its borders is an enormous asset in any consideration of levels of development and utilisation of natural resources of land constitutes the "bedrock of it's economy", the intensity and efficiency of their use determine the level and pace of development. In India, the primary activities under agricultural enterprises is vast : it comprises some 182 million hectares of cropped land (=1.82 million sq. km.) which is again 59.75 per cent of the total geographic area of the country. Again though the share of primary activities in the net national production in India has declined from 58.9 per cent in 1950-51 to 45.3 per cent in 1965-66 and to 40 per cent in 1980-81, it still constitutes the "bedrock of economy" for the country. This is demonstrated in case of food production. In India, about 70 per cent of the country's income is spent of food. Of the total cost of production, about 30 per cent is paid out to the owners of land in form of rent. Thus, approximately 21 per cent of the total national income is spent on land services. Compared to India, these percentage are 12,20 and only 2.[11]

The specific characteristics of land-man relationships lie in the analysis of the links and feedbacks between the various ecological factors and man. Basically, it comprises all traditional and modern methods which in one extreme form is determined by the nature of physical environment of land but, generally, it is looked upon as one economic activity, and is determined by economic forces. In the context of the latter, the land user in land-man relationships is a subspecies of "economic man" who, according to Simon (1957), seeks to maximise or optimise his gain form his land enterprise, but in practice, "he satisfies himself".[12]

Geography which has been concerned with land-man relationships has long subsumed it under the term "landuse". And as Sauer (1919) used it,......" consequence of land as a natural habitat by vegetation and its economic use by man" so that land use = land cover plus land utilisation. Hence, any analysis on land-man relationships is landuse and study of the latter under the theme of natural resource potential, is a part of the "core of geography". Thus, over the decades in the land-man relationship studies, there has been a continual return from the "landschaft" concept of the German school of Humboldt and Ritter throughout the whole of its present century over past 90 years to the environmentalists of the present day in which the focus of study is upon land/landscape/environment which is produced by inter-relationships between a host of direct natural influences and man operating through his social and economic functions/activities as the process of production for development.[14] This concept of land-man relationships permists, or perhaps even enforces close links not only between the fields of physical and human geography but also academic enquiries of our subject and ours role in society as geographers and in solving its "real world" problems. Again, land is so basic and an enormous asset to man to possess, so that land-man relationships is the concern of many disciplines and not the exclusive reserve of geography. Land-man relationships in terms of land use are largely based on geology and economics of land, but, geography also recognises that landuse by itself is the product of human decisions operating within social, political and legal frameworks. Hence, land-man relationship studies tend to be a multi-disciplinary or inter-disciplinary area of enquiry in which geographers cooperate fully with researches from many other cognate branches of study. Perhaps, the UNESCO bulletin, "*Nature and Resource*" is the most effective illustration of this.[15]

But it is all too easy to regard land-man relationships from single viewpoint and to forget that it is a multi-faceted subject. The aim of the present approach is to view it as multi-faceted and to develop it on a truly scientifically based ecological/environmental approach to assessment of relationships so that the underlying forces get emphasized and a synthesis in terms of basis inter-relationships could be advanced. In this direction, an attempt is made in this paper to examine the basic landuse statistics of India through a long-run analysis, so that the landuse dynamics of India are properly brought out and the associated ecological implications are understood for determining how the bedrock of our economy determines the level and pace of development.

Landuse Statistics

The concept of dividing the land put into various categories of uses was adopted by the Bureau of Economics and Statistics, Government of India which is presently known as Directorate of Economics & Statistics, Department of Agriculture & Cooperation, Ministry of Agriculture & Irrigation, Government of India. The 9-fold classification of uselanduse was advocated by the Revenue Department and it was incorporated in a standard proforma to be followed right down to the village-wise record. The area of land of the conceived unit of territory (i.e., the reporting area) that falls under standard category of classification has been recorded each year. This has been termed as "basic landuse statistics" by the Land Revenue Department. In this process has been collected a huge load of records that reveal the landuse characteristics of the country in the form of statistics on areas when each category of uses. The 9-fold classification are: (1) Forest (Fr) : This includes all lands classed as forests under any legal enactment dealing with forests or administered as forests, whether state-owned or private, (2) Permanent Pastures and Other Grazing Lands (P) : includes grazing lands both permanent pastures and other grazing lands or meadows (having grass/scrub as land cover), (3) Land under Miscellaneous Tree Crops and Fruit Groves (M) : All lands under bamboo bushes, casurina trees other groves used for fuel which are not included under orchards/net area sown, (4) Barren and Uncultivable Lands (U) : All lands in mountains like icecaps or rocky lands and, in deserts like shifting sands that are really wastelands being non-capable of producing any material in accounting terms, (5) Culturable Waste Lands (W) : land once cultivated but are not cultivated in last five years in succession, (6) Fallow Lands Other than Current Fallows (F) : All lands which are not in cultivation for a period of not less than one year but not more than five years, (7) Current Fallows (Fc) : Cropped areas which are kept fallow during the current year, (8) Net Area Sown (C) : Area sown with crop and orchards, this category includes areas sown more than once in the same year, and (9) Land put to Non-Agricultural

Uses (N) : It includes all lands occupied by settlements with (or, without) trees, rural or urban, mines, roads and railways, water bodies including rivers and lakes. These nine land use units together constitute the "reporting area". Area sown more than once (C2) is also shows separately.

The commendable feature of this above 9-fold basic landuse statistics is the fact that detailed statistics down to the village level on areas under each of the 9 land use classes are available for each year. However, reliability of the data, i.e. area under each category of landuse that are recorded is very doubtful because: (i) figures are entered into record year after year with little checking making them non-dependable, (ii) the inbuilt weakness of ignoring degraded forests, denuded forest and forest lands put to shifting cultivation that have developed over the years within the administered forest areas and lump them together as "forests" based on their administered status (and not on the basis of actual forest canopy or stock), makes the data all the more non-realistic, (iii) the landuse nomenclature adopted is cumbersome, neither very explicit so that a broken and undulating land may be classed as "barren & uncultivable land" by one and as "culturable waste" by another person, in the fields, (iv) there are no means to check as to how far uniformity in this respect are being maintained throughout the country, (v) finally, the reliability of the data, i.e. area under each category of landuse that are recorded is very doubtful. It is virtually impossible to establish a total correlation between the data on record and situation in the field.

However, in spite of all the above weaknesses in the basic land-use statistics, it would be still more difficult to arrive at a reliable statistics on area under different landuses by making use of the classifications adopted by other government institutions like National Atlas and Thematic Mapping Organisation, All India Soil and Landuse Survey, Wasteland Survey and Reclamation (1959, report in 1963), National Remote Sensing Agency (1983, mapping of forest cover through visual interpretation of satellite data), Forest Survey of India (1986, national vegetation mapping through ground truth and satellite imagerys), National Commission on Agriculture (1976), National Wasteland Development Board (established on May 7, 1985). They are detailed and ambitious in their efforts but for correct identification of each of the landuse classes and their areal demarcation they still require to develop suitable methodology to generate their own base line data, independent of the data by Directorate of Economics & Statistics. Thus, the landuse classification nomenclature in favour of the above 9-fold classification is that it is simple, straightforward and realistic so that the average *Patwari* or village level worker can comprehend the implications of this nomenclature throughout the country and be able to identify and assess their spatial extent.[16]

This sectiva examines the landuse dynamics and its ecological implications by budgeting landuse changes in India for the period of 36 years from 1950-51 to 1985-86. In the present compilation of the time series data on land use classes, besides the above 9 classes, another 2 landuse units are taken. They are : the area sown more than once (C2) and the total cropped area (CA), the latter equals to C plus C2, i.e. CA = C + C2.

Landuse Dynamics Model and a Priori Information

Existing landuses, because of their strong influence on how land will be used in the future, are a crucial element in the development of a landuse or land development plan. Perhaps the most important function the existing landuse data represents is that it serves a base against which the landuse changes can be measured. Land use change data are necessary both for developing landuse policies, for carrying on various land conservation activities and for monitoring their effectiveness.[17] In the present case, a landuse dynamics model is attempted. It was adopted by Pandey and Tiwari (1987)[18] and by Das (1988)[19]

This landuse dynamics model assumes that since the total geographical area is constant so the changes in landuse (or, landuse dynamisms) can occur through inter-class transfers. Thus, starting the landuse accounting with the base equation that the Reporting Area (R) is:

$$R = Fr + P + M + U + W + F + Fc + C + N \quad \text{.......... (Eq. 1).}$$

Now, differentiating over time, the change in the Reporting Area (dR) will be the sum of the proportional growth rates of the above nine landuse classes. Thus, the differentiation equation will be simply additive and, hence, the accounting identity of the landuse dynamics can be expressed as :

$$dR = dFr + dP + dM + dU + dW + dF + dFc + dC + dN \ldots\ldots\ (Eq.\ 2)$$

i.e. the rate of change in landuse will be equal to the aggregate rate of change in the nine units.

But the accounting identity in Eq. 2 is too cumbersome to facilitate any meaningful inference about the direction and magnitude of the inter-class transfers. Therefore, it is needed that some a priori knowledge about the possible land flows among the landuse classes is added. For this purpose, the total land endowment (i.e. the reporting area) can be divided into 3 broad sectors : (a) Ecological or the ECO-SECTOR comprising Fr, P, M and U and (b) Agricultural or the AGRO-SECTOR comprising W, F, Fc and C and (c) Non-Agricultural or the NON-AGRO comprising N only.

In terms of a-priori knowledge, the possible landuse shifts (or, changes) within and across the three sectors can be taken up now. In the ECO-SECTOR, the landuse may shift from M and P to Fr which has no adverse ecological implications and the shift from U to Fr gives rise to favourable ecological consequences rather in mitigation of desertification and flooding problems. But the shift from Fr to P which may possibly occur in the hills, does have some adverse implications. And more serious implications will be if the landuse shift is in the reverse direction, i.e. from Fr and M to U. Keeping this dynamics in view, it will be possible to divide the ECO-SECTOR into 2 sub-sectors, viz, (i) the DESIRABLE ECOLOGY (E1) comprising Fr, P and M and (ii) the UNDESIRABLE ECOLOGY (E2) comprising U. Thus, the net change in the ECO-SECTOR (dE) can be budgeted as—

$$dE = (dFr + dP + dM) + dU = dE1 + dE2 \ldots\ldots\ldots\ (Eq.\ 3)$$

Similarly, the net change in the AGRO-SECTOR (dA) can be budgeted as—

$$dA = dW + dF + dFc + dC \ldots\ldots\ldots\ldots\ (Eq.\ 4)$$

In the AGRO-SECTOR, there is no possibility of landuse shift from NON-AGRO SECTOR to AGRO SECTOR. Hence, the net change in the AGRO-SECTOR will have serious implications. This net change, if positive (+dA), will be at the expense of the ECO-SECTOR. If this change is negative (-dA), the landuse shift may go to ECO-SECTOR or to NON-AGRO SECTOR, or to both the sectors, but, definitely at the expense of the ECO-SECTOR. If this occurs, the country requires to lay down its social preference function for the trade of between ecological balance, agricultural as well as non-agricultural growth in an explicit manner. This is required to be donc within the available technological options by adopting suitable landuse planning so that laws are enforced to prevent undesirable change in landuse like shift of Fr to N or C to N unduly.

Similarly, the changes (+dA) within the AGRO-SECTOR will have some implications also towards ecological situation and/or agricultural growth. For example, with the addition of land to Net Sown Area (+dC), the situation will be favourable at least for agricultural growth, though it will be at the cost of the ECO-SECTOR. However, if there is no increase in the Net Sown Area (dC=0) as well as in the Area Sown More Than Once (dC2=0), but there has been an increase in the net area in AGRO-SECTOR (dA), it would suggest an adverse phenomenon, like that of the vicious circle, indicating a depletion of Net Sown Area (-dC), by the way of landuse shifts to Fc, F and W, on the other one hand and depletion of area under Fr (-Fr) to keep the food production on. This phenomenon is the worst of all, because, first, the cultivated land gets degraded into wastelands (i.e., the Culturable Wastes, W) the reclamation of the latter back to the former requires large investments and, second, the conversion of the land under ECO-SECTOR for cultivation to meet the self-sufficiency for rising population is not a feasible one from the point of view of long-term ecological security.

Now, the overall inter-sectoral landuse transfers can be budgted as :

$$dR = dE1 + dE2 + dA + dN \text{ ....} \qquad \text{(Eq. 5)}$$

The identity in the Eq. 5 is the same as the identity in Eq. 2, except, that being represented in aggregate sectoral terms, the former facilitates quick glance at their sectoral transfers.

Landuse is generally seen from the point of view of land economics rather than from land. Thus, conversion of land from both ECO- and/or AGRO-SECTORs to NON-AGRO SECTOR is very common, but, in case there has been no executive directives to prevent "overbounding" of urban settlements in the name of urbanisation and/or industrialisation, this may result in a disjointed urban spread of mammoth population to become highly uneconomic for being provided with the urban infrastructure investments as well as on the public utility services by the local municipality (Pal, 1989)[20]. Figure 22.1 provides the flow chart of dynamics of land use changes.

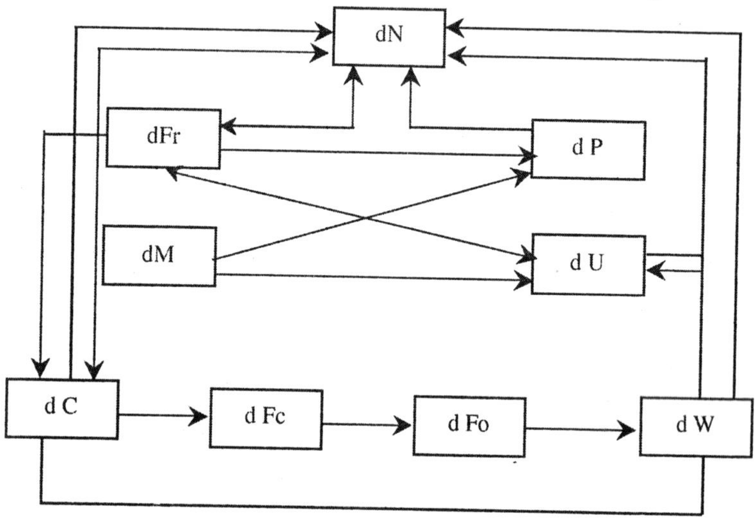

Fig. 22.1: Flow Chart of Dynamics of Landuse Changes.

Data Base

The Annexures 1(a), (b) and (c) provide the gross statistics of land-use under the above three sectors : the ECO-SECTOR, AGRO-SECTOR and NON-AGRO SECTOR by their respective landuse classes in a time series frame. The Annexure 1(d) provide the classification of the Reporting Area (R) under the three sectors collectively. The total time period taken is for 36 annual years spread from 1950-51 to 1985-86. This time period of 36 years cover the six Five Year Plan, three Annual Plans as well as the 1st year of the 7th Five Year Plan (the 6th plan being ended on 1984-85) and hence this is desirable that this time series data on basic landuse is disagregated into three sub-periods, namely : the Sub-Period 1 from 1950-51 to 1965-66, the Sub-Period II from 1966-67 to 1973-74 and the Sub-Period III from 1974-75 to 1985-86. The Sub-Period I includes the first three Five Year Plans (the First Five Year Plan being started in 1951-52), the year 1966-67 heralded the Green Revolution breaking the long continued stagnation in Indian agriculture; the second Sub-Period II which is taken from this year is spread over three successive Annual Plans and the Fourth Five Year Plan. Finally, the third Sub-Period III is counted from the year 1974-75 and it spreads over the Fifth and the Sixth Five Year Plans and the first year of the Seventh Five Year Plan, 1985-86. Accordingly, in the Annexure I(a) to I(d), the data is given in these three sub-periods. These three sub-periods, it is to be noted, are having a span of uneven number of years : 16, 8 and 12 years respectively. Of these three sub-periods, during the second sub-period, the high rate of agricultural growth through Green Revolution, on the one hand, and rapid pace of urbanisation through spread of urban areas, on other hand, were witnessed in the Indian scene. But, since the middle of the seventies all these developments and increasing human activities make also clear that there is the importance of incorporating environmental issues into evaluation of the merit of any landshift programme (by landuse) and of putting the demand on conserving natural resources in an ecologically oriented agricultural technology so that the soils are used more rationally for increasing the agricultural production rather than on encroaching on forestry land. Similarly, there was the demand that the existing urban areas were developed for increasing the role of the tertiary share in the G.N.P. of the country, but, they need not grab the neighbouring agricultural land in the name of urbanisation and/or industrialisation.[21] For this purpose in the Annexure I(a), for analysing the desirable intra-sectoral transfer (dEI) in the ECO-SECTOR, the separate totals are given by adding the landuse data for Fr, P and M together. Hence, the landuse data for U is kept separate to treat it as the undesirable eco-sub-sector, E2. Again, since irrigation plays an important role in supporting both the increase in agricultural output as well as making the increase reliable, so for the Agro-Sector data given in Annexure I(b), the Area Sown More Than Once (C2) is shown separately and it is added with the Net Sown Area (C) to have a separate Total Cropped Area (CA) which is in addition to the total for the Agro-Sector.

Methodology and Analysis

The above three sub-period data for the various landuse classes and the total for the three sectors (as well as for their sub-sectors) are now subjected to trend analysis. The word "trend" refers to the general direction in which the graph of the data collected regarding a theme in successive intervals appears to be going over a period of time. For finding the trends, the linear and log-linear time trend equations are fitted into the time series data of the annexures I(a) to I(d). The linear trend equation is Y = a + bX where Y is the landuse class/sector, X is the number of years in the time series data, a is the intercept of the trend line on the ordinate and b is the regression coefficient of Y on X (or, trend) value. The analytical solution to the trend fitting of the time series data is an exponential curve in the form of $Y = ae^{bX}$ which on taking natural logarithm, In on both sides we have : $Log_e{}^Y = Log_e$ a + bX, which on assuming $Log_e Y = Y_{ln}$ and $Log_e a = a_{ln}$ we have the logarithmic transformation of the exponential identity $Y = ae^{bX}$ in its linear form of $Y_{ln} = a_{ln+}$ bX and we can have the computation of the constants a and b in form analogous to those in the linear trend equation like in

$$b = \frac{\sum xy - n(\bar{x})(\bar{y})}{\sum x^2 - n(\bar{x})^2} \; \ldots\ldots \text{(Eq. 6) and } a = \bar{y} - b\bar{x} \ldots \text{(Eq. 7)}$$

Further, the growth rate per annum in the linear trend equation has been defined as :

$$\text{Growth, Gr} = \frac{b}{\bar{y}} \times 100 \text{ (in per cent)} \ldots\ldots\ldots \quad \text{(Eq. 8)}$$

This growth in log-linear equation is a compound (or, exponential) growth rate defined as :

Compound Growth Rate,

$$\text{CGr} = (\text{Antilog } b_{ln} - 1) \times 100 \text{ (in per cent)} \ldots\ldots\ldots \quad \text{(Eq. 9)}$$

Next, is the linearity test of the trend line by Analysis of Variance in which the mean square (or, estimates of variances) of the Total Variability, SST (i.e., the sum of squares of total variance) is divided into a variability accounted for by the regression, SSR (i.e. the sum of squares accounted for by regression) and a variability unaccounted for by the regression, SSU (i.e. the sum squares unaccounted) ; SST is calculated equal to :

$$\text{SST} = \sum Y_i^2 - \frac{\left(\sum Y_i\right)^2}{n} \ldots \quad \text{(Eq. 10)}$$

$$SSR = b(\Sigma XY - n. \overline{X} \cdot \overline{Y} \ldots \quad \text{(Eq. 11)}$$

and, finally, SSU = SST – SSR .... (Eq. 12)

$$\text{Now, the F Test (d.f. = 1, n-2)} = \frac{SSR/I}{SSU/(n-2)} \ldots\ldots \quad \text{(Eq. 13)}$$

In case, the computed value is larger than the critical F value at the same d.f. – 1, n-2 at a particular significance level i.e. the SSR is large enough in comparison to the SSU and the conclusion will be that the linear trend line by the regression equation is due to a real effect rather than a random sampling.

The ratio of SSR/SST .... (Eq. 14) in analysis of variance is the coefficient of determination, R^2 the square root of which is the correlation of coefficient, r. The test of the correlation coefficient is by forming the t-Statistics using the t distribution with d.f. = n-2 in a manner analogous to :

$$t = r.\sqrt{\frac{n-2}{I-r^2}} \ldots \quad \text{(Eq. 15)}$$

Finally, the measure of scatter of the estimated value of Y is the "Standard Error Estimate of the Growth Rate", SE which is :

$$SE = \frac{\sum Y^2 - a\sum Y - b\sum XY}{(n-2)/(n-1)} \ldots \quad \text{(Eq. 16)}$$

Results of the Landuse Dynamics

The above trend analysis of the time series data of landuse classes/ sectors/sub-sectors by following the above linear and the log-linear regression equations have been given in Tables 22.2 to 22.4, for the sub-periods as well as for total period. Table 22.2 provides the regression coefficients and intercepts for the linear and log-linear trends. It will be found that of all the sectors, the AGRO- and the NON-AGRO sectors have the positive beta coefficients in all the sub-periods as well as in the total period. However, the beta coefficients are always higher for the NON-AGRO sector in comparison to those of the AGRO sector. The highest beta coefficient attained for the NON-AGRO sector was 0.27424 in the sub-period I

Table 22.2: Regression Coefficients and Intercepts for Linear and Log-Linear Trends (Y *in million hectares*).

Landuse Classes/ Sectors/ Sub-Sectors	*Sub-Period I*		*Sub-Period II*		*Sub-Period III*		*Total Period*	
	a	*b*	*a*	*b*	*a*	*b*	*a*	*b*
Fr	44.508	1.114	63.228	0.275	66.862	0.016	49.064	0.626
	44.899	1.021	63.333	1.004	66.862	1.000	49.247	1.011
P	8.050	0.491	14.033	-0.176	12.678	-0.072	11.666	0.042
	8.199	1.045	14.044	0.987	12.679	0.994	11.277	1.005
M	8.788	-0.338	3.931	0.057	3.670	-0.014	7.026	-0.122
	8.911	0.948	3.932	1.014	3.696	0.996	6.947	0.996
El	61.346	1.267	81.256	0.140	82.738	0.007	67.188	0.576
	61.799	1.018	80.983	1.002	82.621	1.0005	67.651	1.007
U (or, E2)	36.598	-0.169	34.042	-1.090	21.879	-0.190	40.017	-0.596
	36.548	0.995	34.419	0.963	21.855	0.991	42.093	0.979
W	24.602	-0.024	16.007	0.168	17.835	-0.180	22.114	-0.203
	24.936	0.976	16.012	1.010	17.867	0.989	22.197	0.989
Fc	12.225	-0.042	13.027	-0.058	13.995	0.042	11.202	0.090
	12.182	0.997	13.032	0.994	13.894	1.003	11.259	1.007
F\`	15.655	-0.425	9.151	-0.006	9.215	0.050	13.135	-0.142
	15.951	0.966	9.151	0.994	9.214	1.005	12.982	0.988
C2	14.005	0.476	20.455	0.189	26.371	0.906	12.385	0.635
	13.265	1.030	20.585	1.035	26.679	1.029	14.032	1.027

(*Contd.*)

Table 22.2 (*Contd.*)

Landuse Classes/ Sectors/ Sub-Sectors	*Sub-Period I*		*Sub-Period II*		*Sub-Period III*		*Total Period*	
	a	*b*	*a*	*b*	*a*	*b*	*a*	*b*
CA	133.888	1.702	157.868	1.193	164.068	1.068	138.070	1.181
	134.156	1.124	157.928	1.035	164.824	1.006	138.756	1.008
C	120.155	1.252	137.413	0.373	139.837	0.163	126.241	0.528
	120.264	1.009	137.319	1.003	139.396	1.002	126.179	1.004
ECO-SECTOR	97.945	1.098	115.328	-0.955	105.116	-0.260	107.773	-0.260
	98.235	1.010	115.388	0.991	105.121	0.997	107.621	0.997
AGRO-SECTOR	172.637	0.295	175.598	0.434	180.806	0.099	172.646	0.278
	172.630	1.002	175.090	1.002	180.805	1.0006	172.687	1.002
NON-AGRO SECTOR (or. N)	11.623	0.274	11.130	0.245	13.719	0.190	14.448	0.249
	11.587	1.021	12.259	1.015	14.487	1.010	12.027	1.016

Note: The regression coefficients and intercepts for log-linear trends are underlined.

Table 22.3: Correlation Coefficient, its Significance and Linearity Test.

Landuse Classes/Sectors/Sub-Sectors	*Sub-Period I*		*Sub-Period II*		*Sub-Period III*		*Total Period*	
	r^o indicates the F test values for linearity test							
	r	*r^o*	*r*	*r^o*	*r*	*r^o*	*r*	*r^o*
Fr	0.925	0.910	0.741[§]	0.686[§]	0.111[§§]	0.112[§§]	0.917	0.897
P	0.943	0.549[§]	0.912	0.913	0.873	0.887	0.260[§§]	0.285[§§]
M	0.823	0.115[§§]	0.623[§§]	0.626[§§]	0.901	0.674[§]	0.771	0.363[§§]
E1	0.928	0.671	0.044[§§]	0.243[§§]	0.000[§§]	0.138[§§]	0.755	0.671
U (or, E2)	0.507[§§]	0.496[§§]	0.976	0.973	0.714	0.717	0.946	0.939
W	0.985	0.986	0.613[§§]	0.609[§§]	0.915	0.916	0.858	0.877
Fc	0.232[§§]	0.226[§§]	0.030[§§]	0.130[§§]	0.118[§§]	0.016[§§]	0.375	0.162
F	0.933	0.957	0.325[§§]	0.329[§§]	0.506	0.509	0.741	0.749
C2	0.854	0.893	0.888	0.881	0.937	0.967	0.983	0.982
CA	0.947	0.942	0.754	0.754	0.827	0.826	0.969	0.962
C	0.953	0.613[§]	0.494[§§]	0.509[§§]	0.359[§§]	0.428[§§]	0.868	0.681
ECO-SECTOR	0.917	0.917	0.962	0.963	0.919	0.922	0.111[§§]	0.106[§§]
AGRO-SECTOR	0.691	0.691	0.916	0.912	0.616[§]	0.621	0.928	0.926
NON-AGRO SECTOR (or, N)	0.853	0.811	0.935	0.938	0.991	0.991	0.970	0.950

Note: §§ indicates non-significance of the correlation coefficient at both 0.05 and 0.01 significance level. This also indicates the non-linearity of the trend at these two levels.

§ indicate the significance of the correlation coefficient as well as linearity of the trend at 0.05 level only.

decreasing to 0.189868 in the Sub-Period III in the linear trend which in the exponential provides a in value of 0.0210 and 0.00974 respectively. During these sub-periods, the AGRO-SECTOR was having an uniform beta coefficient of 1.002 in about all the sub-periods as well as in the total time period (it decreases to 1.0006 in the sub-period III). It should be noted that the beta coefficient in the log-linear trend is a multiple of the intercept 'a' on the ordinate. On other hand, the ECO-SECTOR is characterised by a negative beta coefficient from the sub-period II onwards so much that for the total time period also it reveals a negative beta value and the intercept value also decreased from a high of 115.328 million hectares in the Sub-Period II to a low of 105 million hectares in the Sub-Period III giving an overall low of 107.621 million hectares for the total period. In the intra-ECO-SECTOR, the area under permanent pastures and other grazing land (P) shows a negative beta coefficient from the Sub-Period II onwards but the decrease is low, another land use variable is M which shows a negative beta coefficient in the Sub-Periods I and III. The regional pattern of sectoral landuse shown under E1 reveals a Positive beta coefficient in all the three sub-periods which is due to a surprising positive beta coefficient in the land recorded under forest (Fr).

Table 22.3 provides the correlation coefficient as well as a linearity test for the landuse classes, sectors and the sub-sectors. Both the tests reveal the identical responses. It is found that for the Sub-Period II which is of the smallest of all the Sub-Periods reveal non-linearity (and hence no significance of linear correlation) for the maximum number of land classes like M, W, Fc, F and C, in the sub-sector level, E1 shows the non-linearity at the Sub-Sectors II and III. In the intra-AGRO-SECTOR, the land use class Fc shows the non-linearity in all the three Sub-Periods, the other land use class which comes close to Fc in non-linearity is the Total Cropped Area, C. At sector level, the ECO-SECTOR shows the non-linearity in the Total Period.

The computation of the per annum growth rate, both linear and log-linear (or, compound) is an important part of the trend analysis in context of the landuse dynamics model. Table 22.4 provides the necessary computation. As regards the growth rates, this table shows higher values for most of the landuse classes/sectors. The positive trend has been singularly for the AGRO- and NON-AGRO sectors for all the time periods. Among the two SUB-SECTORS of the ECO-SECTOR, the desirable ecology, E1 has been characterised by positive growth, though

Table 22.4: Growth Rates : Linear and Compound.

Landuse Classes/ Sectors/Sub-Sectors	*Sub-Period I*		*Sub-Period II*		*Sub-Period III*		*Total Period*	
	Gr	*Gr°*	*Gr*	*Gr°*	*Gr*	*Gr°*	*Gr*	*Gr°*
Fr	2.064	2.123	0.427	0.394	0.024	0.025	1.032	1.090
P	4.016	4.514	-1.329	-1.313	-0.593	-0.585	0.338	0.474
M	-5.705	-1.548	1.361	1.376	-0.392	-0.388	-2.569	-1.552
E1	1.757	1.790	0.172	0.171	0.009	0.046	0.021	0.742
U (or, E2)	-0.481	-0.465	-3.742	-3.720	-0.919	-0.886	-2.058	-2.144
W	-2.395	-2.385	1.004	1.009	-1.078	-1.077	-1.103	-1.063
Fc	-0.357	-0.338	-0.456	-0.573	0.294	0.352	0.699	0.687
F	-3.529	-3.416	-0.550	-0.559	0.519	0.521	-1.346	-1.215
C2	2.638	2.974	3.394	3.515	2.807	2.878	2.632	2.744
CA	1.147	0.233	0.731	0.733	0.617	0.621	0.739	0.753
C	0.957	0.976	0.268	0.293	0.115	0.163	0.388	0.399
ECO-SECTOR	1.023	1.025	-0.860	-0.856	-0.251	-0.251	-0.048	-0.051
AGRO-SECTOR	0.169	0.170	0.245	0.245	0.054	0.055	0.156	0.156
NON-AGRO SECTOR	1.965	2.131	1.518	1.534	1.175	0.979	1.534	1.576

Note: The growth rates with ° for example Gr° indicate compound growth rate.

the growth is becoming minimal in the Sub-Period III but there has been a declining trend in the undesirable ecology E2 which shows that land otherwise of no use is also put to some marginal use in meeting the demands of the swelling population which has increased from 361 million in 1951 to 844 million in 1991. In the landuse class only the net sown area (C) has been able to maintain a positive growth but in view of the fact that "there will be no possibility of adding more land for agricultural production" so this growth also can not be guaranteed in future.

Development Prospects

Development should be based on 'total systems' view, this view is not axiomatic with economic growth on overall production basis nor it is with modernisation but development should be through distributive or social justice. Against this background concept of development there should be a transformation of the modes of production which refer to the elements, activities and social relationships of the production. The elements refer to resources, infrastructure and human labour. Activities refer to the process of bringing these elements together in the production of goods and services and in the context of available technology. Social relationships define the social basis for coordinating the productive activities of people involved in these processes. Basic shifts in any aspects of the mode of production can trigger off wide-ranging changes which may culminate not only in the transformation of mode but also changes in the relative importance of social classes. It is such a socio-economic transformation that really constitutes development. Conceptualization of development concentrating on articulation of the processes by which the entire society is mobilised restructure and reconstruct the total national space to facilitate increased social and economic productivity. Since land is the primary source of income and wealth and the basis of social relations for majority of the population, development cannot be said to have begun without the four key words : productivity, stability, sustainability and equitability. There can be trade-offs between productivity and stability, on the one hand, and sustainability and equitability, on the other, in the process of development. These four key words bring us back to the 'total systems' view, in which the strategy of growth through he process of incremental change and development as an intense concentration of activities requiring "synchronization" leading to transformation of space and society, are required to fitted in view of the constraints of natural resources and swelling of population in India.

NOTES

1. E. Huntington (1915), *Civilization and Climate*, N. York.
2. H.H. Burrows (1923), "Geography as Human Ecology", in, *AAAG*, Vol. 13, New York.
3. Ellen C. Sample (1911), *Influences of Geographic Environment*, New York.
4. Vidal De La blache and Brunhes, *Human Geography*, Paris.
5. K.G.T. Clard (1950), "Certain Underpinings of Our Arguments in Human Geography", Transactions of the Institute of *Br. Geographers*, Vol. 16, London.
6. W.W. Firey (1960), *Man, Mind and Land*, Illinois.
7. Table 22.1 modified after "Statement 2, Series I, Paper I, Census of India, 1991.
8. R. Brinkman, A.J. Smith (1973).
9. The World Resources Institute (1990), World Resources, 1990-91, New York.
10. L.D. Stamp (1952), *Land for Tomorrow*, London.
11. *Indian Agriculture in Brief*, Bombay, 2nd edition, December, 1981.
12. H.A. Simon (1957), *Models of Man*, New York.
13. C.O. Sauer (1919), Mapping the Utilisation of Land, *Geography Review*, Vol. 9, New York.
14. R.J. Chorley (ed. 1973), *Directions in Geography*, London.
15. C.M. Embleton & J.J. Cappock (eds., 1968), *Landuse and Resources Studies in Applied Geography*, Institute of Br. Geographers, Spl. Publication No. 1, London.
16. G.K. Dutt (1988), "Wastelands in India and Their Assessment and Mapping" in K.A. Sankarnarayan (ed.), *Wasteland Development and Their Utilisation*, Jodhpur.
17. David T. Lindgren (1985), *Landuse Planning and Remote Sensing*, Boston.
18. V.K. Pandey and S.K. Tewari (1987), Ecological Implications of Landuse Dynamics in Uttar Pradesh, *Indian Journal of Agriculture Economics*, Delhi.
19. H.C. Das (1988), Spatial and Temporal Variations in Area, Productivity and Production of Forests in Bihar, 1957-58 to 1984-85, *Indian Journal of Agriculture Economics*, Delhi.
20. Saroj K. Pal (1989), Changing Jurisdictions of Urban Centres in India and Related Aspects, Proceed. National Seminar on Patterns of Urban Changes, Varanasi, mimeo.
21. K.V. Sundaram (1988), Rural Development and Environmental Conservation, Keynote Speech in Intern. Symp. On the Development of Regional Agriculture and Environmental Conservation in South-East Asian Countries, Manila, mimeo.
22. Anupam Verma (1988), Manpower Development and Utilisation in Agriculture Research, Indian Science Congress Association, 7th Session, mimeo.

Annexure I(a)

Basic Landuse Statistics on Ecological Sector : India

The Ecological Sector Comprises four broad land use classes, namely : Forest (Fr), Permanent Pastures & Grazing Lands (P), Miscellaneous Tree Crops & Groves (M) and Barren & Uncultivable Land (U). Following is the time series data on these four landuse classes of the Ecological Sector. The data is given in million hectares and it is generally limited to three decimal places for the fraction of the million hectare. The time period used is from 1950-51 to 1985-86 which is again divided into three sub-periods : 1950-51 to 1965-66, 1966-67 to 1973-74 and 1974-75 to 1985-86. The total of all these four landuse classes is given in the next column as the 'total for the Eco-sector'. Finally, the total is given for the 'Desirable Eco-Sub-sector', E1.

The Sub-Period I : 1950-51 to 1965-66

Year	*Fr*	*P*	*M*	*U*	*Total*	*EI*
1950-51	40.482	6.675	11.828	38.160	97.145	58.985
1951-52	49.989	8.592	7.881	37.484	102.846	65.362
1952-53	51.154	8.634	7.757	37.420	104.965	67.545
1953-54	51.079	10.881	5.800	36.379	104.139	67.760
1954-55	50.431	11.218	5.770	34.517	101.935	67.418
1955-56	51.343	11.473	5.885	34.475	103.176	68.701
1956-57	52.178	10.186	5.784	33.387	103.535	70.148
1957-58	51.406	12.831	6.087	33.232	103.556	70.324
1958-59	54.021	13.090	6.000	33.155	106.266	73.111
1959-60	54.052	13.679	5.818	33.434	106.983	73.549
1960-61	54.189	13.966	4.459	35.911	108.525	72.614
1961-62	60.692	14.082	4.500	35.921	115.195	79.274
1962-63	60.351	14.104	4.558	35.164	114.177	79.013
1963-64	60.700	14.594	4.378	34.811	114.483	79.672
1964-65	61.100	14.743	4.113	34.795	114.751	79.956
1965-66	61.543	14.810	4.076	34.327	114.756	80.429
Mean	53.975	12.225	5.918	35.161	107.277	72.116

The Sub-Period II : 1966-67 to 1973-74

Year	*Fr*	*P*	*M*	*U*	*Total*	*EI*
1966-67	63.458	14.017	4.018	32.159	113.652	81.433
1967-68	64.217	13.795	3.969	31.830	113.811	81.981
1968-69	64.790	13.316	3.878	31.591	113.575	81.984
1969-70	63.980	12.992	4.379	30.311	111.662	81.351
1970-71	63.917	13.361	4.299	28.161	109.738	81.577
1971-72	63.771	12.960	4.284	27.996	109.011	81.015
1972-73	65.729	12.707	4.531	25.821	108.788	82.967
1973-74	65.878	12.781	4.138	25.217	108.014	82.797
Mean	64.218	13.241	4.187	29.136	111.031	81.888

The Sub-Period III : 1974-75 to 1985-86

Year	*Fr*	*P*	*M*	*U*	*Total*	*EI*
1974-75	65.778	12.856	3.771	23.160	105.565	82.405
1975-76	66.699	12.542	3.630	21.578	104.449	82.871
1976-77	67.159	12.532	3.628	21.522	104.841	83.319
1977-78	67.138	12.353	3.691	20.219	103.401	83.182
1978-79	67.465	12.136	3.627	20.143	103.371	83.228
1979-80	67.501	12.118	3.527	20.156	103.302	83.146
1980-81	67.433	11.991	3.512	20.051	102.987	82.936
1981-82	67.365	12.035	3.603	20.092	103.095	83.003
1982-83	67.330	11.930	3.545	20.104	102.909	82.805
1983-84	66.599	12.034	3.628	20.337	102.598	82.261
1984-85	66.391	12.000	3.569	20.239	102.199	81.960
1985-86	66.756	11.967	3.528	20.151	102.402	82.251
Mean	66.968	12.208	3.605	20.646	103.426	82.781

Annexure I(b)

Basic Landuse Statistics on Agricultural Sector : India

The Agricultural Sector comprises four broad landuse classes, namely Culturable Wastes (W), Fallows other than Current Fallows (Fc), Current Fallows (F) and Net Area Sown (C). The following Table gives the time series data on these landuse classes of the Agricultural Sector as well as the total for the latter. In addition, it is having the year-wise data on 'Area Sown More than Once' (C2) which gives rise to another corresponding column of the 'Total Cropped Area' (CA). The data is given in million hectares and it is generally limited to three decimal places for the fraction of the million hectare. The time period used is 1950-51 to 1985-86 which is again divided into three sub-period used is 1950-51 to 1985-86 which is again divided into three sub-periods : 1950-51 to 1965-66, 1966-67 to 1973-74 and 1974-75 to 1985-86.

The Sub-Period I : 1950-51 to 1965-66

Year	*W*	*Fc*	*F*	*C*	*Total*	*C2*	*CA*
1950-51	22.943	17.445	10.679	118.746	169.813	13.147	131.893
1951-52	23.929	15.154	13.808	119.400	172.291	13.834	133.234
1952-53	23.680	13.480	12.898	123.442	173.501	14.233	137.675
1953-54	22.898	12.750	12.025	126.806	174.479	15.674	142.480
1954-55	22.805	13.045	11.963	127.845	175.658	16.242	144.087
1955-56	21.537	12.544	11.583	129.156	174.820	18.155	147.311
1956-57	21.276	11.968	11.358	131.828	176.430	18.644	149.492
1957-58	20.602	12.620	12.700	129.080	175.002	16.752	145.832
1958-59	20.175	12.255	11.452	132.939	176.821	19.801	151.629
1959-60	19.463	11.076	11.952	133.199	175.690	19.885	152.824
1960-61	19.212	11.180	11.639	135.399	177.430	19.573	152.772
1961-62	18.632	10.478	11.155	136.341	176.606	20.810	157.141
1962-63	17.908	10.195	11.058	136.483	175.644	20.419	156.902
1963-64	17.653	10.059	11.229	138.120	177.061	20.480	158.600
1964-65	17.366	9.166	11.156	136.199	173.887	21.109	157.308
1965-66	16.965	9.262	13.184	137.810	177.221	19.078	156.888
Mean	20.128	12.042	11.865	122.183	175.147	17.990	148.354

The Sub-Period II : 1966-67 to 1973-74

Year	*W*	*Fc*	*F*	*C*	*Total*	*C2*	*CA*
1966-67	16.671	9.046	13.529	137.232	176.478	20.123	157.355
1967-68	16.235	8.800	11.924	139.876	176.835	23.860	163.736
1968-69	16.064	9.189	14.096	137.313	176.662	23.216	160.529
1969-70	15.788	9.483	12.323	138.772	176.366	23.493	162.265
1970-71	17.500	8.759	11.116	140.267	177.642	25.524	165.791
1971-72	17.456	8.312	12.669	139.721	178.158	25.465	165.186
1972-73	17.332	9.193	15.176	137.144	178.845	25.006	162.150
1973-74	17.066	8.655	11.292	142.416	179.429	27.456	169.872
Mean	16.764	8.930	12.766	139.093	177.551	24.268	163.235

The Sub-Period III : 1974-75 to 1985-86

Year	*W*	*Fc*	*F*	*C*	*Total*	*C2*	*CA*
1974-75	17.032	8.969	16.307	137.791	180.099	26.400	164.191
1975-76	17.743	9.229	12.546	141.652	181.170	29.644	171.296
1976-77	17.474	9.385	14.570	139.476	180.905	27.858	167.334
1977-78	17.183	9.567	13.028	141.953	181.731	30.279	172.232
1978-79	17.292	9.366	12.470	142.981	182.109	31.821	174.802
1979-80	16.658	10.026	15.698	138.903	181.285	30.686	169.589
1980-81	16.698	9.992	14.821	140.013	181.524	32.675	172.638
1981-82	16.424	9.657	13.365	142.121	181.567	34.980	177.101
1982-83	16.332	9.357	14.817	140.813	181.319	32.959	173.772
1983-84	16.565	9.271	13.308	143.211	182.355	37.557	180.768
1984-85	15.882	9.512	15.882	140.901	181.653	35.513	176.414
1985-86	15.718	10.117	15.718	140.908	181.664	36.711	177.619
Mean	16.750	9.537	14.267	140.894	181.448	32.257	173.146

Annexure 1(c)

Basic Landuse Statistics for Non-Agricultural Sector : India

The Non-Agricultural Sector refers to the land put to Non-Agricultural Uses (N) only. Thus it comprises only one out of the possible nine basis landuse units to which an area is generally reported. Following is the time series data on the Area put to Non-Agricultural Uses which itself is synonymous with Non-Agricultural Sector. The total time period 1950-51 to 1985-86 is split in three sub-periods : 1950-51 to 1965-66, 1966-67 to 1973-74 and 1974-75 to 1985-86. The data is given in million hectares and it is generally limited to three decimal places for the fraction of the million hectare.

The Sub-Period I : 1950-51 to 1965-66

Year	*N*
1950-51	9.357
1951-52	12.690
1952-53	12.321
1953-54	13.283
1954-55	13.784
1955-56	13.921
1956-57	13.981
1957-58	14.105
1958-59	14.300
1959-60	14.899
1960-61	14.840
1961-62	14.795
1962-63	15.111
1963-64	15.270
1964-65	15.442
1965-66	15.170
Mean	13.954

The Sub-Period II : 1966-67 to 1973-74

Year	*N*
1966-67	15.357
1967-68	15.474
1968-69	15.648
1969-70	15.868
1970-71	16.478
1971-72	16.972
1972-73	16.658
1973-74	16.799
Mean	16.157

The Sub-Period III : 1974-75 to 1985-86

Year	*N*
1974-75	18.377
1975-76	18.660
1976-77	18.934
1977-78	19.047
1978-79	19.201
1979-80	19.543
1980-81	19.658
1981-82	19.630
1982-83	19.865
1983-84	20.237
1984-85	20.458
1985-86	20.514
Mean	19.510

Annexure I(d)

Sector-wise Basic Landuse Statistics of India (For 1950-51 to 1985-86)

Following is the three sector-wise data on Basic Landuse Statistics of India for the period 1950-51 to 1985-86, the latter as the total time period of study is disaggregated into three sub-periods : 1950-51 to 1965-66, 1966-67 to 1973-74 and 1974-75 to 1985-86. The data is in million hectares and it is generally limited to three decimal places for the fraction of the million hectare. The total of these three sectors is the 'Total Reporting Area' and this is given in the last column of the table.

Year	*Agro-Sector*	*Eco-Sector*	*Non-Agro-Sector*	*Total Reporting Area*
1950-51	169.813	97.145	9.357	276.315
1951-52	172.291	102.846	12.690	287.827
1952-53	173.501	104.965	12.321	290.787
1953-54	174.479	104.139	13.283	291.901
1954-55	175.658	101.935	13.784	291.377
1955-56	174.820	103.176	13.921	291.917
1956-57	176.430	103.535	13.981	293.946
1957-58	175.002	103.556	14.105	292.663
1958-59	176.821	106.266	14.300	297.387
1959-60	175.690	106.983	14.899	297.577
1960-61	177.430	108.525	14.840	300.795
1961-62	176.606	115.195	14.795	306.596
1962-63	175.649	114.177	15.111	304.937
1963-64	177.061	114.483	15.270	306.814
1964-65	173.886	114.751	15.442	304.079
1965-66	177.221	114.756	15.170	307.147

(The above data is for the Sub-Period I : 1950-51 to 1965-66)

Year	*Agro-Sector*	*Eco-Sector*	*Non-Agro -Sector*	*Total Reporting Area*
1966-67	176.478	113.652	15.357	305.487
1967-68	176.835	113.811	15.474	306.120
1968-69	176.662	113.575	15.648	305.885
1969-70	176.366	111.662	15.868	303.896
1970-71	177.642	109.738	16.478	303.858
1971-72	178.158	109.011	16.972	304.141
1972-73	178.845	108.788	16.658	304.291
1973-74	179.429	108.014	16.799	304.242

(The above data is for the Sub-Period II : 1966-67 to 1974-75)

Year	*Agro-Sector*	*Eco-Sector*	*Non-Agro-Sector*	*Total Reporting Area*
1974-75	180.099	105.565	18.377	304.041
1975-76	181.170	104.449	18.660	304.279
1976-77	180.905	104.841	18.934	304.680
1977-78	181.731	103.401	19.047	304.179
1978-79	182.109	103.371	19.201	304.681
1979-80	181.285	103.302	19.543	304.130
1980-81	181.524	102.987	19.658	304.169
1981-82	181.567	103.095	19.630	304.292
1982-83	181.319	102.909	19.865	304.093
1983-84	182.355	102.598	20.237	305.190
1984-85	181.653	102.199	20.458	304.310
1985-86	181.664	102.402	20.514	304.580

(The above data is for the Sub-Period III : 1975-76 to 1985-86)

Following is given the mean of the above three Sectors for the above three sub-periods.

Sub-Period I : 175.147 (for Agro) 107.277 (for Eco) 13.954 (for Non-Agro).

Sub-Period II : 177.552 (for Agro) 111.031 (for Eco) 16.157 (for Non-Agro).

Sub-Period III : 181.448 (for Agro) 103.426 (for Eco) 19.510 (for Non-Agro).

Index